T. Minami (Editor in Chief)
M. Tatsumisago, M. Wakihara
C. Iwakura, S. Kohjiya, I. Tanaka (Editors)

Solid State Ionics for Batteries

T. Minami (Editor in Chief)
M. Tatsumisago, M. Wakihara
C. Iwakura, S. Kohjiya, I. Tanaka (Editors)

Solid State Ionics for Batteries

With 325 Figures

Editor in Chief
Tsutomu Minami, Ph.D.
President, Osaka Prefecture University
1-1 Gakuen-cho, Sakai, Osaka 599-8531, Japan

Editors
Masahiro Tatsumisago, Ph.D.
Department of Applied Materials Science, Osaka Prefecture University
1-1 Gakuen-cho, Sakai, Osaka 599-8531, Japan

Masataka Wakihara, Ph.D.
Department of Applied Chemistry, Tokyo Institute of Technology
2-12-1 O-okayama, Meguro-ku, Tokyo 152-8550, Japan

Chiaki Iwakura, Ph.D.
Department of Applied Chemistry, Osaka Prefecture University
1-1 Gakuen-cho, Sakai, Osaka 599-8531, Japan

Shinzo Kohjiya, Ph.D.
Institute for Chemical Resarch, Kyoto University
Gokasho Uji, Kyoto 611-0011, Japan

Isao Tanaka, Ph.D.
Department of Materials Science and Engineering, Kyoto University
Yoshida Honmachi, Sakyo-ku, Kyoto, 606-8501, Japan

Library of Congress Control Number: 2005923754

ISBN 4-431-24974-5 Springer-Verlag Tokyo Berlin Heidelberg New York

Printed on acid-free paper

© Springer-Verlag Tokyo 2005

Printed in Japan
This work is subject to copyright. All rights are reserved, whether the whole or part of the material is concerned, specifically the rights of translation, reprinting, reuse of illustrations, recitation, broadcasting, reproduction on microfilms or in other ways, and storage in data banks.
The use of registered names, trademarks, etc. in this publication does not imply, even in the absence of a specific statement, that such names are exempt from the relevant protective laws and regulations and therefore free for general use.

Springer is a part of Springer Science+Business Media
springeronline.com

Printing and binding: Hicom, Japan
SPIN: 11394129

Preface

In recent years, cordless electronic devices such as personal computers, cellular phones, and digital cameras have become very popular. Faster development of electric vehicles and fuel-cell cars is eagerly anticipated. For these devices and machines, one of the essential parts is the power source. As the power source, many types of batteries are used; batteries are usually composed of anode, cathode, and electrolyte. For the electrolyte, flammable liquids are generally used, which involves the issues of leakage and combustion of the liquids. If excellent solid electrolytes were developed and replaced the liquid electrolytes, these issues could be solved at once.

In the twentieth century our daily life was drastically changed by the innovative development of many electronic devices; the key to the development was the invention of the transistor. In other words, the invention of the solid device transistor resulted in a kind of revolution in our daily life. The transistor is, of course, a solid device composed of semiconductors. The functions of transistors are controlled with the movement of electrons and holes in semiconductors.

As is well known, there are two types of electric conduction: electronic and ionic. In contrast to the transistor in electronic conduction, solid devices are very rare in the field of ionic conduction. Liquid electrolytes are still used in ionic devices such as batteries, because ions such as cations and anions usually can move fast in liquids but not in solids because of their large size and heavy mass.

The field of science and technology where the movement of electrons (and holes) plays a key role is called electronics. Similarly, the field where the movement of ions such as cations and anions plays a key role is called ionics; and where the movement of ions is done in solids, it is called solid state ionics. In this Preface the words "ion", "anode", and "cathode" have been used a priori; these words were first used by M. Faraday in 1834, and "ionics" or "solid state ionics" by T. Takahashi in the 1960s (See E. Tsagaratis, *Ionics*, vol 9, pp. VII–XII, 2003).

With the expectation of solving the issues mentioned above in currently used batteries, scientific research and technological development for solid state ionics have been carried out extensively, especially in North America, Europe, and Asia, including China, India, Korea, and Japan. The Ministry of Education, Culture,

Sports, Science and Technology of Japan supports scientific research of various kinds; one type of support for large scientific research projects is the Grant-in-Aid for Scientific Research in a Priority Area. As one of the priority area programs a group of about 30 members, represented by T. Minami, was supported for a period of 5 years, from the fiscal year 1999 to 2003, for the project "Fundamental Studies for Fabrication of All Solid State Ionics Devices." Based on the 5-year research project, this review book, *Solid State Ionics for Batteries*, has been published. There are many kinds of solid state ionics devices, including solid state batteries, capacitors, sensors, and electrochromic devices. Among these, batteries must be the most important, so for that reason, they are the focus of our attention in this book.

The book is composed of seven chapters. Chapter 1 is a general introduction. Chapter 2 describes the recent development and present status of research and development of batteries by the three leading battery companies in Japan: Sony, Panasonic, and Sanyo Electric, which we hope will be of great interest to readers. The main topics are, of course, the reviews of the results related to the research program, which are described in Chapters 3–7.

November 2004
Tsutomu Minami
Editor in Chief
President of Osaka Prefecture University

Contents

1	**Introduction**		1
2	**Recent progress in batteries and future problems**		5
	2.1	Recent technologies of secondary batteries in Sony Corporation	5
		2.1.1 Introduction	5
		2.1.2 Lithium ion secondary battery	5
		2.1.3 Lithium polymer battery	7
		2.1.4 Future of LIB and LPB	11
	2.2	Recent technologies on materials for advanced lithium rechargeable batteries [Panasonic]	12
		2.2.1 Introduction	12
		2.2.2 Active materials for positive electrodes	12
		2.2.3 Active materials for negative electrodes	13
		2.2.4 Electrolytes	14
		2.2.5 Concluding remarks	17
	2.3	Development of secondary battery electrode materials toward high energy and power density [Sanyo]	19
		2.3.1 Requirements for high performance of the batteries from evolving functions of the mobile devices	19
		2.3.2 Approach to high power density of nickel-metal hydride batteries	20
		2.3.3 Approach to high energy density of lithium secondary batteries	24
		2.3.4 Conclusions	28
3	**Recent development of amorphous solid electrolytes and their application to solid-state batteries**		31
	3.1	Introduction	31
	3.2	Lithium ion conductors	32
		3.2.1 Solid electrolytes	32
		3.2.2 Electrode materials	42

		3.2.3	All-solid-state lithium secondary batteries	53
		3.2.4	Thin film batteries	64
	3.3	Proton conductors		73
		3.3.1	Sol-gel ionics	73
		3.3.2	Proton-conducting composite materials	80
		3.3.3	Proton-conducting hybrid materials	87
	3.4	Conclusions		93
4	**Recent development of electrode materials in lithium ion batteries**			95
	4.1	Candidate cathode material with high voltage for lithium ion batteries: Structural study of 5V class $LiNi_{0.5}Mn_{1.5}O_4$ and $LiNi_{0.5}Mn_{1.5}O_{4-\delta}$		95
		4.1.1	Introduction	95
		4.1.2	Crystal structure of spinel type phase	97
		4.1.3	Electrochemical properties of $LiNi_{0.5}Mn_{1.5}O_4$ and $LiNi_{0.5}Mn_{1.5}O_{4-\delta}$	99
		4.1.4	Coulombic potential calculation of diffusion path for ordered ($P4_332$) and disordered ($Fd\bar{3}m$) spinels	100
		4.1.5	Conclusions	102
	4.2	Electrode/electrolyte interfaces in all-solid-state lithium ion batteries		104
		4.2.1	Introduction	104
		4.2.2	Control and charge transfer process at electrode/electrolyte interface	104
		4.2.3	Characterization of electrode/solid electrolyte interface in solid-state batteries	126
5	**Construction of solid/solid interface between hydrogen storage alloy electrode and solid electrolyte for battery application**			133
	5.1	Introduction		133
	5.2	Hydrogen storage alloy electrode/polymer hydrogel electrolyte interface		136
		5.2.1	Preparation and characterization of polymer hydrogel electrolyte	136
		5.2.2	Application of polymer hydrogel electrolyte to nickel-metal hydride batteries	146
		5.2.3	Application of polymer hydrogel electrolyte to electric double-layer capacitors	159
		5.2.4	Preparation and characterization of proton-conducting polymeric gel electrolytes	162
	5.3	Construction of hydrogen storage alloy electrode/inorganic solid electrolyte interface		171
		5.3.1	Preparation, characterization and application of phosphoric acid-doped silica gel electrolytes	171
		5.3.2	Preparation, characterization and application of inorganic oxide solid electrolytes	177

	5.4 Conclusions	185
6	**Polymer solid electrolytes for lithium-ion conduction**	**187**
	6.1 Introduction: Role of rubbery state for ionic conduction	187
	6.2 Branched poly(oxyethylene)s as polymer solid electrolytes	191
	6.2.1 Copolymers of ethylene oxide	191
	6.2.2 High molar mass poly(oxyethylene)s	192
	6.2.3 Poly(oxyethylene) networks	196
	6.3 Viscosity behaviors of branched poly(oxyethylene)s	199
	6.4 Composite electrolytes based on branched poly(oxyethylene)	200
	6.4.1 Introduction: Composite-type polymer electrolytes	200
	6.4.2 Hybrid solid electrolytes from oxysulfide glass and branched poly(oxyethylene)	201
	6.5 Effect of elongation of elastomer electrolytes on conductivity	206
	6.6 Ionene elastomers for polymer solid electrolytes	209
	6.6.1 Introduction: Ionene elastomers	209
	6.6.2 Polymer solid electrolytes prepared from poly(oxytetramethylene)s	211
	6.6.3 Viologen-type poly(oxytetramethylene) ionene elastomers	212
	6.6.4 Aliphatic poly(oxytetramethylene) ionene elastomer	216
	6.7 Further usefulness of rubbery matrix	221
	6.8 Concluding remarks	223
7	**First principles calculations of lithium battery materials**	**225**
	7.1 Introduction	225
	7.2 Changes in chemical bondings by lithium insertion/extraction in $LiMO_2$ (M = V, Cr, Co and Ni)	230
	7.2.1 Introduction	230
	7.2.2 Computational procedure	232
	7.2.3 Electronic and bonding states of $LiCoO_2$ and CoO_2	233
	7.2.4 Differences in bonding states among $LiMO_2/MO_2$	237
	7.2.5 Conclusion	238
	7.3 First principles study on factors determining voltages of layered $LiMO_2$ (M = Ti - Ni)	240
	7.3.1 Introduction	240
	7.3.2 Computational procedure	241
	7.3.3 Molecular orbital calculations using model clusters	242
	7.3.4 FLAPW band-structure calculations for $LiMO_2$ and MO_2	245
	7.3.5 Conclusion	248
	7.4 New fluorides electrode materials for advanced lithium batteries	250
	7.4.1 Introduction	250
	7.4.2 PW-PP calculation for structural optimization	252
	7.4.3 FLAPW calculation for electronic structure and voltage	252
	7.4.4 Lattice parameters and relaxation by delithiation	254
	7.4.5 Electronic structure	254

		7.4.6	Average voltage	256
		7.4.7	Conclusion	256
	7.5	\multicolumn{2}{l}{First principles calculations of formation energies and electronic structures of defects in oxygen-deficient $LiMn_2O_4$}	258

- 7.5 First principles calculations of formation energies and electronic structures of defects in oxygen-deficient $LiMn_2O_4$ 258
  - 7.5.1 Introduction 258
  - 7.5.2 Computational procedure 259
  - 7.5.3 Defects of oxygen-vacancy type 260
  - 7.5.4 Defects of metal-interstitial type (I): simple interstitial atoms 262
  - 7.5.5 Defects of metal-interstitial type (II): with occupation of Mn at the 8a position 264
  - 7.5.6 Local electronic structures around defects 265
  - 7.5.7 Discussion 267
  - 7.5.8 Conclusion 268
- 7.6 Summary and conclusions 269

Index 273

List of Authors

Preface: **Tsutomu Minami**
Osaka Prefecture University

Chapter 1: **Tsutomu Minami**
Osaka Prefecture University
Masahiro Tatsumisago
Department of Applied Materials Science, Osaka Prefecture University

Chapter 2:
2.1 **Yoshio Nishi**
Materials Laboratories, Sony Corporation
2.2 **Tetsuo Nanno**
Technology Development Center, Matsushita Battery Industrial Co., Ltd.
Hiroshi Yoshizawa
Technology Development Center, Matsushita Battery Industrial Co., Ltd.
Munehisa Ikoma
Technology Development Center, Matsushita Battery Industrial Co., Ltd.
2.3 **Shin Fujitani**
Mobile Energy Company, R&D Business Unit, Sanyo Electric Co., Ltd.
Toshiyuki Nohma
Mobile Energy Company, R&D Business Unit, Sanyo Electric Co., Ltd.

Hisashi Tarui
Mobile Energy Company, R&D Business Unit, Sanyo Electric Co., Ltd.
Ikuo Yonezu
Mobile Energy Company, R&D Business Unit, Sanyo Electric Co., Ltd.

Chapter 3:
3.1, 3.4   **Masahiro Tatsumisago**
           Department of Applied Materials Science, Osaka Prefecture University
3.2.1, 3.2.3 **Masahiro Tatsumisago**
           Department of Applied Materials Science, Osaka Prefecture University
           Akitoshi Hayashi
           Department of Applied Materials Science, Osaka Prefecture University
3.2.2   **Nobuya Machida**
           Department of Chemistry, Konan University
3.2.4   **Junichi Kawamura**
           Institute of Multidisciplinary Research for Advanced Materials, Tohoku University
3.3.1, 3.3.2 **Atsunori Matsuda**
           Department of Materials Science, Toyohashi University of Technology
3.3.3   **Kiyoharu Tadanaga**
           Department of Applied Materials Science, Osaka Prefecture University

Chapter 4:
4.1   **Masataka Wakihara**
           Department of Applied Chemistry, Tokyo Institute of Technology
           Takeshi Miki
           Department of Applied Chemistry, Tokyo Institute of Technology
           Masanobu Nakayama
           Department of Applied Chemistry, Tokyo Institute of Technology

	Hiromasa Ikuta
	Department of Applied Chemistry, Tokyo Institute of Technology
4.2	**Yoshiharu Uchimoto**,
	Department of Applied Chemistry, Tokyo Institute of Technology
	Masataka Wakihara
	Department of Applied Chemistry, Tokyo Institute of Technology
Chapter 5:	**Chiaki Iwakura**
	Department of Applied Chemistry, Osaka Prefecture University
	Hiroshi Inoue
	Department of Applied Chemistry, Osaka Prefecture University
	Naoji Furukawa
	Department of Applied Chemistry, Osaka Prefecture University
	Shinji Nohara
	Department of Applied Chemistry, Osaka Prefecture University
	Hiroki Sakaguchi
	Department of Materials Science, Tottori University
	Masayuki Morita
	Department of Applied Chemistry and Chemical Engineering, Yamaguchi University
Chapter 6:	
6.1, 6.3, 6.7, 6.8	**Shinzo Kohjiya**
	Institute for Chemical Research, Kyoto University
6.2	**Shinzo Kohjiya**
	Institute for Chemical Research, Kyoto University
	Yuko Ikeda
	Faculty of Engineering and Design, Kyoto Institute of Technology
6.4, 6.6	**Yuko Ikeda**
	Faculty of Engineering and Design, Kyoto Institute of Technology
6.5	**Shinzo Kohjiya**
	Institute for Chemical Research, Kyoto University
	Kazunobu Senoo
	Institute for Chemical Research, Kyoto University

Chapter 7: **Yukinori Koyama**
Department of Materials Science and Engineering, Kyoto University
Isao Tanaka
Department of Materials Science and Engineering, Kyoto University

1

Introduction

Since the discovery of electricity, electrical properties of materials have supported our convenient everyday life and "electrical conduction" has been one of the most important properties of materials. There are two types of electrical conduction in materials: one is electronic conduction in which charge carriers are electrons or holes and the other is ionic conduction in which charge carriers are ions. The field "electronics" which is usually based on the devices using electronic conduction has extensively developed since the "transistor" was invented. In other words, the big innovation was carried out by the appearance of the all-solid-state device of transistor in the field of electronics. On the other hand, liquid electrolytes are still used in the field "ionics." If the liquid electrolytes can be replaced by the solid electrolytes, we can obtain all-solid-state ionic devices, which might realize a big innovation also in the field "ionics." All-solid-state batteries, solid fuel cells, all-solid capacitors are strongly required as advanced energy sources because of their high safety and high reliability. In particular, microionic devices like thin-film batteries and power ionic devices like large-scaled batteries for vehicles are most promising field of all-solid-state ionics.

Batteries are known as the most important ionic devices. In recent years, the need for portable power sources has accelerated due to the miniaturization of electronic appliances where in some cases the battery system is as much as half the weight and volume of the powered devices. From another point of view it has been possible to operate some small electric devices even by such small power sources like thin film batteries because low power devices have been used due to the recent advancement of semiconductor technology. The rapid developments of micro-electric devices such as wearable computers, radio frequency integrated circuits (RF-IC) tags, and micro-machines also strongly demand micro-power sources. The development of thin film batteries should be based on the solid state ionics and the all-solid-state battery technology. On the other hand, large-scaled batteries with no safety problems have also been strongly desired for utilization in automobiles. Not only electric vehicles and hybrid electric vehicles, but also fuel cell vehicles demand rechargeable power sources with high safety and high reliability. All-solid-state rechargeable batteries using nonflammable solid electrolytes with single ion conduction are one of the ultimate goals of vehicle-use batteries.

This book start with overviews of recent development of advanced batteries and prospects of this field in relation with all-solid-state systems from a standpoint of industries. Chapter 2 was written by three industry researchers who are active in the front line in SONY, Matsushita Battery, and Sanyo Electric. Those three are the leading companies representing the Japanese battery industry.

Chapter 3 describes recent development of amorphous solid electrolytes and their application to all-solid-state batteries. As already mentioned, solid electrolytes with high ionic conductivity and high performance have been desired for all-solid-state ionic devices with high safety and high reliability. Many efforts were devoted to the preparation of solid electrolytes made of various materials, *e.g.* single crystals, ceramics, glasses, organic polymers, and so on. In this chapter, we concentrate on the amorphous-based inorganic materials, glass, glass-ceramics, and inorganic gels which have several advantages for battery use. We also focus on the lithium ion conductors and proton conductors from a viewpoint of charge carriers because those carrier ions are most important in attractive application like lithium ion batteries, metal hydride batteries, and fuel cells.

Electrode materials are of course most important in battery materials. Recent development of electrode materials in lithium ion secondary batteries is described in Chapter 4. Lithium ion secondary batteries have been rapidly popularized as a power source for portable electric devices such as cellular phones, notebook-type personal computers and video cameras, because these batteries have superior properties like high voltage, high energy density and light weight. However, those battery systems usually use flammable organic electrolytes, so that all solidification is most desired in the systems. In this Chapter two topics are included: advanced electrode materials for lithium ion batteries like modified spinel compounds and electrode/electrolyte interface in lithium ion batteries.

Metal hydride batteries are also very important commercialized secondary batteries. They are environment-friendly and have excellent performance as shown in the lithium ion batteries. Since the battery systems are basically safe, they are used in commercially available hybrid electric vehicles. However, the systems are also desired to be solidified because the aqueous electrolyte solution used in the systems can freeze in cold climates and dry out on the way of charge-discharge cycles. Chapter 5 describes the construction of solid / solid interface between hydrogen storage alloy electrode and solid electrolyte for battery application. The key materials for electrolytes in solid-state metal hydride batteries are polymer hydrogels, inorganic silica-based gels, and inorganic oxide polycrystals. Characterization of the interface between hydrogen storage alloy electrode and those key materials and application of those materials to all-solid-state metal hydride batteries are reported in this Chapter.

Chapter 6 describes the polymer solid electrolytes for batteries. There are big gaps between the present commercialized battery system and the future all-solid-state battery system because the solid electrode / solid electrolyte contact in the latter is essentially different from the solid electrode / liquid electrolyte contact in the former. The smooth solid / solid interface is basically very difficult to be realized. In such a standpoint, flexible materials like organic polymers can play an important role in producing an excellent interface between solid electrode and solid electrolyte. In this Chapter, various elastomers such as elastomeric branched poly(ethylene oxide) and ionene elastomers are reported as the polymer electrolytes.

Theoretical studies for batteries are introduced in the last Chapter, which describes the first principles calculations of battery materials. The importance of theoretical treatment on the basis of the first principles calculations has been growing larger and larger in the field of materials science. Chapter 7 introduces the success story in application of the calculations to the electrode and electrolyte materials for lithium battery systems.

2
Recent progress in batteries and future problems

2.1 Recent technologies of secondary batteries in Sony Corporation

2.1.1 Introduction

In 1990 Sony succeeded in the development of a new kind of electrochemical cells with non-aqueous electrolytes, lithiated carbon anodes and $LiCoO_2$ cathodes. They were named lithium ion secondary batteries (LIBs) and were introduced into the market for the first time in 1991 by Sony [1].

LIBs have outstanding properties in comparison with conventional secondary batteries such as nickel / cadmium, nickel / metal hydride and lead-acid batteries. Several among them are:

1) High operating voltage,
2) High energy density (both gravimetric and volumetric),
3) No memory effect,
4) Low self-discharge rate (less than 10% per month),
5) Operation over a wide temperature range.

Much effort has been done incessantly to improve further the performances of LIBs, which boost LIBs into the most suitable power sources for portable equipment like cellular phones and notebook computers.

The latest status of Sony's LIBs will be described in the pages that follow.

2.1.2 Lithium ion secondary battery

Electrolyte solutions based on organic solvents have been used in LIBs. Due to the lower ionic conductivity of non-aqueous electrolytes than aqueous ones, it was misunderstood at the start of LIBs manufacturing (in the first half of 1990's) that LIBs could not have excellent performances, especially regarding to low temperature characteristics and drain capability.

Table 2.1.1. Performances of LIB

Size (DxWxH)	18.3 X 65.1 mm
Weight	44.1 g
Capacity	2400mAh (0.2C, 3.78V)
Average Voltage	3.7V
Charge Voltage	4.2V
Charging Time	2.5hrs
Vol. Energy Density	530 Wh/dm^3
Gr. Energy Density	205 Wh/kg
Cyclability	80% @ 300th cycle
Temperature Range of Operation	Charge 0°C~45°C Discharge -20°C~60°C Storage -20°C~45°C
Cathode	LiCoO$_2$
Anode	Graphite

We have succeeded, however, in the improvement of electrolytes, electrode active compounds and other cell materials as well as cell structures, and according to the up-to-date data, LIBs have sufficient power even at low temperatures and can be discharged at significantly high load current. Table 2.1.1 shows the performances of the latest model of so-called 18650 cylindrical cells.

LIBs of the first generation, which was introduced in 1991 into the market, had volumetric and gravimetric energy densities of only 200 Wh/dm^3 and 80 Wh/kg respectively. It can be recognized from Table 2.1.1 that the latest LIB has very high volumetric and gravimetric energy densities in comparison with the 1991 model. They are almost 2.5 times larger than those of the old model.

The first generation LIBs were applied to cellular phones and 2 cells were required for each phone due to poor output power of LIBs of those days. Moreover, it was very difficult to use LIBs of the first generation under the cold circumstances like ski runs. As described above, the effort to improve the performances has enabled LIBs to be used in various conditions and at a large discharge current. Drain capability of the recent LIB models is illustrated in Fig.2.1.1 and the temperature dependence of discharge capacity is shown in Fig.2.1.2.

It can be recognized from Fig.2.1.2 that capacity retention at −20°C is greater than 80% of the capacity at room temperature, and drain capability is almost constant within the discharge current measured. Thus, LIB performances at low temperatures and the drain capability are by no means inferior to aqueous electrolyte cells including nickel-cadmium and nickel-metal hydride secondary cells. At present, specially designed LIBs can be applied even to equipment of high power consumption like power tools.

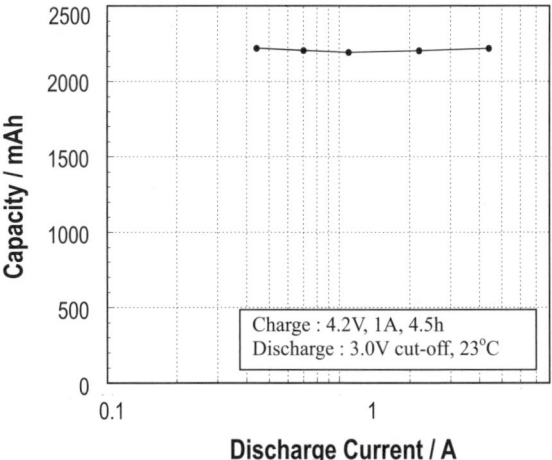

Fig. 2.1.1. Drain Capability of LIB

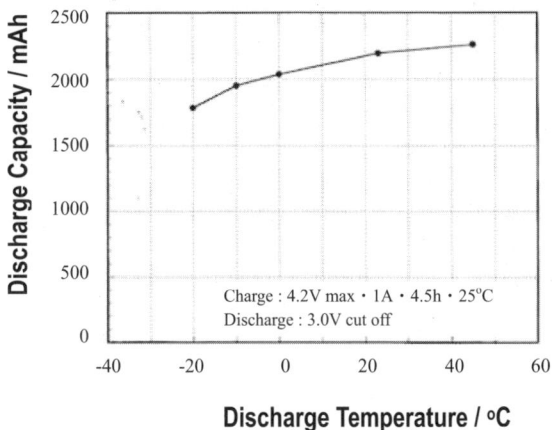

Fig. 2.1.2. Temperature dependence of discharge capacity of LIB

2.1.3 Lithium polymer battery

Electrolyte leakage has been one of the most annoying problems in primary and secondary cells regardless of aqueous or non-aqueous electrolytes. The immobilization of electrolyte has been considered effective against the leakage.

Genuine polymer electrolytes, which consist of matrix polymers and Li salts, have only low ionic conductivity, and gelled polymer electrolytes (GPE*s*) have been investigated intensively by many researchers to improve conductivity. GPE*s* are composed of polymer matrices and solvents or plasticizers and they have so high ionic conductivity that they can be used as solidified electrolytes in place of genuine polymer electrolytes.

Sony developed lithium polymer batteries (LPB*s*), namely lithium ion secondary batteries with polymer gel electrolytes and introduced them into the market in 1998.

It has been widely said that the drain capability and low temperature performances of LPB*s* compare unfavorably with LIB*s* due to the poorer ionic conductivity of GPE*s* than liquid electrolytes and that they cannot have sufficient performances as power sources for portable electronic equipments.

Keeping these anxieties in mind, we fixed our concepts and targets as follows when we started our developmental program of LPB*s*:

1) Low vapor pressure of solvents from gelled electrolytes,
2) Good adhesiveness of gelled electrolytes to active cathode and anode materials,
3) Adoption of aluminum-laminated polymer film as a casing for a cell,
4) Anti-leakage properties,
5) Comparable performances to conventional LIB*s*.

It is well known that there are 2 kinds of GPE*s*, namely heterogeneous and homogeneous ones. In case of the former, micropores of matrix polymers are filled with electrolyte solutions. In the latter, on the other hand, electrolyte solutions and matrix polymers are mixed completely to form the homogeneous phase. We adopt the latter type because it has high ability to hold solvents tightly within the polymer matrices and no free solvent is present, resulting in the leakage-proof characteristics of LPB*s*.

GPE*s* need to have 2 incompatible properties. They are required to have ability to hold as much solvent as possible in order to increase ionic conductivity. Furthermore, high mechanical strength is also essential. Too much solvent, however, causes fragile gels.

We have developed gel electrolytes which consist of block copolymer of polyvinylidene fluoride and polyhexafluoropropylene as matrix resin with ethylene carbonate (EC) / propylene carbonate (PC) / $LiPF_6$ as electrolyte solutions and it has been proven that they have sufficiently high ionic conductivity and high mechanical strength to realize those concepts listed above [2].

It is widely recognized that PC decomposes spontaneously during charge of LIB*s* (i.e. during lithium insertion into graphite anodes). Consequently, EC based electrolytes are commonly utilized in LIB*s* with graphite anodes. PC based electrolytes, however, show higher ionic conductivity than EC based ones and thus it is favorable to use PC as plasticizers to raise conductivity of resulting GPE*s*.

In our LPB*s*, surfaces of graphite particles are modified with amorphous carbonaceous materials to avoid PC decomposition and our GPE*s* have comparable ionic conductivity to liquid non-aqueous electrolytes and also exhibit excellent compatibility with graphite anodes.

As shown in Table 2.1.2, Sony's LPB*s* have almost equal performances with LIB*s*. Please note that the size of LPB*s* in Table 2.1.2 is much smaller than LIB*s* in

Table 2.1.2. Performances of LPB

Size (DXWXH)	3.8 x 35 x 62 mm
Weight	16 g
Capacity	900mAh
Average Voltage	3.75 V
Charge Voltage	4.2 V
Charging Time	150 min.
Vol. Energy Density	410 Wh/dm^3
Gr. Energy Density	210 Wh/kg
Cyclability	85% @ 1000th Cycle
Temp. Range of Operation	-20°C ~ 60°C
Cathode	$LiCoO_2$
Anode	Graphite

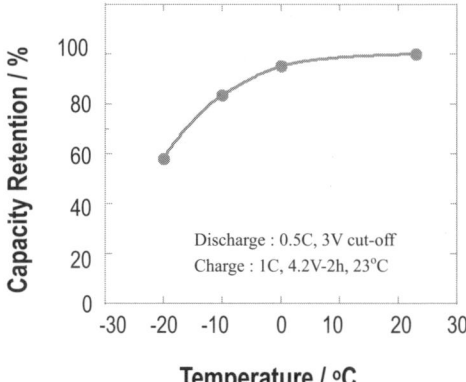

Fig. 2.1.3. Dependence of capacity retention of LPB on discharge temperature

Table 2.1.1 (volume is almost half) and the gravimetric energy density is higher than LIB*s* in spite of this handicap. Light weight is achieved by the adoption of light-weight aluminum-laminated polymer film as a casing material instead of a heavy metal can.

Dependence of capacity retention of LPB*s* on temperature is shown in Fig.2.1.3. The capacity retention of 60% (vs. capacity at room temperature) at −20°C is obtained and this value is sufficient for the outdoor use in winter, though it is somewhat smaller than that of LIB*s*.

In some countries, especially in Europe, cellular phones are operated by so-called GSM (Global System for Mobile Communications) mode. The power consumption for GSM is alternate between 2 A for 6 ms and 0.15 A for 4 ms and this value is larger than the power for other systems like NTT DoCoMo, Japan. This pulse

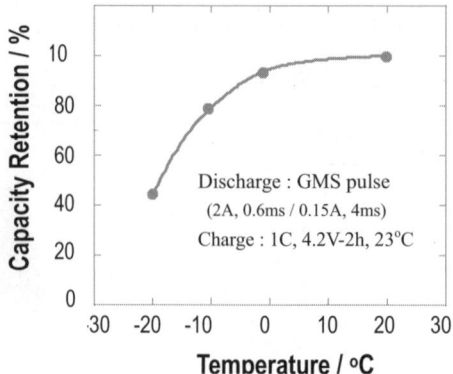

Fig. 2.1.4. Dependence of capacity retention of LPB on temperature in case of GSM pulse discharge

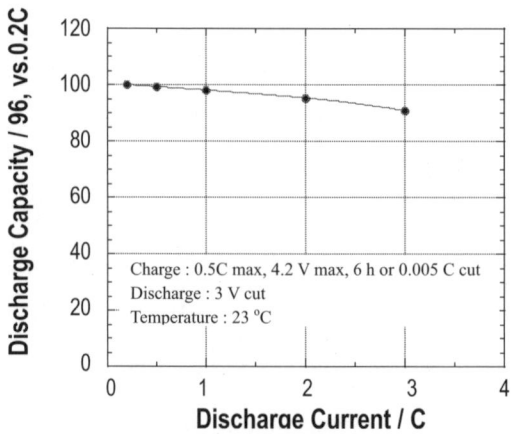

Fig. 2.1.5. Drain capability of LPB

discharge mode of cellular phones requires excellent drain capability to batteries even at low temperatures up to −20°C. The effect of ambient temperatures on the capacity retention of Sony's LPBs in case of GSM pulse discharge mode is shown in Fig.2.1.4, and it can be seen from this figure that the retention is satisfactory even at −20°C.

Drain capability is shown in Fig.2.1.5 and the discharge capacity retention (capacity at 0.2 C rate = 100%) is plotted against the discharge rate and the retention of 90% is obtained even at the 3 C rate, being comparable to LIBs.

Figure 2.1.6 shows the cycle performances of LPBs. Cells are charged with the constant current / constant voltage method (1 C mA and 4.2 V) following discharge at a constant current of 1 C mA to the cut-off voltage of 3.0 V. The capacity fade during this cycling course is shown in this figure. The performance is excellent, and

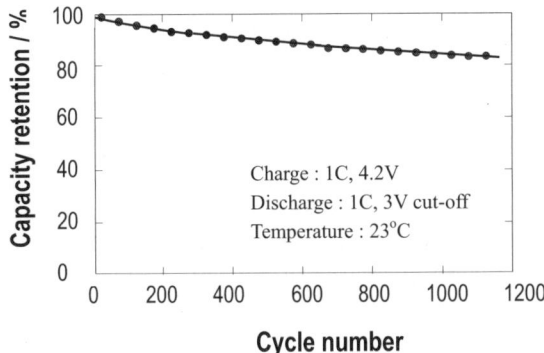

Fig. 2.1.6. Cycle Performance of LPB

the capacity retention at 1000th cycle is around 85%. This value is much better than that of LIB*s* (cf. Table 2.1.1).

2.1.4 Future of LIB and LPB

LIB chemistry consists of $LiCoO_2$ (or $LiNiO_2$ and $LiMn_2O_4$) cathodes and graphite (or amorphous carbon) anodes at present. The energy density of LIB (and LPB) with this chemistry is believed to be nearing its technological limits.

It is essential to develop new active electrode materials which have potentially higher energy densities. Metals or metal alloys including Sn and Si might be candidates of anode materials. Operation of cells at higher voltages around 4.5 V is assumed also effective to raise energy densities.

Lithium metal anode has high energy density and it has been being mentioned as the ultimate anode for lithium secondary cells. Dendritic formation during charge / discharge cycles is the most difficult hurdle to be overcome. Utilization of genuine polymer electrolytes could be an answer to this issue. We have been investigating dimethylsiloxane polymers with ethylene oxide side chains as solvent-free electrolytes. Ionic conductivity of our genuine polymer electrolytes has reached to 0.2 mScm^{-1} at 25°C and the mechanical strength is so tough that they can serve as separators as well. Further extensive effort, however, is needed before cells with lithium metal anodes are put to practical use.

References

1. Y. Nishi, *Chem. Rec.*, **1** (2001) 406.
2. Y. Nishi, *Advances in Lithium-Ion Batteries* ed. by W. A. van Schalkwijik and B. Scrosati, Kluwer Academic / Plenum Publishers, New York, NY, p.233 (2002).

2.2 Recent technologies on materials for advanced lithium rechargeable batteries [Panasonic]

2.2.1 Introduction

Lithium ion batteries were introduced to the consumer market in 1991. Although nickel metal hydride batteries were used as popular secondary batteries of high capacity at that time, the advantage of lightweight of lithium ion batteries was widely accepted and the lithium ion batteries have been applied to cellular phones and laptop personal computers. As an active material for positive electrodes, $LiCoO_2$ was used [1–7], and carbon was mainly adopted for negative electrodes. The evolution in energy density of various secondary batteries is shown in Fig.2.2.1. The energy density of lithium ion batteries has been improved twice as much in the last decade. Because of this improvement in energy density, lithium ion batteries are now used most extensively as the main power supply for mobile equipment such as cellular phones and laptop PCs. Despite this marked improvement in energy density, the chemistry adopted in the batteries has no change since their introduction to the market; $LiCoO_2$ and carbon are used in positive and negative electrodes, respectively, even today. Here, we will report some well-known and promising works out of numbers of research for advanced lithium ion batteries from the viewpoint of material.

2.2.2 Active materials for positive electrodes

As for active materials for positive electrodes, $LiNiO_2$ was evaluated as an alternative candidate of $LiCoO_2$ because $LiNiO_2$ was expected having a high capacity. The

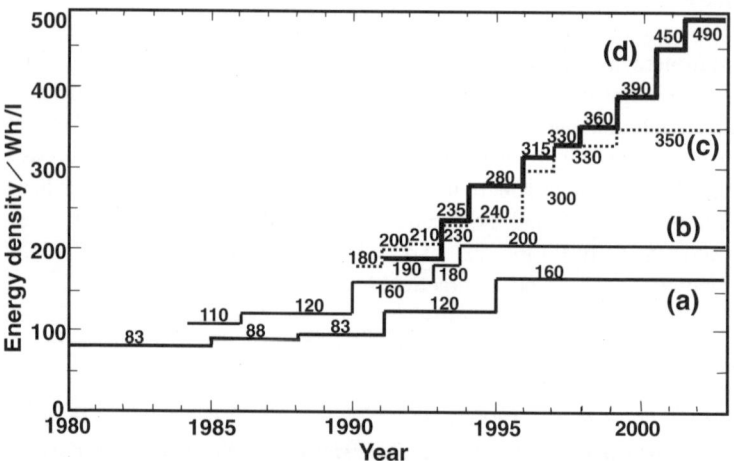

Fig. 2.2.1. Change in energy density of Panasonic rechargeable batteries ('80–'03) The energy density of Li-ion improved the twice of ten years ago. (a) Sintered electrode Ni-Cd battery, (b) SME electrode Ni-Cd battery, (c) Ni-MH battery, and (d) Li-ion battery.

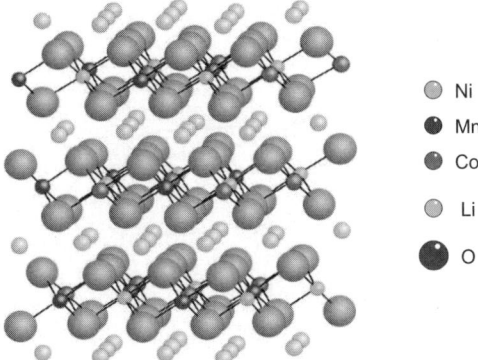

Fig. 2.2.2. Schematic illustration on the in-plane super-lattice $LiNi_{1/3}Mn_{1/3}Co_{1/3}O_2$ for positive active material of advanced lithium ion batteries.

drawbacks of $LiNiO_2$ were short cycle life and safety problem originating in its unstable crystal structure. To overcome these drawbacks, substitutions of nickel to various elements were examined. It was reported that addition of cobalt suppressed the crystal structure change during charge and discharge, and that safety was improved by aluminum addition. More improvements are, however, necessary for $LiNiO_2$ to displace $LiCoO_2$. Recently, a new material was found in which equimolar of nickel, manganese and cobalt was arranged in the transition metal layers [8–10]. This material, $LiNi_{1/3}Mn_{1/3}Co_{1/3}O_2$, was a layered oxide with in-plane superlattice structure, as shown in Fig.2.2.2. The crystal structure was stabilized with such arrangement of nickel, manganese and cobalt. The material was able to retain its structure during charge and discharge as well as to improve safety problems.

A discharge characteristic of $LiNi_{1/3}Mn_{1/3}Co_{1/3}O_2$ is shown in Fig.2.2.3 in comparison with $LiCoO_2$. While $LiCoO_2$ has a discharge capacity of 140 mAhg^{-1}, $LiNi_{1/3}Mn_{1/3}Co_{1/3}O_2$ is expected to have 170 mAhg^{-1}. This highly stabilized crystal structure enabled $LiNi_{1/3}Mn_{1/3}Co_{1/3}O_2$ to be charged up to 4.45 V, at which $LiCoO_2$ was very unstable.

2.2.3 Active materials for negative electrodes

Active materials such as Si and Sn for negative electrodes that can store large quantity of lithium are attracting attention as candidate materials for the next generation. These materials accompany large volumetric changes during charge and discharge. The large volumetric changes and stresses associated with result in cracking and crumbling of the materials, which cut current flow path between active materials and current collectors. These materials were developed aiming at retainment of current collectivity by high electric conductive matrix, and at suppression of cracking and crumbling of the material by adoption of amorphous silicon.

Figure 2.2.4 shows discharge characteristics of Si alloy and graphite. More than twice discharge capacity was obtained with Si alloy materials comparing with

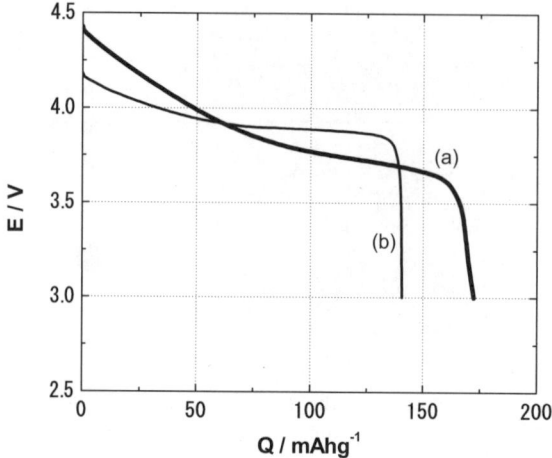

Fig. 2.2.3. Discharge curves of (a) a Li / LiNi$_{1/3}$Mn$_{1/3}$Co$_{1/3}$O$_2$ cell operated in voltage of 3.0–4.45 V and (b) a Li / LiCoO$_2$ cell operated in voltage of 3.0–4.2 V.

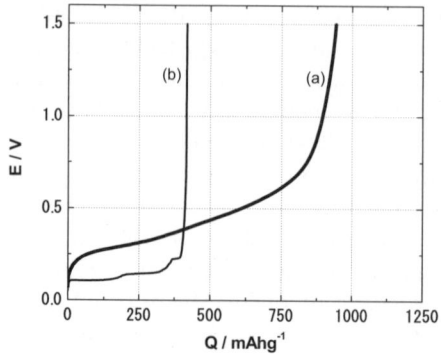

Fig. 2.2.4. Discharge curves of (a) a Li / Si alloy cell and (b) a Li / Graphite cell operated in voltage of 0.0–1.5 V.

conventional graphite. Charge-discharge cycle life of this material is considerably better than that of simple silicon powders, though it requires more improvement with respect to conventional graphite anode.

2.2.4 Electrolytes

The relationship between lithium ion conductivity and temperature of various electrolytes is illustrated in Fig.2.2.5. A conventional organic liquid electrolyte shows lithium ion conductivity of 10^{-2} Scm^{-1} at room temperature, which is one order lower than an alkaline aqueous electrolyte. Gel electrolytes with organic liquid electrolytes impregnated into host polymers are reported to have the conductivity almost as high as conventional organic liquid electrolytes. However, dry polymer

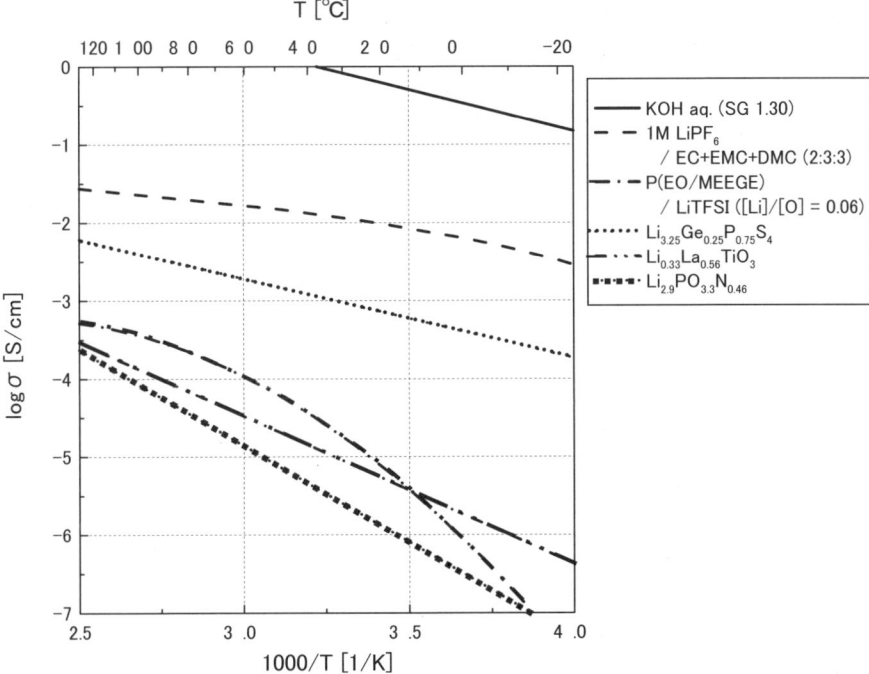

Fig. 2.2.5. The relationship between lithium ion conductivity and temperature of various electrolytes.

electrolytes such as PEO (polyethylene oxide) in which lithium salts are dissolved show only low ionic conductivity around 10^{-7} Scm^{-1} at room temperature. The conductivity of dry polymers is being improved owing to many dedicated works by lowering the glass transition temperature with incorporation of branched structure to the polymers. As for inorganic solid electrolytes, some were reported to show the ionic conductivity of 10^{-3} Scm^{-1}. The typical activation energy of lithium ion conduction is in the range of 0.2 to 0.6 eV.

Properties of some representative solid electrolytes are shown in Table 2.2.1. We classify solid electrolytes as in Table 2.2.1 according to some remarks.

• Gel electrolytes [12]

Gel electrolytes are made by impregnation of organic liquid electrolytes consisting of a solvent mixture, such as ethylene carbonate and propylene carbonate, and lithium salts such as LiPF$_6$, into host polymers such as polyacrylonitrile, poly(vinylidene fluoride) [13] and polyether. The activation energy of lithium ion conduction is around 0.2 eV, which is comparable to that of organic liquid electrolytes. The ionic conductivity depends on the weight ratio of host polymer/organic liquid electrolyte. To increase ionic conductivity, more liquid electrolytes are needed but giving rise to a

2 Recent progress in batteries and future problems

Table 2.2.1. Comparison of Representative Solid Electrolytes

	Organic Material		Inorganic Material				
	Gel	Dry Polymer	Sulfide		Oxide		
			Crystalline	Amorphous	Crystalline		Amorphous
Representative Solid Electrolyte	PVDF-12%HFP /60vol%(1M LiPF$_6$ EC/DMC)[13]	P(EO/MEEGE) /LiTFSI ([O]/[Li] = 0.06)[20]	Li$_{3.25}$Ge$_{0.25}$P$_{0.75}$S$_4$[19]	Li$_3$PO$_4$-63Li$_2$S-36S iS$_2$[20]	Li$_{0.33}$La$_{0.56}$TiO$_3$[21]	Li$_{1.3}$Al$_{0.3}$Ti$_{1.7}$(PO$_4$)$_3$[22]	Li$_{2.9}$PO$_{3.3}$N$_{0.46}$[23]
Li$^+$ Conductivity at 25 °C [S/cm]	3 x 10^{-3}	1.1 x 10^{-5}	2.2 x 10^{-3}	10^{-3}	1.3 x 10^{-3} (bulk) 2 x 10^{-5} (total)	2 x 10^{-3} (bulk) 7 x 10^{-4} (total)	2.3 x 10^{-6}
Activation Energy of Li$^+$Conduction [eV]			0.21	0.29	0.33	0.38	0.55
Transference Number	0.04 (at 60 °C)	1	1	1	1	1	
Electric Conductivity [S/cm]			1.5 x 10^{-7}		5 x 10^{-10}		< 10^{-14}
Note			H$_2$S is generated by the reaction with H$_2$O	H$_2$S is generated by the reaction with H$_2$O			PH$_3$ is generated by the reaction with H$_2$O[24]

safety problem. The rate capability of battery and the safety problem are in trade-off relationship even in the gel electrolytes system.

• **Dry polymer electrolytes** [12]

Dry polymer electrolytes are prepared by dissolution of lithium salts to polyether that usually contain branched structure induced by the addition of siloxanes, acrylic monomers and polyols [14]. The thermal property of lithium ion conductivity follows Vogel-Tamman-Fulcher (VTF) equation rather than a simple Arrhenius function.

$$\sigma = \sigma_0 \exp[-B/(T - T_0)]$$

For this reason, the ionic conductivity of dry polymer electrolytes drops steeply in the low temperature range as seen in Fig.2.2.5. To increase ionic conductivity, the shifting of glass transition point to lower temperature is required, which renders the mechanical strength of polymer membrane decline. The charge-discharge performance of battery and the reliability are in trade-off relationship in this system. Besides, the PEO system begins to be oxidized at 3.9 V versus Li/Li$^+$, that makes this system incompatible with 4 V cathode materials such as LiCoO$_2$ and LiMn$_2$O$_4$[15, 16].

• **Sulfide solid electrolytes (amorphous and crystalline)**

Amorphous and crystalline sulfide solid electrolytes were reported to have the ionic conductivity of 10^{-3} Scm^{-1} [17–19]. The outstanding feature of this system is the low grain-boundary resistance, so that the mechanically molded pellet of a sulfide electrolyte shows high ionic conductivity without further treatment such as sintering. An active material, a sulfide electrolyte and electric conduction additives are mixed to be used as an electrode material. A cell is fabricated by molding a pellet consisting of a positive electrode layer, a solid electrolyte layer and a negative electrode layer. It was reported that a discharge current density of 1 mAcm^{-2} was attained with a LiCoO$_2$/carbon cell that contained two solid electrolytes [20]. A Li$_2$S-GeS$_2$-P$_2$S$_5$ crystalline material that was stable against oxidation was used for the ionic conductor of the positive electrode while LiI-Li$_2$S-P$_2$S$_5$ glass that was stable against reduction was used for the negative electrode. Due to the high reactivity with water, sulfide solid electrolytes are to be handled with care.

- **Crystalline oxide solid electrolytes**

An oxide with perovskite structure, Li$_{3x}$La$_{2/3x}$TiO$_3$, is well known which had bulk ionic conductivity of 10^{-3} Scm^{-1}. Due to high grain boundary resistance, however, the total conductivity of the sintered pellet is as low as 10^{-5} Scm^{-1} [21]. A phosphate with NASICON structure, Li$_{1.3}$Al$_{0.3}$Ti$_{1.7}$(PO$_4$)$_3$, shows the total ionic conductivity of 10^{-4} Scm^{-1} [22]. Since both materials contain Ti(IV) as a main element, they are easily reduced around 1.8 V versus Li/Li$^+$ for Li$_{3x}$La$_{2/3x}$TiO$_3$[23] and around 2.5 V versus Li/Li$^+$ for Li$_{1.3}$Al$_{0.3}$Ti$_{1.7}$(PO$_4$)$_3$[24], respectively. It was reported that a cell Li$_4$Ti$_5$O$_{12}$|Li$_{1.3}$Al$_{0.3}$Ti$_{1.7}$(PO$_4$)$_3$|LiMn$_2$O$_4$ was constructed by sintering process with LiBO$_2$-LiF as a sintering additive, which demonstrated charge-discharge cycles with 10 μAcm^{-2} successfully[24].

- **Amorphous oxide solid electrolyte**

In general, solid electrolytes of amorphous oxides show poor lithium ion conductivity and it is difficult to construct a cell like the one using sulfides solid electrolyte. Recently, a film designated as LIPON (Li$_{2.9}$PO$_{3.3}$N$_{0.46}$), made by sputtering of Li$_3$PO$_4$ target in nitrogen atmosphere, was reported to have a ionic conductivity of 3.3 × 10^{-6} Scm^{-1}, which was quite high as an amorphous oxide solid electrolyte [25]. As a LIPON film was made by vapor process, the thickness of the electrolyte layer could be reduced to several micrometers, which realized thin film battery of high rate capability. This thin film battery is expected to create a new market. High price due to the process and the packaging technique still remain to be solved.

2.2.5 Concluding remarks

Lithium nickel manganese oxides with or without cobalt have been regarded as promising candidate materials for advanced lithium ion batteries. Of these, LiNi$_{1/3}$Mn$_{1/3}$Co$_{1/3}$O$_2$ is quite interesting in terms of rate capability, cycle life and safety. High capacity materials for a negative electrode such as Si have a problem of deterioration of electrochemical performance caused by large volumetric changes in charge-discharge cycles. New approach such as nanotechnology is being applied to solve the problem. Solid electrolytes are expected to solve the safety problem associated with the use of flammable organic liquid electrolytes in conventional batteries. Breakthroughs are required in lithium ion conductivity and in the methodology to form good interface between electrolyte and active material.

References

1. K. Mizushima, P. C. Jones, P. J. Wiseman and J. B. Googenough, *Solid State Ionics*, **3–4** (1981) 171.
2. K. Mizushima, P. C. Jones, P. J. Wiseman and J. B. Goodenough, *Mater. Res. Bull.* **15** (1980) 783.
3. H. Tukamoto and A. R. West, *J. Electrochem. Soc.*, **144** (1997) 3316.
4. T. Ohzuku and A. Ueda, *J. Electrochem. Soc.*, **141** (1994) 2972.
5. S. Levasseur, M. Menetrier and C. Delmas, *J. Power Sources*, **112** (2002) 419.
6. J. N. Reimers and J. R. Dahn, *J. Electrochem. Soc.*, **139** (1992) 2091.
7. Y. Shao-Horn, S. Levasseur, F. Weil and C. Delmas, *J. Electrochem. Soc.*, **150** (2003) A366.
8. T. Ohzuku and Y. Makimura, *Chem. Lett.*, (2001) 744.
9. N. Yabuuchi and T. Ohzuku, *J. Power Sources*, **119–121** (2003) 171.
10. Y. Koyama, I. Tanaka, H. Adachi, Y. Makimura and T. Ohzuku. *J. Power Sources*, **119–121** (2003) 644.
11. K. M. Abraham and M. Alamgir, *J. Electrochem. Soc.*, **137** (1990) 1657.
12. P. G. Bruce, *Solid State Electrochemistry*, Cambridge University Press, Cambridge, p. 95 (1995).
13. J.-M. Tarascon, A. S. Gozdz, C. Schmutz, F. Shokoohi and P. C. Warren, *Solid State Ionics*, **86–88** (1996) 49.
14. M. Watanabe and A. Nishimoto, *Solid State Ionics*, **79** (1995) 306.
15. Y. Xia, , T. Fujieda, K. Tatsumi, P. P. Prosini and T. Sakai, *J. Power Sources*, **92** (2001) 234.
16. S. Matsui, T. Muranaga, H. Higobashi, S. Inoue and T. Sakai, *J. Power Sources*, **97–98** (2001) 772.
17. R. Mercier, J.-P.. Malugani, B. Fahys and G. Robert, *Solid State Ionics*, **5** (1981) 663.
18. N. Aotani, K. Iwamoto, K. Takada and S. Kondo, *Solid State Ionics*, **68** (1994) 35.
19. R. Kanno and M. Murayama, *J. Electrochem. Soc.*, **148** (2001) A742.
20. K. Takada, T. Inada, A. Kajiyama, H. Sasaki, S. Kondo, M. Watanabe, M. Murayama and R. Kanno, *Solid State Ionics*, **158** (2003) 269.
21. O. Bohnke, C. Bohnke and J. L. Fourquet, *Solid State Ionics*, **91** (1996) 21.
22. H. Aono, E. Sugimoto, Y. Sadaoka, N. Imanaka and G. Adachi, *J. Electrochem. Soc.*, **140** (1993) 1827.
23. C. H. Chen and K. Amine, *Solid State Ionics*, **144** (2001) 51.
24. P. Birke, F. Salam, S. Doering and W. Weppner, *Solid State Ionics*, **118** (1999) 149.
25. X. Yu, J. B. Bates, G. E. Jellison, Jr. and F. X. Hart, *J. Electrochem. Soc.*, **144** (1997) 524.

2.3 Development of secondary battery electrode materials toward high energy and power density [Sanyo]

Secondary chemical batteries used for driving mobile electronic devices like notebook-type personal computers (PCs) and cellular phones are one of the most significant energy sources with high power density. In this section, the prospect of the battery materials with high performance in connection with the evolution of the mobile electronic devices in the future is discussed; several examples of developing new electrode materials are mainly presented.

2.3.1 Requirements for high performance of the batteries from evolving functions of the mobile devices

Since the function of the mobile devices like PCs and cellular phones has been evolving with increasing the power consumption of the devices, the secondary batteries with higher performance as the main power source of the devices are more and more strongly demanded. The first priority for the function of the batteries should be placed on output power so as to cover the power consumption. Requirement for larger capacity, which allows the devices to work for longer time in one charge, and for longer cycle life is the second point.

The performance is shown in Fig.2.3.1 with power density on the side axis and energy density on the bottom axis in volumetric scale for nickel-cadmium, nickel-metal hydride (Ni-MH) and lithium ion batteries, which are used as main power sources for driving the mobile devices. This figure is derived from the specification data of each battery model on the market described in catalogues [1,2]. The power density is usually calculated on the basis of so-called C-rate method, in which 1C gives 1h for the time spent for discharging the nominal capacity, and accordingly, XC gives 1/X h. Different C-rates were used for each type of the batteries as almost guarantees the nominal capacity, i.e. 8C for nickel-cadmium batteries, 3C for nickel-metal hydride batteries, 2C for lithium ion batteries, and 0.2C for alkaline manganese batteries. The figure illustrates that the most significant feature of the secondary batteries exists not in energy density, but in power density. Alkaline manganese batteries with limitation of use up to around 0.2C are much inferior in power density to the secondary batteries with much more extended limitation so far as 2-8C, though the energy density is competitive to that of lithium ion batteries. This difference comes mainly from a difference in the internal structure. Alkaline manganese batteries with the inside-out structure [3] have much less reaction area than the secondary batteries with the wound structure [4]. Similar case is true for a small size direct methanol fuel cell (DMFC) [5,6] actively involved in the R&D of new power sources for the mobile devices. From this point of view, the low power density is one of the biggest issues for DMFC to be developed into practical use for the mobile devices.

A battery is actually selected for use in a device based on cycle life, temperature performance, *etc.*, as well as on the matching of the power consumption. Lithium ion batteries, however, which have the best balance between power and energy density

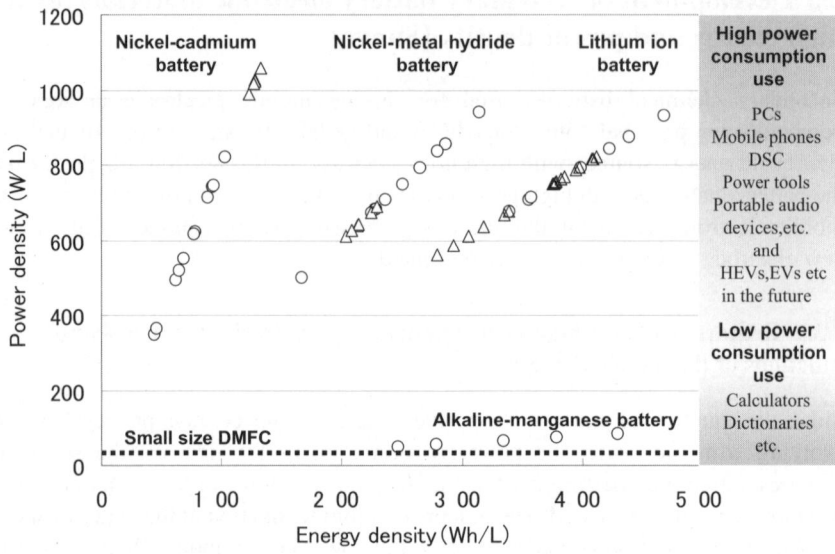

Fig. 2.3.1. Power density and energy density of chemical batteries and small size fuel cell in volumetric scale ; circle for cylindrical type, triangle for prismatic type.

in volumetric scale, will remain dominant in the market of the mobile devices. Nevertheless, some nickel-cadmium batteries still have such an advantage in the balance between cost and power density that they can remain in the market of power tools. Nickel-metal hydride batteries can be positioned in power, energy density and cost between lithium ion and nickel-cadmium batteries, and therefore, they suit digital still cameras (DSCs) and portable audio devices like compact disks (CDs), mini disks (MDs), *etc.* on the currently existing market, and are going to be employed for the power to drive hybrid electric vehicles (HEVs) in the near future.

Batteries must be developed according to requirements in performance specific to each device. High power and high energy density are the most important points for battery developments since the power consumption of mobile devices becomes larger and larger. In the following sections, the development of electrode materials are described for nickel metal-hydride and lithium ion batteries for high power or high energy density.

2.3.2 Approach to high power density of nickel-metal hydride batteries [7]

Recently, Ni-MH batteries become much important due to the market growth of DSCs and power tools since their performances are getting better in balance between energy density and power density. For the other applications, HEVs have been receiving much attention from both environmental and economical points of view, and almost all the commercialized HEVs employ Ni-MH batteries because of their better combination of output power, capacity, life, reliability and cost [8–15].

Ni-MH batteries thus need to match with a variety of applications, and in particular they are required to work in a wide temperature range by improving both high- and low-temperature characteristics. The discharge capacity of a Ni-MH battery charged at high temperature is usually lower than that charged at room temperatures [15–19], which can cause some problems particularly to the applications to HEVs and power tools since high temperature working conditions at high rate charge and discharge are more likely to happen in the batteries.

This is considered to be due to the oxygen evolution reaction or a side reaction (Eq. 2.3.2) of the charging reaction (Eq. 2.3.1) of the positive electrode at high temperature, either in using a sintered nickel electrode or in using a non-sintered nickel electrode:

Charge reaction

$$Ni(OH)_2 + OH^- \rightarrow NiOOH + H_2O + e^- \quad (2.3.1)$$

Side reaction of oxygen evolution

$$4OH^- \rightarrow O_2 + 2H_2O + 4e^- \quad (2.3.2)$$

In the past, the electrolyte composition [16] and co-precipitating elements into nickel hydroxide such as cobalt and calcium [17–21] were investigated in order to improve the high-temperature charge efficiency. Powder additives such as calcium compounds, barium compounds and rare-earth compounds were used for this purpose for a non-sintered nickel electrode [22–26]. However, it is difficult from the manufacturing point of view to apply such powder additives to a sintered nickel electrode which possesses better power characteristics than a non-sintered nickel electrode and is suitable for use in HEVs.

Described below is an example of improving the positive electrode material to better high temperature performance done from the above point of view through the modification of a sintered positive electrode by rare-earth compounds [27]. Our strategy here for improving high-temperature charge efficiency is illustrated in Fig.2.3.2. The sintered nickel electrode has a three-dimensional porous structure consisting of a sintered nickel body filled with nickel hydroxide as the active material. When the cell temperature is high, the side reaction in Eq (2.3.2) occurs, generating oxygen and water. Cobalt co-precipitation in nickel hydroxide is often used in order to improve the charge acceptance characteristics. Instead, we focused on the modification of the active material surface by coating the nickel positive electrode with hydroxides of other elements as this side reaction is considered to occur on the active material surface.

Yttrium from the rare-earth elements was selected as a coating material. Calcium from the alkaline earth metals and cobalt which is commonly used as an element to be co-precipitated in nickel hydroxide were also selected. After filling nickel hydroxide into a porous sintering body by using a chemical impregnation method, the sintering body with nickel hydroxide was immersed in an yttrium nitrate solution, calcium nitrate solution or cobalt nitrate solution of a specific gravity of 1.2. By immersion in a sodium hydroxide solution, the surface of sintered nickel

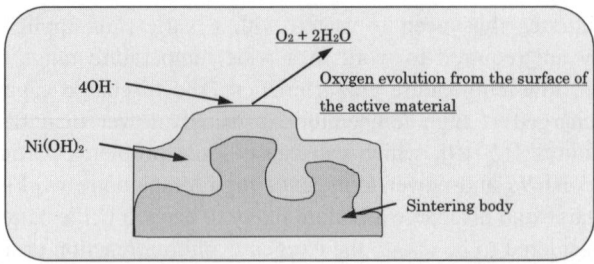

Fig. 2.3.2. Oxygen evolution reaction at the surface of the sintered positive nickel electrode.

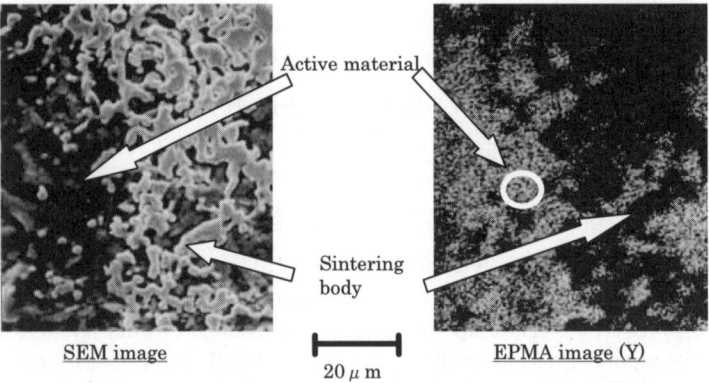

Fig. 2.3.3. Surface SEM and EPMA images of a nickel positive electrode coated with yttrium hydroxide.

positive electrodes was coated with yttrium hydroxide, calcium hydroxide and cobalt hydroxide, respectively. The amount of the coating materials is approximately 2 wt%. SEM and EPMA images of the surface of the nickel positive electrode coated with yttrium hydroxide are shown in Fig.2.3.3. The white dots in the EPMA image indicate the presence of yttrium, showing that the surface of the nickel hydroxide which is the active oxygen evolution site is coated with yttrium hydroxide.

Test cells using the coated positive electrode as the working electrode and a sintered cadmium electrode as the counter electrode were assembled in order to examine the high-temperature characteristics. A three-ingredient solution of potassium hydroxide, sodium hydroxide and lithium hydroxide was used as the electrolyte. The condition testing the high-temperature charge characteristics was as follows: (1) Charge the positive electrode up to 160% of the theoretical capacity for a one-electron reaction at 1/5C within an environmental temperature range from 25°C to 60°C. (2) Set the environmental temperature at 25°C. (3) Discharge the positive electrode up to a cell voltage of 0.8V at 1/3C.

The effect of hydroxide coating on the high-temperature charge efficiency of the test cells is shown in Fig.2.3.4 in terms of the discharge capacity ratio versus the charging temperature. The charge efficiency at high temperatures has been improved

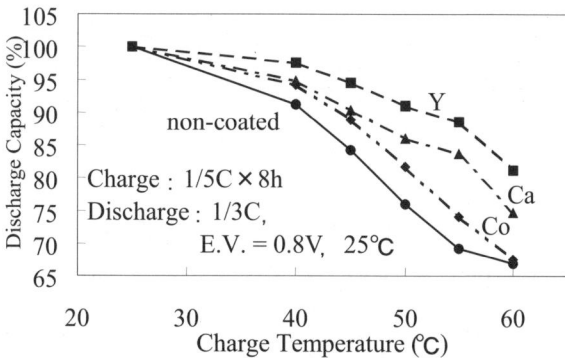

Fig. 2.3.4. Effect of hydroxide coating on high-temperature charge efficiency of test cells.

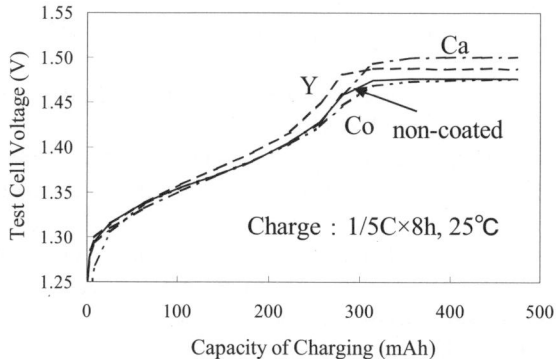

Fig. 2.3.5. Charge characteristics of test cells at 25°C.

for all the three hydroxides. In particular, the electrodes using yttrium hydroxide and calcium hydroxide show a large discharge capacity for charging even at 60 °C. The cobalt hydroxide shows the same characteristic as the non-coated electrode for charging at 60°C while it shows a superior capacity compared to the non-coated electrode up to 55°C.

In order to interpret the difference depending on the coating materials, we analyzed the voltage behavior during charging. Fig.2.3.5 shows the charge characteristics of the test cells at 25°C. The potential difference between oxygen evolution and nickel hydroxide oxidation is large when using the electrodes coated with yttrium hydroxide and calcium hydroxide compared to the non-coated electrode. Coating with yttrium hydroxide and calcium hydroxide suppresses the oxygen evolution reaction. The timing of the oxygen evolution reaction is delayed for the electrode coated with cobalt hydroxide although the oxygen evolution overvoltage does not change.

Figure 2.3.6 shows the charge characteristics of the test cells at 60°C. Among the three elements, yttrium hydroxide and calcium hydroxide suppress the oxygen

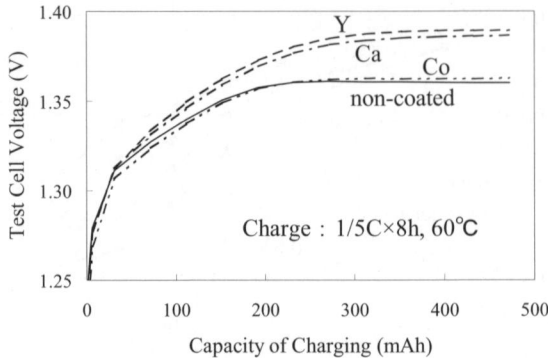

Fig. 2.3.6. Charge characteristics of test cells at 60°C.

evolution reaction more effectively, and high charge efficiency is achieved even at high temperatures.

In contrast to this, the electrode with cobalt hydroxide exhibits the same characteristic as the non-coated electrode because the drop in oxygen evolution potential is large. Thus, high-temperature characteristics of a sintered nickel positive electrode are greatly enhanced when coated with yttrium hydroxide and calcium hydroxide.

Cylindrical sealed-type AA-size cells using the coated nickel positive electrode, a hydrogen-absorbing alloy negative electrode and a three-ingredient electrolyte consisting of potassium hydroxide, sodium hydroxide and lithium hydroxide were also prepared. The composition of the hydrogen-absorbing alloy was $MmNi_{3.6}Co_{0.6}Al_{0.3}Mn_{0.5}$ (Mm: misch metal). The conditions testing the high-temperature charge characteristics were as follows: (1) Charge the cell up to 120% of the cell capacity at 1/2C at an environmental temperature ranging from 25°C to 60°C. (2) Set the environmental temperature at 25°C. (3) Discharge the cell up to a cell voltage of 1.0V at 1/2C.

The effect of the hydroxide coating on the high-temperature charge efficiency of AA cells is shown in Fig.2.3.7. Similarly to the results for the test cells, the AA cells using the coated positive electrodes show a large discharge capacity at the charge temperature above 40°C. In particular, the electrodes using yttrium hydroxide and calcium hydroxide give a large discharge capacity even when charged at 60°C compared to the non-coated electrode.

It is concluded hereby that the high-temperature charge efficiency of a sintered nickel positive electrode is greatly improved by coating the positive electrode with hydroxides including yttrium hydroxide, which leads to increasing the oxygen overvoltage at active oxygen evolution sites.

2.3.3 Approach to high energy density of lithium secondary batteries [28]

The requirement for raising their volume-base energy density is growing stronger, while the market of lithium ion batteries as a power source for portable electric

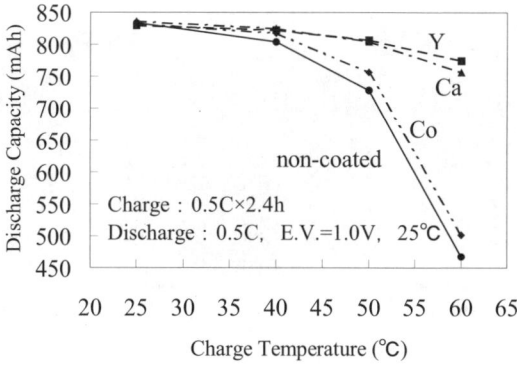

Fig. 2.3.7. Effect of hydroxide coating on high-temperature charge efficiency of AA cells.

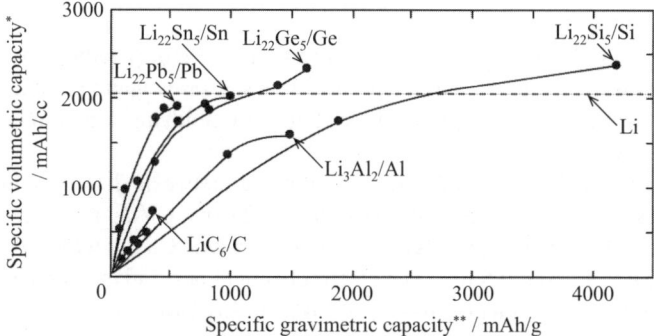

Fig. 2.3.8. Specific capacity of some lithium alloys theoretically calculated in a charged, or lithiated state (∗) and discharged, or non-lithiated state (∗∗). Based on the non-lithiated state, the theoretical gravimetric capacity of lithium is infinite (dashed line).

devices is steadily growing. Graphite is now widely used for the anode of commercially available lithium ion batteries, but the capacity has already approached the theoretical limit of C_6Li.

At this background, other elements like tin and silicon, which can be electrochemically alloyed with lithium, have been gathering a considerable attention from many researchers and summarized in some reviews[29,30], because they are expected to react with far more lithium than carbonaceous materials like graphite. Figure 2.3.8 shows the comparison of these elements in specific capacity, calculated on the stoichiometry for intermetallic compounds with Li in the binary alloy phase diagrams [31]. It should be noted that providing an anode from these chemistries is combined with a lithium compound like $LiCoO_2$ as cathode to construct a battery system like the currently commercialized lithium ion batteries, the non-lithiated state must be taken to calculate the gravimetric capacity while the lithiated state to calculate the volumetric capacity. A similar result is presented by Winter et al[29].

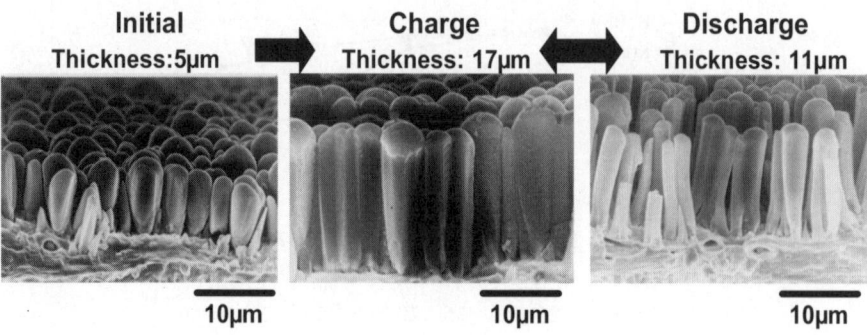

Fig. 2.3.9. Cross-sectional SEM image of a-Si thin film electrode at initial, charged and discharged state.

Since $Li_{22}Si_5$/Si anode system is the most attractive in specific capacity in both volumetric and gravimetric scales, we have focused investigation about the lithium alloy anodes on the $Li_{22}Si_5$/Si anode system by employing a new method[32,33] described hereafter.

In this method, amorphous Si thin films of between 2-10 μm in thickness were deposited to be a test electrode by a plasma CVD or a RF magnetron sputtering method on the current collector of electrolytic Cu foil with average surface roughness, Ra of about 0.5 μm. The electrochemical performance was examined by using a tri-polar test cell with 1 moll^{-1} of $LiPF_6$ in EC/DEC=3:7 electrolyte in which the Si thin film test electrode as a working electrode. The Si thin film electrode (5 μm) exhibited an initial discharge capacity of 3590 mAhg^{-1}[32]. The initial charge-discharge efficiency was 96% and almost equal to that of graphite. There was no capacity decay in 10 cycles.

The structure was observed by using SEM, TEM and Raman spectra analysis. a-Si films of about 5 μm as deposited uniformly covered the surface of copper along with the irregularly-shape. It is observed that the a-Si thin film was extended to 17 μm thickness after first charge, as shown in Fig.2.3.9, where a-Si thin films were fully charged. After the first discharge, the Si thin film was divided into micro columns. However, each column was not peeled off from the copper current collector. This is partly because the adhesion of the Si was enhanced by expanding the contact area for the roughened surface of the copper current collector and partly because an intermixed layer of active mass and underlying material was formed during a vacuum process. The stress relaxation mechanism in the self-organized micro-columnar structure is explained as follows. Microscopic anisotropy of film structure along with irregularities of the current collector surface causes the concentration of the stress from the volume expansion at the valley of irregularities followed by forming of the Si columns at the first charge-discharge. On the contrary, the stress is dispersed at the rest part. Thus, the whole active mass is kept adhered to the current collector.

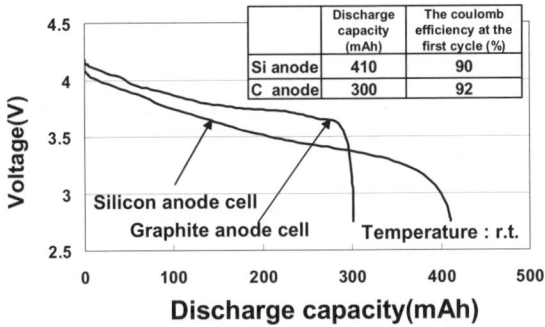

Fig. 2.3.10. Initial charge-discharge curves of test battery Charge and discharge: 1mA/cm², electrode potential: 0-2.0 (V vs Li/Li⁺).

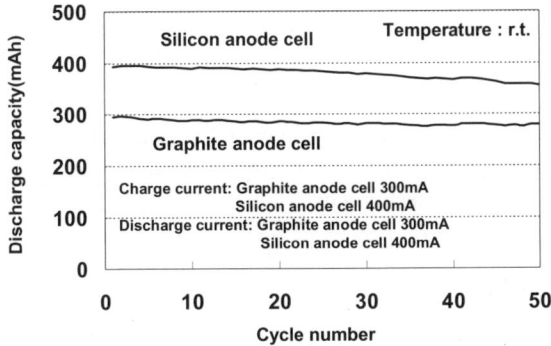

Fig. 2.3.11. Cycle characteristics of the cell using a LiCoO$_2$ cathode and the Si thin film anode.

A test cell of aluminum laminated film casing was fabricated with a LiCoO$_2$ cathode and the Si thin film anode. For comparison, a test cell with a graphite anode was also prepared. They were charged at a constant current (1C) of 400 mA for Si and 300 mA for graphite up to 4.2V followed by constant voltage charge at 4.2V down to 20mAh and then were discharged to 2.75V at a constant current (1C). Initial discharge curves of test cells with the Si anode and the graphite anode are shown in Fig.2.3.10. The volume energy density of the cell using the Si anode was about 1.3 times as large as that using graphite. Cycle performance of these cells is shown in Fig.2.3.11. After 50 cycles, the retained capacity using the Si anode was 92%.

The total variation in thickness of both test cells was almost the same as shown in Fig.2.3.12 in spite of the large volume change shown in Fig.2.3.9. This is attributed to the fact that the active mass thickness of the Si anode can be decreased drastically to around 17 μm at the maximum for the charged state, while that of the carbon anode is about 100 μm for the corresponding capacity per unit electrode area.

A large storage capacity and substantially 100% reversibility for almost fully charge-discharge condition of Li-Si alloy anode are available by employing a-Si thin film electrodes with self-organized micro-structure. The structure illustrated here

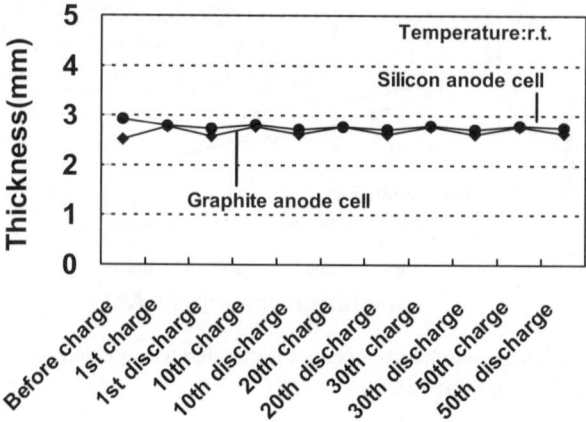

Fig. 2.3.12. The thckness variation of batteries with the thin-film Si and conventional graphite anode.

can overcome the large volume change inevitably incidental to the sorption and desorption of Li during cycling which has remained one of the biggest barriers to progress of Li alloy electrode technology.

2.3.4 Conclusions

In addition to the material improvement mentioned above, improvement of the electrode construction including addition of other materials should be taken into consideration to realize an excellent electrodes with higher electronic and ionic conductivity. Furthermore, if the load imposed on the batteries is heavier, the battery technologies should be requested to solve the safety and reliability problems which would be caused by side reactions other than the charge and discharge reactions.

The future development of secondary batteries with higher power and energy density depends on how to improve the electrochemical reactivity of the electrode materials with electrolytes.

It is convinced that the knowledge resulting from "ionics" field like nature of ion diffusion and chemical reactivity of ions is becoming more instrumental to the development of secondary battery in the future.

References

1. *Catalogs of Sanyo Batteries* (Nickel-Cadmium Batteries, Nickel-Metal Hydride Batteries and Lithium Ion Batteries), Sanyo Electric Co. Ltd. (2003).
2. http://www.tbcl.co.jp/primary/lr/tokusei/index.html.
3. D. Linden and T. B. Reddy ed., *Handbook of Batteries*, McGraw-Hill, New York, NY, (2001) p. 10.11.

2.3 Development of secondary battery electrode materials toward high energy and power density [Sanyo]   29

4. D. Linden and T. B. Reddy ed., *Handbook of Batteries*, McGraw-Hill, New York, NY, (2001) p. 28.4, p. 29.5, p. 35.32.
5. D. Linden and T. B. Reddy ed., *Handbook of Batteries*, McGraw-Hill, New York, NY, (2001) p. 42.4.
6. Y. Si, J.-C. Lin, H. R. Kunz and J. M. Fenton, *J. Electrochem. Soc.*, **151** (2004) A463.
7. K.Shinyama, Y.Magari, A.Funahashi, T.Nohma and I.Yonezu, *Electrochemistry*, **71**, No.8 (2003) 686.
8. I. Yonezu, *Second International Advanced Automotive battery Conference, Proceedings*, (2002) 26.
9. M. Verbrugge and E. Tate, *J. Power Sources*, **126** (2004) 236.
10. Yi-Fu Yang, *J. Power sources*, **75** (1998) 19.
11. A. Taniguchi, N. Fujioka, M. Ikoma and A. Ohta, *J. Power Sources*, **100** (2001) 117.
12. P. Gofford, J. Adams, D. Corrigan and S. Venkatesan, *J. Power Sources*, **80** (1999) 157.
13. T. Sakai, I. Uehara and H. Ishikawa, *J. Alloys Compd.*, **293-295** (1999) 762.
14. R. F. Nelson, *J. Power Sources*, **91** (2000) 2.
15. K. Shinyama, R. Maeda, Y. Matsuura, I. Yonezu and K. Nishio, *65th Annual Meeting of the Electrochemical Society of Japan, Abstr.*, (1998) 162.
16. M. Oshitani, M. Yamane and S. Hattori, *J. Power Sources*, **8** (1980) 471.
17. M. Oshitani, Y. Sasaki and K. Yakashima, *J. Power Sources*, **12** (1984) 219.
18. D. M. Constantin, E. M. Rus, L. Oniciu and L. Ghergari, *J. Power Sources*, **74** (1998) 188.
19. E. J. Casey, A. R. Dubois, P. E. Lake and W. Z. Moroz, *J. Electrochem. Soc.*, **112** (1965) 371.
20. M. E. Folquer, J. R. Vilche and A. J. Arvia, *J. Electroanal. Chem.*, **172** (1984) 235.
21. D. F. Pickett and J. T. Maloy, *J. Electrochem. Soc.*, **125** (1978) 1026.
22. H. Nakahara, H. Sasaki, T. Murata and M. Yamachi, *GS News Tech. Rep.*, **57 (1)** (1998) 26.
23. H. Matsuda and M. Ikoma, *Denki Kagaku* (presently *Electrochemistry*), **65** (1997) 96.
24. A. Yuan, S. Cheng, J. Zhang, and C. Cao, *J. Power Sources*, **76** (1998) 36.
25. C. Shaoan, Y. Anbao, L. Hong, Z. Jianqing and C. Chunan, *J. Power Sources*, **76** (1998) 215.
26. K. Ohta, K. Hayashi, H. Matsuda and Y. Toyoguchi, *186th Annual Meeting of the Electrochemical Society, Abstr.*, (1994) 98.
27. K. Shinyama, Y. Magari, A. Funahashi, T. Nohma and I. Yonezu, *Electrochemistry*, **71** (2003) 686.
28. H.Tarui, H.Yagi, K.Sayama, M.Fujimoto and S.Fujitani, *Abstract of the 43th Battery Symposium in Japan*, Kyusyu, Japan, (2002) 24.
29. M. Winter and J. O. Besenhard, *Electrochimica Acta*, **45** (1999) 31.
30. R. A. Huggins, *J. Power Sources*, **81-82** (1999) 13.

31. Desk Handbook: *Phase Diagrams for Binary Alloys*, ed. H. Okamoto, American Society for Materials, Metals Park, OH (2000).
32. H. Ikeda, M Fujimoto, S. Fujitani, Y. Domoto, H. Yagi, H. Tarui, N. Tamura, R. Ohshita, M. Kamino and I. Yonezu, *Abstract of the 42nd Battery Symposium in Japan*, Yokohama, Japan, (2001) 282.
33. T. Yoshida, T. Fujihara, H. Fujimoto, R. Ohshita, M. Kamino and S. Fujitani, *The 11th IMLB*, Monterey, CA USA, Abstract No.48 (2002).

3

Recent development of amorphous solid electrolytes and their application to solid-state batteries

3.1 Introduction

With recent popularization of various kinds of portable electronic devices and motor-drive vehicles, the importance of energy devices like secondary batteries, fuel cells, capacitors, and so on has become larger and larger. It has been widely understood that all-solid-state energy devices are most promising for improving safety and reliability of the devices. Solid electrolytes as a key material of all-solid-state energy devices have been extensively studied in the field of materials science, polymer science, electrochemistry, and solid state chemistry. Many efforts have been devoted to the preparation of solid electrolytes made of various materials, e.g. single crystals, ceramics, glasses, organic polymers, and so on. Most investigations have been carried out in the individual field of materials.

Amorphous or glassy materials have several advantages from a viewpoint of ion conduction in comparison with the crystalline ones: a wide range selection of compositions, isotropic properties, no grain boundaries, easy film formation, and so on. Because of the so-called open structure, the ionic conductivity of amorphous materials is generally higher than that of crystalline ones. In addition, single ion conduction can be realized because glassy materials belong to the so-called decoupled systems in which the mode of ion conduction relaxation is decoupled from the mode of structural relaxation. Amorphous or glassy materials are thus the most promising candidates of solid electrolytes with single ion conduction and high ionic conductivities.

A number of amorphous-based solid electrolytes from the idea that they are especially favorable for ion conductors have been developed using several unique preparation techniques. This Chapter reports lithium ion conductors of glasses and glass-ceramics for all-solid-state lithium secondary batteries prepared by melt-quenching and mechanical milling techniques. Also reported are proton conductors of inorganic gel-based materials mainly for fuel cell application prepared by the sol-gel process.

3.2 Lithium ion conductors

3.2.1 Solid electrolytes

3.2.1.1 Background

Solid-state ionic materials for use in all-solid-state batteries have been extensively developed during the last three decades. Lithium ion conducting solid electrolytes have been widely investigated on organic polymer and inorganic material fields [1,2]. In this section, inorganic glassy and crystalline solid electrolytes are discussed. The inorganic solid electrolytes offer advantages over liquid and organic polymer electrolytes such as free of hazards of leakage and nonflammability. Single ion conduction is also an important feature of inorganic materials. Only Li ions are mobile while the counter anions and other cations form a rigid network, suggesting that undesirable side reactions or decomposition of the electrolytes due to anionic concentration gradient across the electrolytes would be suppressed in electrochemical cells.

Figure 3.2.1 shows the temperature dependence of electrical conductivities for a variety of glassy and crystalline lithium ion conductors [3–11]. On the basis of the primary interest for all-solid-state lithium secondary batteries, the targeted solid electrolytes should possess high conductivity in the ambient temperature region. Oxide materials [3–7] show ambient temperature conductivities in the range from 10^{-6} to 10^{-4} S cm^{-1} while sulfide materials [9–11] exhibit conductivity as high as 10^{-3} S cm^{-1}, indicating that the replacement of oxide anions by sulfide anions with larger polarizability is effective to improve conductivity [12].

In this section, we focus on sulfide-based lithium conductors, in particular glassy materials. Glasses have several advantages such as isotropic properties for ion migration and easy control of properties with changing chemical compositions. First, electrical and electrochemical properties and local structure of the sulfide-based glasses are reported. We have recently developed the mechanochemical synthesis as a new preparation route of glassy solid electrolytes instead of the well-known melt-quenching method. The usefulness of mechanochemistry and the properties of the obtained glasses are demonstrated. Finally, the effects of crystallization on conductivity of glassy materials are discussed.

3.2.1.2 Li_2S-SiS_2-Li_xMO_y oxysulfide glasses

Ionic conducting glasses are commonly prepared by the melt-quenching method. A large number of studies revealed that an increase in lithium ion concentration in glassy materials is a key point to achieve higher electrical conductivity and lower activation energy for Li$^+$ transport. A rapid quenching technique using a twin-roller apparatus is useful to expand glass-forming region, and thereby oxide and sulfide glasses with large amounts of lithium ions have been prepared. Oxysulfide glasses with high conductivity and high electrochemical stability were prepared by the twin-roller quenching technique [13–15]. Figure 3.2.2 shows the composition dependence

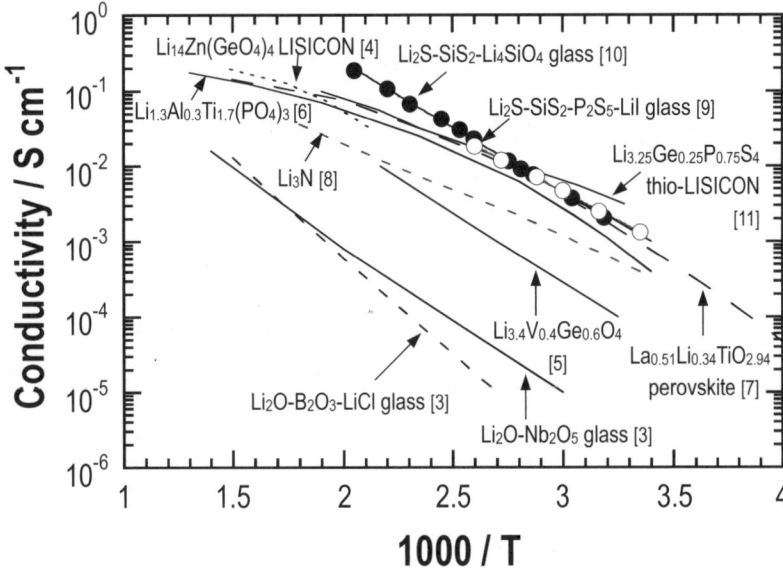

Fig. 3.2.1. Temperature dependence of ionic conductivities for a variety of inorganic lithium ion conductors.

of electrical conductivities at 25 °C (σ_{25}) and activation energies for conduction (E_a) of the oxysulfide glasses in the systems (100-x)(0.6Li$_2$S·0.4SiS$_2$)·xLi$_x$MO$_y$ (M=Si, P, Ge, B, Al, Ga and In). The dotted lines are a guide to the eye. The addition of 5 mol% of Li$_x$MO$_y$ (M=Si, P, Ge, B, and Al) to the 60Li$_2$S·40SiS$_2$ sulfide system keeps the high conductivity of 10^{-3} S cm^{-1}, while the addition of Li$_3$GaO$_3$ and Li$_3$InO$_3$ monotonically decreases the conductivity. Further addition of Li$_x$MO$_y$ decreases the conductivity in all the systems. The composition dependence of Ea corresponds to that of conductivity. The addition of oxide to sulfide is anticipated to monotonically decrease conductivity because oxide glasses show much lower conductivity than sulfide glasses. It is noteworthy that the oxysulfide glasses with 5 mol% of Li$_x$MO$_y$ maintained high conductivity of 10^{-3} S cm^{-1}.

The electrochemical stability of the oxysulfide glasses against Li electrode was examined by using cyclic voltammetry. Figure 3.2.3 shows the cyclic voltammogram of the 95(0.6Li$_2$S·0.4SiS$_2$)·5Li$_4$SiO$_4$ oxysulfide glass. A platinum plate, a lithium foil, and a silver wire are used as working, counter, and reference electrodes, respectively. A cathodic current due to lithium deposition reaction (Li$^+$ + e$^-$ → Li) and an anodic one due to the dissolution of lithium (Li → Li$^+$+e$^-$) are only observed at $-3 \sim -1$ V vs. Ag/Ag$^+$. Another anodic current due to electrolyte decomposition is not observed in the potential up to almost +10 V vs. Ag/Ag$^+$, suggesting that this oxysulfide glass shows a wide electrochemical window at least more than 10 V. On the other hand, the addition of 20 mol% Li$_4$SiO$_4$ lowered the electrochemical stability of the oxysulfide glass [16].

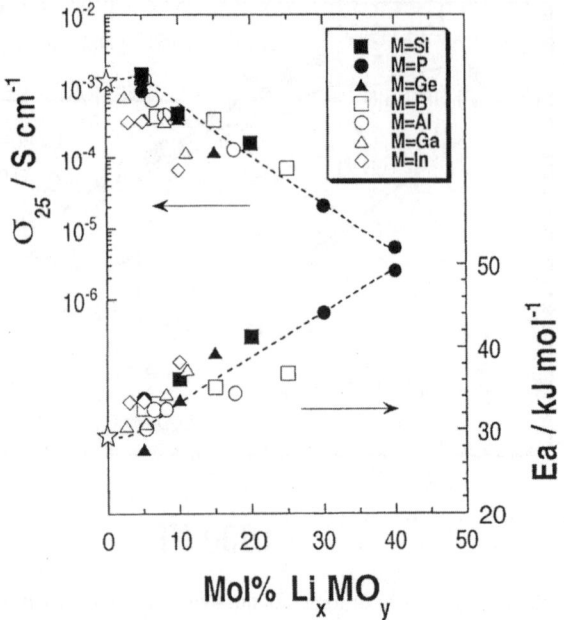

Fig. 3.2.2. Composition dependence of the σ_{25} and Ea of the oxysulfide glasses in the $(100-x)(0.6Li_2S\cdot 0.4SiS_2)\cdot xLi_xMO_y$ ($Li_xMO_y=Li_4SiO_4$, Li_3PO_4, Li_4GeO_4, Li_3BO_3, Li_3AlO_3, Li_3GaO_3 and Li_3InO_3) systems.

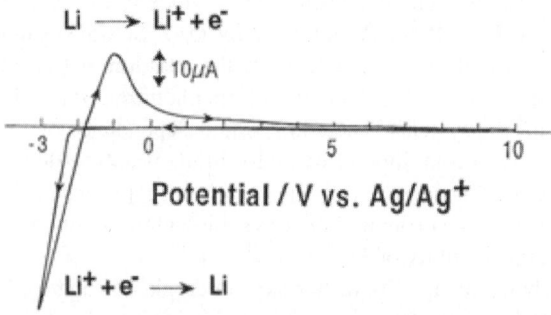

Fig. 3.2.3. Cyclic voltammogram for the $95(0.6Li_2S\cdot 0.4SiS_2)\cdot 5Li_4SiO_4$ oxysulfide glass.

The oxysulfide glasses with small amounts of Li_xMO_y exhibited high conductivity and electrochemical stability. Local structure of the glasses was investigated by several spectroscopic techniques. The main structural unit estimated by solid-state NMR and XPS in the oxysulfide glasses with small amounts of Li_xMO_y is shown in Fig.3.2.4 [17,18]. In the dimer unit, silicon atoms are coordinated with nonbridging sulfur atoms and a bridging oxygen (BO) atom, which

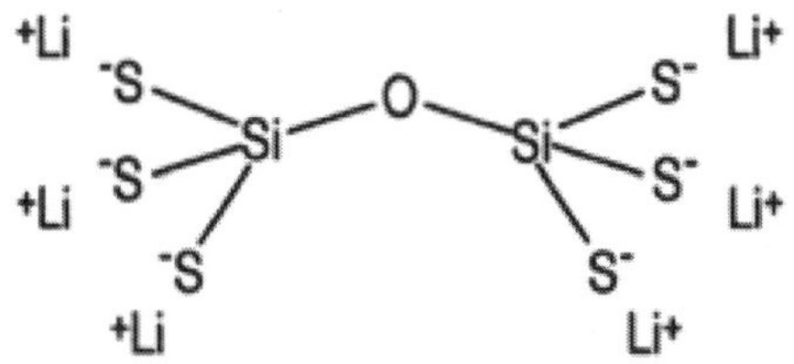

Fig. 3.2.4. Main structural unit expected to be present in the 95(0.6Li$_2$S·0.4SiS$_2$)·5Li$_4$SiO$_4$ oxysulfide glass.

would work as a weak trap of lithium ions. The reason of low conductivity of lithium oxide glasses is based on strong trap of nonbridging oxygen (NBO) atoms to lithium ions. NBO is not mainly present in the oxysulfide glasses added with 5 mol% of Li$_x$MO$_y$, while further addition of Li$_x$MO$_y$ increased the NBO content in the oxysulfide glasses. The doping of oxygen as BO is responsible for high conductivity and wide electrochemical window of the oxysulfide glasses.

3.2.1.3 Li$_2$S-based sulfide glasses prepared by mechanical milling

Glassy solid electrolytes are commonly prepared by melt-quenching. Mechanochemical synthesis using a ball mill apparatus has the following advantages as a new preparation technique of glasses; the whole process is performed at room temperature, and fine electrolyte powders, which can be directly applied to lithium secondary batteries, are obtained without an additional pulverizing procedure. The Li$_2$S-SiS$_2$-Li$_4$SiO$_4$ oxysulfide glasses were prepared by mechanical milling of crystalline starting materials Li$_2$S, SiS$_2$, and Li$_4$SiO$_4$ [19,20]. Mechanical milling was performed for powder mixtures in an alumina pot with alumina balls by a planetary ball mill apparatus (Fritsch Pulverisette 7). Figure 3.2.5 shows the temperature dependence of electrical conductivity of the 95(0.6Li$_2$S·0.4SiS$_2$)·5Li$_4$SiO$_4$ oxysulfide samples prepared by mechanical milling. Numbers in the figure denotes milling periods and the conductivity data were obtained on compressed pellets. In all milling periods, the conductivities of the samples follow the Arrhenius type equation. The conductivity of the as-mixed sample is much less than the order of 10^{-9} S cm^{-1}. The conductivities of mechanically milled samples dramatically increase with an increase in the milling periods. The conductivity of the sample milled for more than 5 h is higher than 10^{-4} S cm^{-1} at room temperature, which is comparable to the conductivity of the corresponding quenched glass in the compressed powder pellet sample. Activation energies for conduction decrease from 78 to 32 kJ mol^{-1} with increasing the milling periods. It is noteworthy that the milling for only 1 h drastically increases the conductivity of the oxysulfide samples by more than three orders of magnitude.

In order to clarify the formation process of the lithium ion conducting phase during mechanical milling, the structural change of the samples was investigated by the solid-state NMR technique [20]. Figure 3.2.6 shows the ^{29}Si MAS-NMR spectra

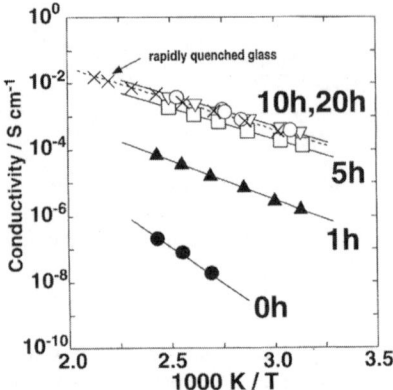

Fig. 3.2.5. Temperature dependence of electrical conductivity of the 95(0.6Li$_2$S·0.4SiS$_2$)·5Li$_4$SiO$_4$ oxysulfide samples prepared by mechanical milling.

of the 95(0.6Li$_2$S·0.4SiS$_2$)·5Li$_4$SiO$_4$ powder samples prepared with different milling periods. The peak at around -20 ppm observed in the as-mixed powder (0 h) is due to the SiS$_4$ tetrahedral units with two edge-sharing (E(2)), which corresponds to the presence of SiS$_2$ crystals. After milling for only 1 h, the peak at around -20 ppm almost disappears and two peaks at around -3 and 5 ppm appear. These peaks are attributable to the SiS$_4$ tetrahedral units with one edge sharing (E(1)) and no edge sharing (E(0)), respectively [21]. The peak due to SiOS$_3$ tetrahedral units overlaps the peak at around 5 ppm [22]. The spectra of powders milled for more than 5 h have three new peaks at around -25, -60, and -110 ppm. These peaks are attributable to SiO$_2$S$_2$, SiO$_3$S, and SiO$_4$ units, respectively [22]. The NMR spectrum for the glassy powder milled for 20 h is very similar to that of the corresponding melt-quenched glass, indicating that the local structure of the 95(0.6Li$_2$S·0.4SiS$_2$)·5Li$_4$SiO$_4$ glassy powder prepared by mechanical milling is very close to that of the corresponding quenched glass. The formation of such unique SiO$_n$S$_{4-n}$ (n = 1, 2, 3) structure as shown in Fig.3.2.4 means that the sulfur-oxygen exchange reaction occurs during not only the melting process but also the mechanical milling process of raw materials. The relationship between electrical conductivity and local structure of milled glasses is discussed. The formation of E(0) units after milling for 1 h drastically enhances electrical conductivity because E(0) units are main units in the highly conductive Li$_2$S-SiS$_2$ sulfide glasses [21]. The similarity of conductivity between milled and quenched glasses is explained from the viewpoint of the similarity of their local structures.

It has been considered that the amorphization mechanism by mechanical milling is not due to a local melt-quenching, but to a solid-state interdiffusion reaction [23], suggesting that there is a great possibility of preparing new amorphous materials which could not be prepared by melt-quenching. The Li$_2$S-Al$_2$S$_3$ binary glasses have not been prepared by a conventional melt-quenching technique. The Li$_2$S-P$_2$S$_5$ glasses can be prepared by quenching method while the melting reaction has to be

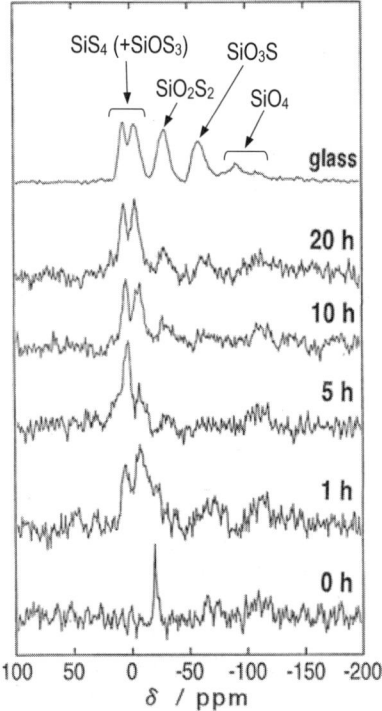

Fig. 3.2.6. ^{29}Si MAS-NMR spectra of the 95(0.6Li$_2$S·0.4SiS$_2$)·5Li$_4$SiO$_4$ powder samples prepared with different milling periods. The spectrum of the melt-quenched glass is also shown.

carried out in sealed quartz tubes because of high vapor pressure of P$_2$S$_5$. Recently, these glasses have been synthesized by mechanical milling at room temperature under normal pressure [24,25]. Figure 3.2.7 shows the composition dependence of electrical conductivities at 25 °C (σ_{25}) and activation energies for conduction (E_a) for the Li$_2$S-M$_x$S$_y$ (M=Al, Si and P) glasses prepared by mechanical milling. In all the systems, the σ_{25} values increase with an increase in the Li$_2$S content and maximize at the composition 60 mol% Li$_2$S. Further increase in the Li$_2$S content decreases the σ_{25} values. It is noteworthy that the conductivities in the Li$_2$S-M$_x$S$_y$ (M=Al, Si and P) systems are maximized at the same Li$_2$S content. The composition dependence of the E_a values corresponds to that of σ_{25} values. The σ_{25} of the 60Li$_2$S·40AlS$_{1.5}$ (mol%) glass is 3.4 ×10^{-5} S cm^{-1}, which is lower than the σ_{25} (= 10^{-4} S cm^{-1}) of the 60Li$_2$S·40SiS$_2$ and 60Li$_2$S·40PS$_{2.5}$ (mol%) glasses. It is revealed that aluminum sulfide based glasses show lower conductivity than silicon sulfide and phosphorus sulfide based glasses.

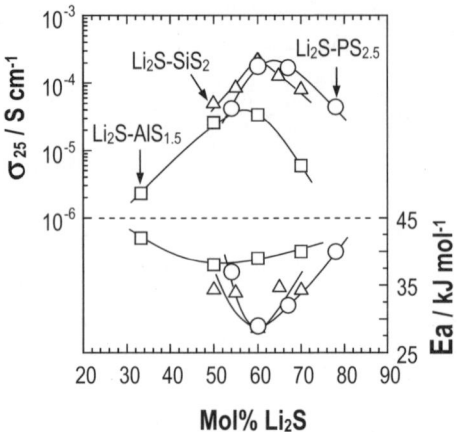

Fig. 3.2.7. composition dependence of electrical conductivities at 25°C (σ_{25}) and activation energies for conduction (Ea) for the Li_2S-M_xS_y (M=Al, Si and P) glasses prepared by mechanical milling.

3.2.1.4 Glass-ceramic solid electrolytes

In general, crystallization of glassy materials is well known to lower conductivities [26]. However, the enhancement of conductivity by crystallization has been discoverd in the Li_2S-P_2S_5 glasses with high Li_2S content [27,28]. The obtained glass-ceramic materials exhibited ambient temperature conductivity of about 10^{-3} S cm^{-1} in compressed pellet so that the glass-ceramics are attractive solid electrolytes for all-solid-state lithium secondary batteries. Figure 3.2.8 shows the temperature dependence of conductivities for the $67Li_2S \cdot 33PS_{2.5}$ glass prepared by mechanical milling. The conductivity measurements were carried out both on the heating and cooling processes and the conductivities on the cooling process are higher than those on the heating process. The ambient temperature conductivity on the cooling process is 7.2×10^{-4} S cm^{-1}, which is almost four times larger than the conductivity of the as-prepared material. Because the highest temperature on the heating process is 230 °C, which is beyond the crystallization temperature (see Fig.3.2.9), the crystallization occurs during the heating process. The inset figure shows the conductivity data for the $95(0.6Li_2S \cdot 0.4SiS_2) \cdot 5Li_4SiO_4$ melt-quenched glass [16]. The crystallization of the Li_2S-SiS_2 based glass decreases conductivity and this behavior is commonly observed in glassy ion conductors. The decrease in conductivity of the glass is due to the crystallization of Li_4SiS_4 crystal with low conductivity [29]. On the other hand, the conductivity enhancement is observed in the Li_2S-P_2S_5 system. This unexpected phenomenon is explained by the formation of metastable superionic crystal phases in this system.

Figure 3.2.9 shows the X-ray diffraction (XRD) patterns of the $67Li_2S \cdot 33PS_{2.5}$ samples before and after being heated up to 240 °C in an DTA apparatus. The inset figure shows the DTA curve of the as-prepared sample. The targeted temperature

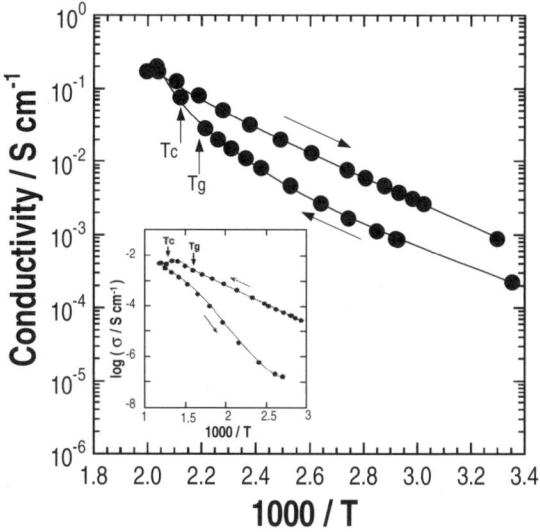

Fig. 3.2.8. Temperature dependence of conductivities for the 67Li$_2$S·33PS$_{2.5}$ glass prepared by mechanical milling. The inset figure shows the conductivity data for the 95(0.6Li$_2$S·0.4SiS$_2$)·5Li$_4$SiO$_4$ melt-quenched glass. Tg and Tc mean glass transition temperature and crystallization temperature, respectively.

of 240 °C is beyond the crystallization temperature (T$_c$). The XRD pattern of the glass-ceramic after crystallization (☆) (except for the peaks due to Li$_2$S crystal (○) which remains a little in the as-prepared glass) is very similar to that of one of the superionic conducting crystalline phases called "thio-LISICON" [11]. This phase, the composition of which is Li$_{4-x}$Ge$_{1-x}$P$_x$S$_4$ (0.6 < x < 0.8), exhibits very high conductivity of 2.2×10^{-3} S cm^{-1} at room temperature [11] as shown in Fig.3.2.1. Although similar crystalline materials, Li$_{3+5x}$P$_{1-x}$S$_4$, were prepared in the binary Li$_2$S-P$_2$S$_5$ system, the crystal structure was found to be very different from that of the Li$_{4-x}$Ge$_{1-x}$P$_x$S$_4$ thio-LISICON and the lower conductivity of 1.5×10^{-4} S cm^{-1} was observed compared to the thio-LISICON and the glass-ceramics [30]. The crystal phases with the same structure as the Li$_{4-x}$Ge$_{1-x}$P$_x$S$_4$ thio-LISICON were not formed by solid phase reaction in the case of Li$_2$S-P$_2$S$_5$ system [29], suggesting that the superionic thio-LISICON analog phase would be precipitated as a metastable phase from the mechanically milled glass. The superionic metastable phase was also confirmed to be formed in the Li-rich 67Li$_2$S·33PS$_{2.5}$ glass while it could not be formed in the 60Li$_2$S·40PS$_{2.5}$ glass; in the latter case, the conductivity enhancement by crystallization was slightly observed [28]. The glass-matrix with high lithium concentration is necessary to form highly conductive metastable phase. The crystallization of the metasable phase with high conductivity is responsible for the marked enhancement of conductivity in the 67Li$_2$S·33PS$_{2.5}$ glass.

Similar glass-ceramics with high lithium ion conductivity were also prepared in the ternary Li$_2$S-P$_2$S$_5$-SiS$_2$ system by the crystallization of mechanochemically

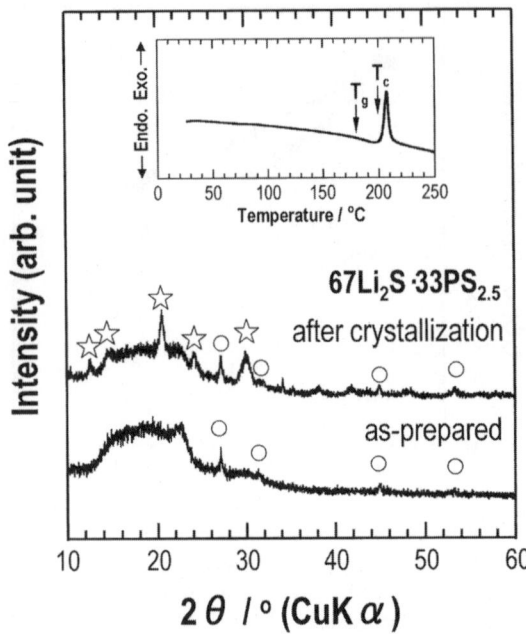

Fig. 3.2.9. XRD patterns of the 67Li$_2$S·33PS$_{2.5}$ samples before and after being heated up to 240°C in an DTA apparatus. The inset figure shows the DTA curve of the as-prepared sample. The targeted temperature of 240°C is beyond the crystallization temperature (Tc). The XRD patterns are assigned to the following crystals: Li$_2$S (○) and thio-LISICON analog (see text in detail, ☆).

prepared sulfide glasses [31]. In the glass-ceramics, the Li$_{4-x}$Ge$_{1-x}$P$_x$S$_4$ thio-LISICON analogs were precipitated by the heat treatment of the glasses. The conductivity of the Li$_2$S-P$_2$S$_5$-SiS$_2$ glass-ceramics was as high as 1×10^{-3} S cm^{-1}, which is higher than the maximum conductivity 6.4×10^{-4} S cm^{-1} of the solid solutions in the Li$_2$S-P$_2$S$_5$-SiS$_2$ system prepared by solid phase reaction [32]. The conductivity enhancement by the heat treatment is also due to the formation of superionic metastable phase of thio-LISICON analogs. It is noteworthy that the highly conductive metastable phases which could not be prepared by the usual solid state reaction formed by the crystallization of mechanochemically prepared glasses. This is obviously a remarkable advantage of glass formation or amorphization in fabrication of highly conductive solid electrolytes.

All-solid-state cells using these glass-ceramic electrolytes worked as lithium secondary batteries at room temperature and exhibited excellent cycling performance. Detailed performance data will be discussed in section 3.2.3.

References

1. C. Julien and G.-A. Nazri, Solid State Batteries: *Materials Design and Optimization*, Kluwer Academic Publishers, Boston, p. 183 (1994).
2. G.-A. Nazri and G. Pistoia, *Lithium Batteries, Science and Technology*, Kluwer Academic Publishers, Boston, p. 623 (2004).
3. H.L. Tuller, D.P. Button and D.R. Uhlmann, *J. Non-Cryst. Solids*, **40** (1980) 93.
4. H. Y-P. Hong, *Mat. Res. Bull.*, **13** (1978) 117.
5. A.R. Rodger, J. Kuwano and A. R. West, *Solid State Ionics*, **15** (1985) 185.
6. H. Aono, E. Sugimoto, Y. Sadaoka, N. Imanaka and G. Adachi, *J. Electrochem. Soc.*, **137** (1990) 1023.
7. M. Itoh, Y. Inaguma, W.-H. Jung, L. Chen and T. Nakamura, *Solid State Ionics*, **70/71** (1994) 203.
8. A. Rabenau, *Solid State Ionics*, **6** (1982) 277.
9. J.H. Kennedy and Z. Zhang, *Solid State Ionics*, **28–30** (1988) 726.
10. K. Hirai, M. Tatsumisago and T. Minami, *Solid State Ionics*, **78** (1995) 269.
11. R. Kanno and M. Murayama, *J. Electrochem. Soc.*, **148** (2001) A742.
12. T. Minami, *Bull. Inst. Chem. Res. Kyoto Univ.*, **72** (1994) 305.
13. S. Kondo, K. Takada and Y. Yamamura, *Solid State Ionics*, **53–56** (1992) 1183.
14. M. Tatsumisago, K. Hirai, T. Minami, K. Takada and S. Kondo, *J. Ceram. Soc. Jpn.*, **101** (1993) 1315.
15. T. Minami, A. Hayashi and M. Tatsumisago, *Solid State Ionics*, **136–137** (2000) 1015.
16. A. Hayashi, M. Tatsumisago and T. Minami, *J. Electrochem. Soc.*, **146** (1999) 3472.
17. A. Hayashi, M. Tatsumisago, T. Minami and Y. Miura, *J. Am. Ceram. Soc.*, **81** (1998) 1305.
18. A. Hayashi, M. Tatsumisago, T. Minami and Y. Miura, *Phys. Chem. Glasses*, **39** (1998) 145.
19. H. Morimoto, H. Yamashita, M. Tatsumisago and T. Minami, *J. Ceram. Soc. Jpn.*, **108** (2000) 128.
20. M. Tatsumisago, H. Yamashita, A. Hayashi, H. Morimoto and T. Minami, *J. Non-Cryst. Solids*, **274** (2000) 30.
21. H. Eckert, Z. Zhang and J.H. Kennedy, *J. Non-Cryst. Solids*, **107** (1989) 271.
22. A. Hayashi, R. Araki, K. Tadanaga, M. Tatsumisago and T. Minami, *Phys. Chem. Glasses*, **40** (1999) 140.
23. R.B. Schwarz and C.C. Koch, *Appl. Phys. Lett.*, **49** (1986) 146.
24. A. Hayashi, S. Hama, H. Morimoto, M. Tatsumisago and T. Minami, *J. Am. Ceram. Soc.*, **84** (2001) 477.
25. A. Hayashi, T. Fukuda, S. Hama, H. Yamashita, H. Morimoto, T. Minami and M. Tatsumisago, *J. Ceram. Soc. Jpn.*, **112** (2004) S695.
26. T. Minami and N. Machida, *Mater. Sci. Eng. B*, **13** (1992) 203.
27. A. Hayashi, S. Hama, H. Morimoto, M. Tatsumisago and T. Minami, *Chem. Lett.*, **2001**, 872.

28. A. Hayashi, S. Hama, T. Minami and M. Tatsumisago, *Electrochem. Commun.*, **5** (2003) 111.
29. A. Hayashi, M. Tatsumisago and T. Minami, *Phys. Chem. Glasses*, **40** (1999) 333.
30. M. Murayama, N. Sonoyama and R. Kanno, *Solid State Ionics*, **170** (2004) 173.
31. A. Hayashi, Y. Ishikawa, S. Hama, T. Minami and M. Tatsumisago, *Electrochem. Solid-State Lett.*, **6** (2003) A47.
32. M. Murayama, R. Kanno, M. Irie, S. Ito, T. Hata, N. Sonoyama and Y. Kawamoto, *J. Solid State Chem.*, **168** (2002) 140.

3.2.2 Electrode materials

3.2.2.1 Background

Development of all-solid-state lithium batteries with high energy density and with high reliability is strongly desired to replace commercially available lithium-ion batteries using conventional electrolyte solutions, because the lithium-ion batteries include flammable electrolyte solutions and then have safety problems [1–13].

The electrode materials are most important in battery systems because the electrode active materials determine the battery performance. A number of electrode materials have been developed for conventional lithium ion secondary battery systems. However, the excellent materials for conventional batteries using liquid electrolytes do not always work in all-solid-state battery systems.

In this section, negative and positive electrode materials suitable for all-solid-state lithium batteries are reported.

3.2.2.2 Negative electrode materials for all-solid-state lithium batteries

Indium metal has been most widely utilized as negative electrode materials of the all-solid-state batteries with the inorganic solid electrolytes [4, 6–12]. The indium metal is, however, unfavorable materials because of its scare resources and high cost.

Graphitic and graphitized carbons have been widely used as negative electrode materials for the commercially available lithium-ion batteries with electrolyte solutions [14–16]. The carbon materials have favorable characteristics as negative electrode materials, such as a highly negative electrode potential of about 0.2 V against the Li / Li$^+$ electrode, a good reversibility for charge-discharge cycling, and a large specific capacity of about 370 mAhg^{-1}. The carbon materials, however, are unavailable for negative electrode materials of all-solid-state lithium batteries with inorganic solid electrolytes. Electrochemical lithium insertion into the carbon materials does not reversibly progress in the all-solid-state cells with the inorganic solid electrolytes such as amorphous 60Li$_2$S·40SiS$_2$ (mol%) electrolyte and lithium oxo-salts doped Li$_2$S-SiS$_2$ electrolytes, which have high lithium-ion conductivities over 10^{-4} Scm^{-1} at room temperature and thus are promising electrolytes for the all-solid-state batteries. Consequently, it is necessary to develop new candidates of negative electrode materials for the all-solid-state batteries.

Lithium silicide ($Li_{4.4}Si$) is one of the favorable candidates as new materials for the negative electrode of the lithium batteries, because the lithium silicide has a large specific capacity (theoretical specific capacity of 2011 $mAhg^{-1}$) and highly negative potentials close to that of the metallic lithium (about 0.3 V vs. Li / Li^+) [17–21]. In addition, the silicide has plentiful resources and contains no harmful elements, and thus it is environment-friendly materials.

This section mainly deals with lithium silicide $Li_{4.4}Si$ that has been prepared by a high-energy ball-milling process and its electrochemical properties as negative electrodes of the all-solid-state lithium batteries with sulfide-based solid electrolytes [22, 23].

The lithium silicide ($Li_{4.4}Si$) has been prepared from silicon and lithium metals by a high-energy ball-milling process and the X-ray diffraction (XRD) spectra of the ball-milled sample are shown in Fig.3.2.10. The XRD spectra of the starting materials, Si and Li, are also shown as (a) and (b) in the figure, respectively. The XRD spectrum (d) of the sample ball-milled for 54 h has no diffraction lines attributed to the starting materials, Li and Si, and shows new diffraction lines with closed circles only. The newly appearing diffraction lines are different from those of the $Li_{4.4}Si$ alloy that is prepared by a conventional melt-quenching method. The XRD spectrum of the melt-quenched $Li_{4.4}Si$ is presented as (e) in the figure for comparison. The diffraction spectrum of the melt-quenched $Li_{4.4}Si$ alloy well agrees with the spectrum reported by Nesper et al. [17] . Four lithium-silicide phases, $Li_{4.4}Si$, $Li_{3.25}Si$, $Li_{2.33}Si$, and $Li_{1.71}Si$ are reported as inter-metallic phases in the Li-Si system. The XRD pattern of the ball-milled sample is completely different from those of the four lithium silicide phases. Thermal analysis on the ball-milled sample has revealed that heat treatment above 500 °C prompts the ball-milled $Li_{4.4}Si$ alloy to transform exothermically into the well-known $Li_{4.4}Si$ phase, which is the same phase as shown in Fig.3.2.10 (e). Those observations confirmed that the $Li_{4.4}Si$ alloy prepared by the ball-milling to be a meta-stable new phase.

Figure 3.2.11 shows the XRD patterns of the ball-milled Li-Si alloys that have some different chemical compositions. In the figure, MM denotes the time for which the samples have been ball-milled. The X-ray diffraction peaks with closed circles are attributed to the meta-stable phase. The meta-stable alloy has been obtained over the wide composition range of $Li_{3.8}Si$ to $Li_{5.2}Si$ by the high-energy ball-milling process. This result indicates that the meta-stable phase has the wide solid-solution range and that the $Li_{4.4}Si$ alloy is located in the middle of the solid-solution range. Thus the alloy is expected to be negative electrode materials without significant structural changes during the electrochemical lithium insertion and/or extraction reactions. Two types of the all-solid-state batteries have been assembled with the meta-stable $Li_{4.4}Si$ alloy using as negative electrode materials. One of them employed $LiCo_{0.3}Ni_{0.7}O_2$ powder as positive electrode materials. This type of batteries was utilized to investigate electrochemical lithium insertion into the alloy. The other type of batteries employed TiS_2 powder as positive electrode materials to investigate lithium extraction from the alloy. Those batteries utilized the sulfide-based amorphous solid electrolytes, which were also prepared by use of the high-energy ball-milling process. The batteries were assembled by successively pressing

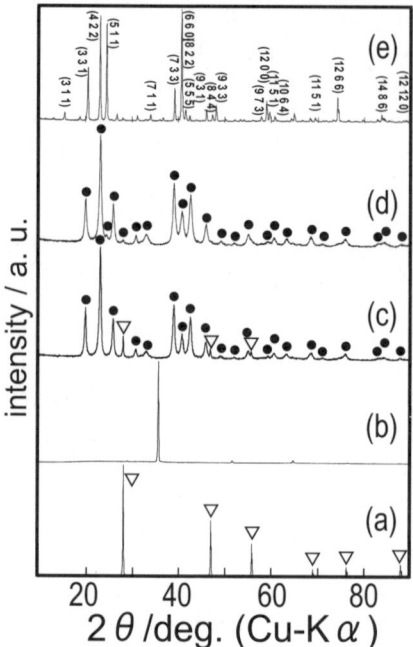

Fig. 3.2.10. XRD patterns of metallic silicon, (b) metallic lithium, (c) Li$_{4.4}$Si ball-milled for 18 h, (d) Li$_{4.4}$Si ball-milled for 54 h and (e) Li$_{4.4}$Si alloy prepared by a melt-quenching method.

the positive electrode materials, the solid electrolytes, and the meta-stable Li$_{4.4}$Si alloy powders at 300 MPa into a pellet of 10 mm in diameter.

Figure 3.2.12 shows the charge-discharge curves of the all-solid-state battery, Li$_{4.4}$Si /a-60Li$_2$S·40SiS$_2$/ LiCo$_{0.3}$Ni$_{0.7}$O$_2$, with testing under a constant current density of 64 μAcm^{-2} in the potential range of 4.1 to 2.3 V; a-60Li$_2$S · 40SiS$_2$ means the 60Li$_2$S · 40SiS$_2$ (mol%) glass prepared by mechanical milling. The positive electrode of the battery is a mixture of the active material LiCo$_{0.3}$Ni$_{0.7}$O$_2$, the solid electrolyte a-60Li$_2$S·40SiS$_2$, and acetylene black powders in the weight ratio of 59, 39, and 2 mass%, respectively. The charge-discharge cycling test of the battery has been started with a charge process, which corresponds to the lithium insertion into the Li$_{4.4}$Si alloy. The abscissa of the figure is the specific capacity of the meta-stable Li$_{4.4}$Si alloy. Additionally, the specific capacity of the LiCo$_{0.3}$Ni$_{0.7}$O$_2$ positive electrode material is plotted in the top abscissa. The capacity of the first charge is about 220 mAhg^{-1} and that of the first discharge is about 160 mAhg^{-1}. While the capacity of the first discharge is smaller than that of the first charge, the charge and discharge curves after the second cycle are almost the same. Thus, the charge-discharge efficiency is almost unity after the second cycle. Those results indicate that the meta-stable Li$_{4.4}$Si alloy has good electrochemical properties as the negative electrode materials of the all-solid-state lithium batteries with sulfide-based inorganic solid electrolytes.

3.2 Lithium ion conductors   45

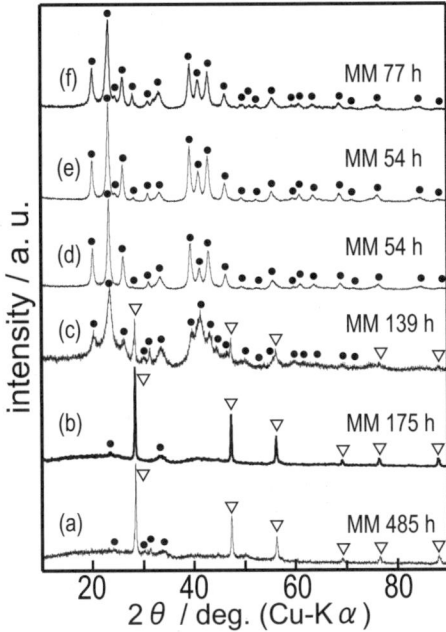

Fig. 3.2.11. XRD spectra of ball-milled Li-Si alloys with various chemical compositions; (a) $Li_{1.71}Si$, (b) $Li_{2.33}Si$, (c) $Li_{3.25}Si$, (d) $Li_{4.2}Si$, (e) $Li_{4.4}Si$, and (f) $Li_{5.2}Si$. MM denotes the milling times for obtaining samples.

In order to investigate structural changes of the meta-stable alloy during the charge-discharge process, XRD measurements have been carried out for the alloys after the first charge and discharge. Figure 3.2.13 shows the XRD patterns of the meta-stable alloys after first charge (Fig.3.2.13 b), and also after first discharge (Fig.3.2.13 c). In the figure, the XRD pattern of the as-ball-milled $Li_{4.4}Si$ alloy powder is also shown as (a) for comparison. The compositions of the charged and discharged alloys are calculated to be $Li_{4.88}Si$ and $Li_{4.53}Si$, respectively, from the charge and discharge capacities with a coulometric equation. In the XRD spectra (b) and (c) weak halo patterns are observed at around $2\theta = 27°$. These halo patterns are attributed to the amorphous $60Li_2S \cdot 40SiS_2$ (mol%) solid electrolyte utilized for the all-solid-state batteries.

In the X-ray pattern of the charged sample (b), the diffraction peaks are slightly shifted to lower angels. These peak shifts are caused by the lithium insertion into the lithium silicide. On the other hand, in the X-ray pattern of the discharged sample (c), the diffraction peaks are shifted back to the higher angles. In both patterns (b) and (c) there are no peaks attributed to the as-ball-milled $Li_{4.4}Si$ sample. Those results indicate that all particles of the negative electrode materials play a part in electrochemical reactions and that the electrochemical reactions progress thoroughly homogeneously in the all-solid-state batteries. In addition, there is no significant structural change of the lithium silicide during the electrochemical lithium insertion

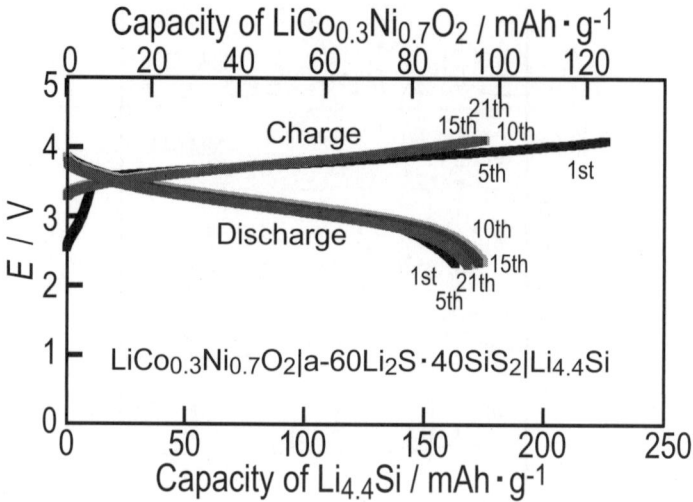

Fig. 3.2.12. Charge-discharge cycle curves of the all-solid-state battery $Li_{4.4}Si$ $|a\text{-}60Li_2S\cdot40SiS_2|\ LiCo_{0.3}Ni_{0.7}O_2$. The current density of the charge-discharge measurements was 64 μAcm^{-2}.

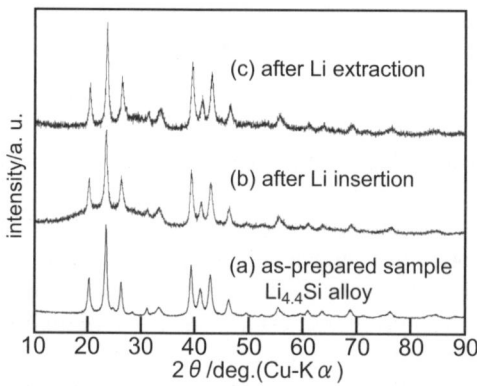

Fig. 3.2.13. XRD spectra of the meta-stable $Li_{4.4}Si$ alloy; (a) as-prepared sample, (b) after Li insertion, and (c) Li extraction.

and extraction. The characteristics of the Li-Si alloy are noteworthy as reversible electrode materials for the all-solid-state lithium batteries.

Figure 3.2.14 shows the discharge-charge curves of the other all-solid-state battery, $Li_{4.4}Si\ /a\text{-}7.5Li_2O\cdot67.5Li_2S\cdot25P_2S_5/\ TiS_2$. The positive electrode of the battery is a mixture of TiS_2 (60 mass%) and a-$7.5Li_2O\cdot67.5Li_2S\cdot25P_2S_5$ (40 mass%). The discharge-charge cycling test has been started with the discharge process of the battery under a constant current density of 64 μAcm^{-2} in the potential range of 1.1 to 2.6 V. The abscissa of the figure is the specific capacity of the meta-stable $Li_{4.4}Si$

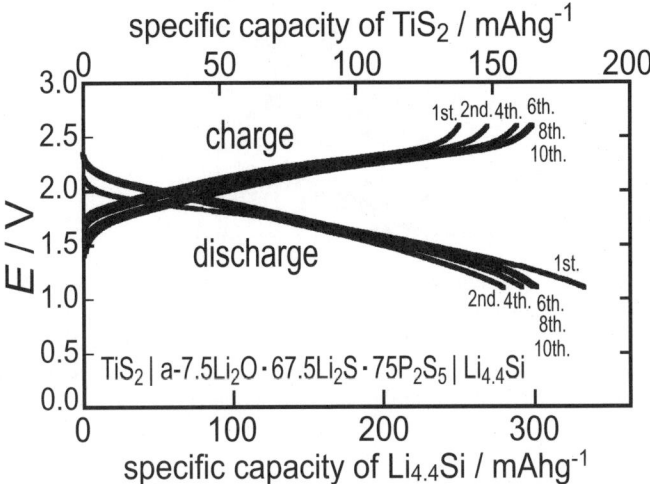

Fig. 3.2.14. Charge-discharge cycle behavior of the all-solid-state lithium battery, $Li_{4.4}Si$ $|a$-$7.5Li_2O \cdot 67.5Li_2S \cdot 25P_2S_5|$ TiS_2. The current density of the charge-discharge measurements was 64 μ Acm^{-2}.

alloy. Additionally, the specific capacity of the TiS_2 positive electrode material is plotted in the top abscissa [24].

The specific capacity of the alloy is 330 $mAhg^{-1}$ for first discharge. While the capacity of the second discharge of 290 $mAhg^{-1}$ is smaller than that of the first discharge, the discharge-charge cycle properties after second cycle are almost the same. Those results indicate that the meta-stable $Li_{4.4}Si$ alloy is also successfully utilized as negative electrode materials for the all-solid-state batteries of which the discharge-charge cycle has been started with the discharge process.

Hence, those results as shown above confirm that the meta-stable $Li_{4.4}Si$ alloy is available for the negative electrode materials for both types of all-solid-state lithium batteries of which the charge-discharge cycles are started with the charge process and/or with the discharge process. This characteristic of the meta-stable $Li_{4.4}Si$ alloy is an advantage over other negative electrode materials for all-solid-state batteries.

3.2.2.3 Positive electrode materials for all-solid-state batteries

In all-solid-state batteries, electrochemical reaction progresses through the solid-solid interface between solid electrolytes and electrode materials. Thus, adequate contacts between these materials are necessary to achieve good charge-discharge properties of the solid-state batteries. To obtain large contact area between solid electrolytes and electrode materials, the particle size of those materials are required to be small, and also the difference between the particle sizes of the solid electrolytes and electrode materials is to be small.

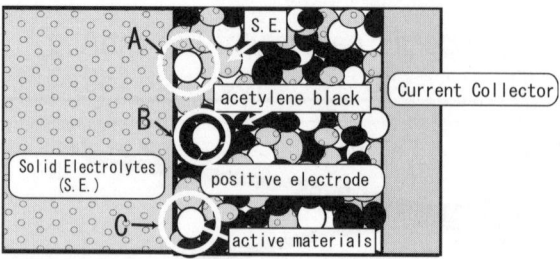

Fig. 3.2.15. Schematic diagram of the positive electrode mixture of all-solid-state batteries with inorganic solid electrolytes. Three examples of different contact situations of active electrode materials are illustrated: the active material particles contacted (A) with only solid electrolytes, (B) with only acetylene black powders, and (C) with both solid electrolytes and acetylene black powders. (See text for detailed descriptions.)

The electrode materials for all-solid-state batteries also require high stability against over charge and over discharges. The electrode of all-solid-state batteries is usually constructed of a mixture of electrode materials, solid electrolytes, and electronic conductors such as acetylene black. Figure 3.2.15 shows a schematic diagram of the composite electrode. In the composite electrode, the particles of the electrode materials must be kept in contact with other additives in order to play a part in electrochemical reactions. If the particles lose the contact with other additives, the particles could not take part in the reaction. Three examples of different contact situations of active electrode materials are illustrated in Fig.3.2.15. The particle of active materials in area A is in contact with only solid electrolytes, and the particle in area B is in contact with only acetylene black powders. The active materials in both areas A and B could not take part in the electrochemical reactions. The particle of active material in area C is in contact with both solid electrolytes and acetylene black powders, and then the particle plays a part in the electrochemical reactions. The difference in the contact situation creates wide variation in the reaction level of the particles. It means that electrochemical reactions proceed progressively on some particles but hardly at all on the other particles. Accordingly, the progressively reacted particles run the danger of being over-reaction. These situations more frequently arise in the electrodes of all-solid-state batteries as compared to the case of the batteries with electrolyte solutions. Hence, high stability against over-reaction is required for the electrode materials.

The $Li_yCo_{0.3}Ni_{0.7}O_2$ solid solutions, which have the same structure as the well-known $LiCoO_2$, have been reported to have high structural stability against overcharge more than the $LiCoO_2$. Thus the $LiCo_{0.3}Ni_{0.7}O_2$ fine powders have been prepared by an oxalate decomposition method. The particles of $LiCo_{0.3}Ni_{0.7}O_2$ sample prepared by the oxalate decomposition method have hexagonal shapes and 1 to 2 μm diameters as the average particle size [10].

The charge-discharge cycling properties of the all-solid-state battery that was assembled with the $LiCo_{0.3}Ni_{0.7}O_2$ fine powders were already shown in Fig.3.2.12.

3.2 Lithium ion conductors   49

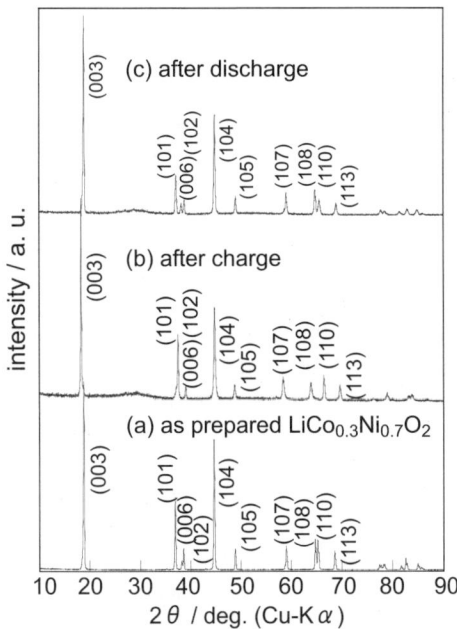

Fig. 3.2.16. XRD patterns of the $LiCo_{0.3}Ni_{0.7}O_2$ powders: (a) before electrochemical measurements, (b) after first charge, and (c) after first discharge.

The $LiCo_{0.3}Ni_{0.7}O_2$ powder showed good charge-discharge cycle properties with a high specific capacity of about 100 mAhg^{-1} at room temperature.

Figure 3.2.16 shows the XRD patterns of the $LiCo_{0.3}Ni_{0.7}O_2$ sample after charge (Fig.3.2.16 b) and after discharge (Fig.3.2.16 c). In the figure the pattern of as-prepared $LiCo_{0.3}Ni_{0.7}O_2$ powder (Fig.3.2.16 a) is also shown for comparison. The compositions of the charged and discharged samples are calculated to be $Li_{0.53}Co_{0.3}Ni_{0.7}O_2$ and $Li_{0.89}Co_{0.3}Ni_{0.7}O_2$, respectively. In the X-ray pattern of the charged sample (Fig.3.2.16 b), the diffraction peak (003) is shifted to a lower angle, and the peaks (108) and (110) are split well compared to the pattern of as-prepared $LiCo_{0.3}Ni_{0.7}O_2$ powder. Those peak shifts are caused by the de-lithiation from the $LiCo_{0.3}Ni_{0.7}O_2$ powder. On the other hand, in the X-ray pattern of the discharged sample (Fig.3.2.16 c) the diffraction peak (003) is shifted back to the higher angle, and the (108) and (110) peaks converge. In both pattern (b) and (c) there are no peaks attributed to the as-prepared $LiCo_{0.3}Ni_{0.7}O_2$ powder. Those results indicate that all particles of the positive electrode materials play a part in the electrochemical reactions and that the mixing of the electrode materials, solid electrolytes, and acetylene black powders in the composite electrode is thoroughly homogeneous. Consequently, the $LiCo_{0.3}Ni_{0.7}O_2$ powder prepared by the oxalate decomposition has sufficiently fine particle size for all-solid-state batteries.

3.2.2.4 All-solid-state battery with sulfur as positive electrode materials

Electrode materials with large specific capacity are strongly desired for developing and upgrading high-energy-density lithium batteries. Elemental sulfur is very attractive as the positive electrode materials for the high-energy-density lithium batteries because of its large theoretical specific capacity of 1670 mAhg^{-1}. Thus many efforts have been devoted to study the electrochemical properties of the sulfur electrode for lithium batteries [24–28]. Sulfur, however, does not work successfully as reversible positive electrode materials in lithium batteries using electrolyte solutions, because the electrochemical capacity of sulfur rapidly fades away on cycling [28]. The capacity fading has been attributed to the facts that the reduction of sulfur produces several lithium poly-sulfides in the batteries and the lithium poly-sulfides can dissolve into the electrolyte solutions of the batteries.

The inorganic solid electrolytes do not dissolve the lithium poly-sulfides that are produced during the discharge process of the sulfur electrode. Thus the elemental sulfur is expected to work successively as positive electrode materials of the all-solid-state batteries. The charge-discharge properties of sulfur electrode in the all-solid-state lithium batteries have investigated and have been compared with those in a conventional lithium battery with an electrolyte solution [13, 30].

The sulfur electrode was prepared by mixing of sulfur, metallic copper and acetylene black powder by a ball-milling process. XRD measurements revealed that the metallic copper reacted with the sulfur during the ball milling and consequently the CuS phase precipitated in the ball-milled composite. The XRD peaks of the elemental sulfur remained in the XRD spectrum of the composite. Assuming that all the metallic copper used as raw materials changed into CuS during the milling process, the mole ratio of CuS:S was calculated to be 1:1.5 in the composite electrode.

Figure 3.2.17 shows discharge-charge cycling curves of the battery with the sulfur composite electrode and with the electrolyte solution of 1M LiClO$_4$-PC at a constant current density of 100 µAcm^{-2}. The abscissa of figure is the specific capacity of sulfur. The first discharge curve shows three plateaus at 2.4, 2.0, and 1.6 V. The total capacity of the first discharge is about 430mAhg^{-1}. After the first cycle, the voltage plateau at 2.4 V disappears in the discharge curves and the discharge capacity of the battery rapidly fades away on cycling. This result indicates that the positive electrode composite containing sulfur does not work reversibly in the cell with the electrolyte solution. After the discharge-charge cycling test, the color of the electrolyte solution turned to yellow. The yellow color was attributed to the soluble poly-sulfide ions [31]. Thus, the irreversibility of the battery was caused by the poly-sulfide ions as mentioned above.

Figure 3.2.18 shows discharge-charge cycle behavior of the all-solid-state battery, Li$_{4.4}$Ge /a-60Li$_2$S·40SiS$_2$/ S, CuS. The abscissa of figure is the specific capacity of sulfur. Two plateaus are observed at 1.5 and 1.0 V in the first discharge curve. The total capacity of the first discharge is about 980 mAhg^{-1}. In the first charge, two plateaus are also observed at 1.3 and 2.0 V. The total capacity of the first charge is about 810 mAhg^{-1}. After the first cycle, the discharge and charge curves

3.2 Lithium ion conductors   51

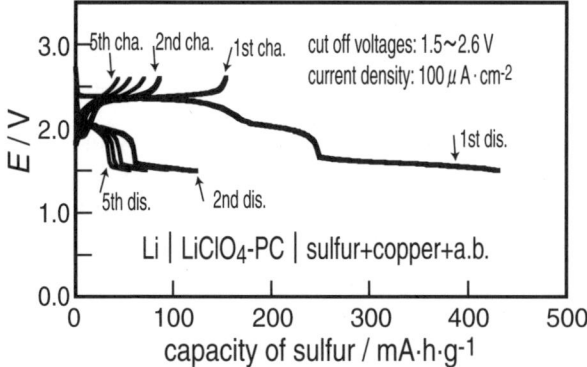

Fig. 3.2.17. Discharge-charge cycle behavior of the conventional lithium battery with the ball-milled composite of sulfur, copper, and acetylene black powders as positive electrode materials and with the electrolyte solution of 1M LiClO$_4$-PC.

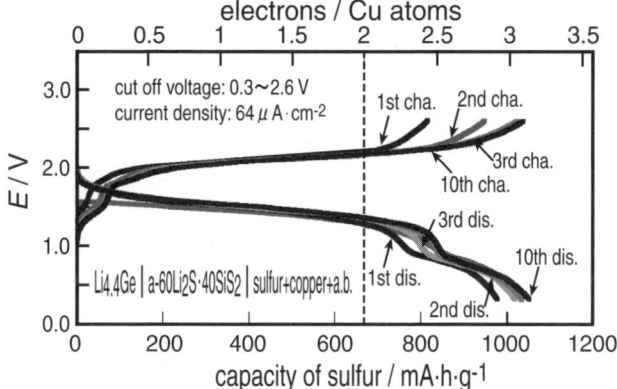

Fig. 3.2.18. Discharge-charge cycle behavior of the conventional lithium battery with the ball-milled composite of sulfur, copper, and acetylene black powders as positive electrode materials and with the solid electrolyte of a-60Li$_2$S·40SiS$_2$(mol%)

show almost the same properties and rechargeable specific capacities are about 1080 mAhg^{-1}. The all-solid-state battery shows good discharge-charge reversibility with the large specific capacity.

On the discharge and charge mechanism of the all-solid-state battery, there is a question whether only copper sulfide in the composite electrode works as active materials or not sulfur. The mole ratio of the reacted electrons to the total copper atoms included in the composite electrode, e$^-$ / Cu, is additionally plotted in the top abscissa of the figure. If it is supposed that the metallic copper was completely converted into the stoichiometric CuS phase by the milling process, CuS is theoretically able to accept up to two electrons. The theoretical value is shown as a broken line in the figure. The discharge and charge capacities of the all-solid-state batteries are beyond the theoretical value. The results confirm that not only CuS but also the elemental sulfur contributes to the electrochemical reactions of the composite electrode. The observations indicate that the sulfur electrode is available for the all-solid-state batteries with high energy density.

In conclusion, the new Li-Si alloy and the sulfur composite are available for the negative and positive electrode materials with large specific capacities for all-solid-state batteries, respectively. Those observations should contribute to the development of all-solid-state batteries with high energy density.

Referemces

1. K. Kanehori, K. Masumoto, K. Miyauchi and T. Kudo, *Solid State Ionics*, **9–10** (1983) 1445.
2. H. Otsuka, S. Okada and J. Yamaki, *Solid State Ionics*, **40–41** (1990) 964.
3. S.D. Jones and J.R. Akridge, *Solid State Ionics*, **53–56** (1992) 628.
4. S. Kondo, K. Takada and Y. Yamamura, *Solid State Ionics*, **53–56** (1992) 1183.
5. J.B. Bates, N.J. Dudney, G.R. Gruzalski, R.A. Zuhr, A. Choudhury, C.F. Luck and J.D. Robertson, *J. Power Sources*, **43–44** (1993) 103.
6. K. Takada, N. Aotani and S. Kondo, *J. Power Sources*, **43–44** (1993) 135.
7. K. Iwamoto, N. Aotani, K. Takada and S. Kondo, *Solid State Ionics*, **79** (1995) 288.
8. K. Takada, N. Aotani, K. Iwamoto and S. Kondo, *Solid State Ionics*, **79** (1995) 284.
9. R. Komiya, A. Hayashi, H. Moromoto, M. Tatsumisago and T. Minami, *Solid State Ionics*, **140** (2001) 83.
10. N. Machida, H. Maeda, H. Peng and T. Shigematsu, *J. Electrochem. Soc.*, **149** (2002) A688.
11. M. Tatsumisago, S. Hama, A. Hayashi, H. Morimoto and T. Minami, *Solid State Ionics*, **154–155** (2002) 635.
12. F. Mizuno, A. Hayashi, K. Tadanaga, T. Minami and M. Tatsumisago, *Electrochemistry*, **71** (2003) 1196.
13. N. Machida and T. Shigematsu, *Chem. Lett.*, **33** (2004) 376.
14. M. Winter and J.O. Besenhard 5. Lithiated Carbons, In: J.O. Besenhard (ed.) *Handbook of Battery Materials*, p.383. Weinheim: Wiley-VCH (1999),
15. R. Fong, U.von Sacken and J.R. Dahn, *J. Electrochem. Soc.*, **137** (1990) 2009.
16. K. Tatumi, K. Zaghib, Y. Sawada, H. Abe and T. Ohsaki, *J. Electrochem. Soc.*, **142** (1990) 1090.
17. R. Nesper and H.G.V. Schnering, *J. Solid State Chem.*, **70** (1987) 48.
18. R.N. Seefurth and R.A. Sharma, *J. Electrochem. Soc.*, **128** (1977) 1207.
19. C.J. Wen and R.A. Huggins, *J. Solid State Chem.*, **37** (1981) 271.
20. C.S. Wang, G.T. Wu, X.B. Zhang, Z.F. Qi and W.Z. Li, *J. Electrochem. Soc.*, **145** (1998) 2751.
21. J. Niu and J.Y. Lee, *Electrochem. Solid-State Lett.*, **5** (2002) A107.
22. R. Tamori, N. Machida and T. Shigematsu, *J. Jpn. Soc. Powder and Powder Metallurgy*, **48** (2001) 267.

23. Y. Hashimoto, N. Machida and T. Shigematsu, *Solid State Ionics*, **175** (2004) 177.
24. N. Machida, Y. Yoneda and T. Shigematsu, *J. Jpn. Soc. Powder and Powder Metallurgy*, **51** (2004) 91.
25. H. Yamin, J. Penciner, A. Gorenshtain, M. Elam and E. Peled, *J. Power Sources*, **14** (1985) 129.
26. H. Yamin, A. Gorenshtein, J. Penciner, Y. Sternberg and E. Peled, *J. Electrochem. Soc.*, **135** (1988) 1045.
27. S.-E. Cheon, J.-H. Cho, K.-S. Ko, C.-W. Kwon, D.-R. Chang, H.-T. Kim and S.-W. Kim, *J. Electrochem. Soc.*, **149** (2002) A1437.
28. Y.V. Mikhaylik and J.R. Akridge, *J. Electrochem. Soc.*, **150** (2003) A306.
29. S.-E. Cheon, K.-S. Ko, J.-H. Cho, S.-W. Kim, E.-Y. Chin and H.-T. Kim, *J. Electrochem. Soc.*, **150** (2003) A796.
30. N. Machida, K. Kobayashi, Y. Nishikawa and T. Shigematsu, *Solid State Ionics*, **175** (2004) 247.
31. S. Tobishima, H. Yamamoto and M. Matsuda, *Electrochim. Acta*, **42** (1997) 1019.

3.2.3 All-solid-state lithium secondary batteries

3.2.3.1 Background

All-solid-state lithium secondary batteries have attracted much attention because the replacement of conventional liquid electrolyte with inorganic solid electrolyte essentially improves safety and reliability of lithium batteries [1,2]. There are two approaches of developing all-solid-state batteries; one is a thin-film micro-battery prepared by RF sputtering and laser ablation, and the other is a bulk-type battery constructed of electrolyte and electrode powders. Several all-solid-state thin film batteries with inorganic glassy electrolytes such as LiPON have been reported to show excellent long-cycling performances (over 50,000 times) at room temperature [3,4] and the related film batteries will be discussed in the next section. The bulk-type battery has an advantage of enhancing cell capacity by the addition of large amounts of active materials to the cell. Both utilizing highly conductive solid electrolytes and achieving a close contact between electrolyte and electrode powders are key to improve the cell performance. As mentioned in the section 3.2.1, Li_2S-based sulfide materials are promising solid electrolyte for all-solid-state batteries. The electrochemical performance of all-solid-state In / $LiCoO_2$ cells with the Li_2S-SiS_2-Li_3PO_4 glasses was firstly reported in 1994 by Kondo *et al.* [5] and then these cells with sulfide solid electrolytes have been developed [6–11]. In order to build lithium ion and electron conduction pathways in the electrode for all-solid-state batteries, the design of the composite electrode with active material, solid electrolyte, and conductive additive powders is important as shown in Fig. 3.2.15 in the previous section 3.2.2. The suitable composite electrodes would improve the properties of all-solid-state batteries, especially high-rate performance.

In this section, we focus on three components constructing composite electrodes; (i) solid electrolyte, (ii) electrode material, and (iii) conductive additive. The effects of these components on the electrochemical performance of the all-solid-state cells with sulfide solid electrolytes were investigated.

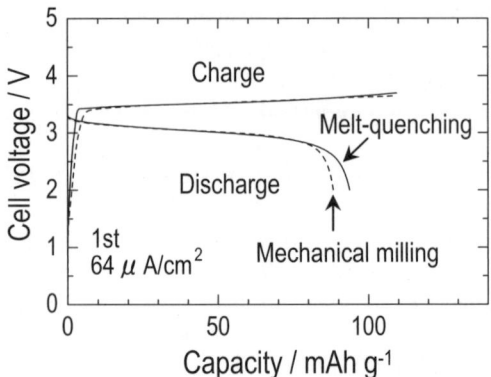

Fig. 3.2.19. Charge-discharge curves at the first cycle for the In / LiCoO$_2$ cells with the 95(0.6Li$_2$S·0.4SiS$_2$)·5Li$_4$SiO$_4$ oxysulfide glasses prepared by melt quenching and mechanical milling. The current density was 64 μAcm^{-2}.

3.2.3.2 Solid electrolytes in all-solid-state cells

The 95(0.6Li$_2$S·0.4SiS$_2$)·5Li$_4$SiO$_4$ oxysulfide glass with ambient temperature conductivity of 10^{-3} S cm^{-1} was used as a solid electrolyte [12]. Electrochemical measurements were carried out for all-solid-state cells, which were three-layered powder compressed pellets. The first layer was an indium foil as a negative electrode. The second layer was the glassy powder as an electrolyte. The third layer was a composite as a positive electrode. In order to achieve smooth electrochemical reaction in the cell, we prepared the composite positive electrode composed of three kinds of powders: the active material, the solid electrolyte powder providing lithium ion conduction path, and the acetylene black providing electron conduction path. The composite positive electrode consisting of LiCoO$_2$, the oxysulfide glass and acetylene-black powders with a weight ratio of 20 : 30 : 3 was used for all-solid-state cells [7,8]. These cells were charged and discharged under a constant current density of 64 μA cm^{-2} at room temperature in Ar atmosphere.

Figure 3.2.19 shows the first charge-discharge curves for the In / LiCoO$_2$ cells with the 95(0.6Li$_2$S·0.4SiS$_2$)·5Li$_4$SiO$_4$ oxysulfide glasses prepared by melt quenching and mechanical milling [13]. Both cells with the quenched and milled glasses were charged up to the capacity of 110 mAh g^{-1}, which corresponds to the composition of x=0.4 in Li$_{1-x}$CoO$_2$, and discharged to 2.0 V; these cells work as lithium secondary batteries. The charge-discharge curves of the cell with the mechanically milled glass are similar to those of the cell with the melt quenched glass. The average cell voltages on charge and discharge are 3.6 V and 3.1 V, respectively. The discharge capacity of the cell with the mechanically milled glass is about 90 mAh g^{-1}, which is just smaller than that of the quenched glass. The mechanically milled glasses as well as the melt quenched glasses can be applied as solid electrolytes to all-solid-state lithium secondary batteries.

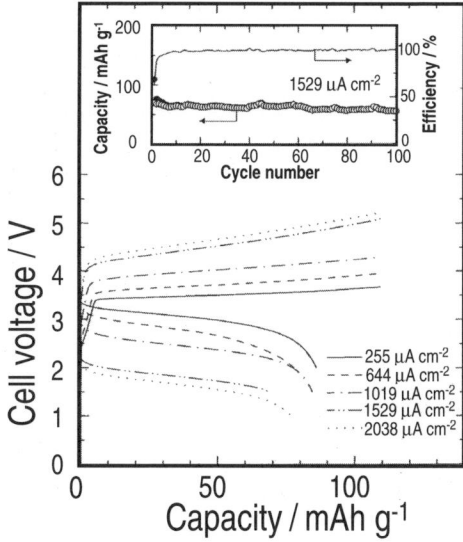

Fig. 3.2.20. Charge-discharge curves at the first cycle of In / $LiCoO_2$ cells with the $95(0.6Li_2S \cdot 0.4SiS_2) \cdot 5Li_3BO_3$ oxysulfide glass prepared by melt quenching under various current densities. The inset shows cycling performance of the cell under the current density of 1529 µAcm^{-2}.

The rate performance of the cells was also investigated. Figure 3.2.20 shows the first charge-discharge curves of In / $LiCoO_2$ cells with the $95(0.6Li_2S \cdot 0.4SiS_2) \cdot 5Li_3BO_3$ oxysulfide glass prepared by melt quenching. The measurements were examined under various current densities from 255 to 2038 µA cm^{-2}. Although the polarization increases with an increase in current density, the cell can be charged and discharged even under the high current density over 1mA cm^{-2}. The inset shows cycling performance of the cell under the current density of 1529 µA cm^{-2}. The cell capacities decrease at the first few cycles and then keep a constant value of about 60 mAh g^{-1} up to 100 cycles. Although the charging voltage reached over 5 V under the current density, the cell exhibits excellent cycling performance over 100 times and the charge-discharge efficiency is almost 100 %, suggesting that the cell works as a lithium secondary battery without the decomposition of the glassy electrolyte.

The all-solid-state In / $LiCoO_2$ cell using Li_2S-P_2S_5 glass-ceramics as a solid electrolyte was also constructed in a similar way [10,14]. The glass-ceramics with ambient temperature conductivity of about 10^{-3} S cm^{-1} were prepared by crystallization of the mechanically milled $80Li_2S \cdot 20P_2S_5$ glass [15]. Figure 3.2.21 shows the cell capacities and charge-discharge efficiencies under the constant current density of 64 µA cm^{-2} as a function of cycle number. Although an irreversible capacity is initially observed at the first few cycles, the all-solid-state cell maintains the reversible capacity of about 100 mAh g^{-1} and the charge-discharge efficiency of 100% (no irreversible capacity) for 500 cycles. The charge-discharge profile of

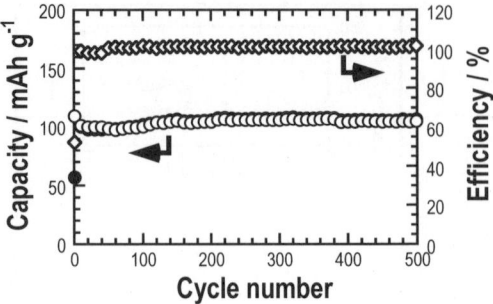

Fig. 3.2.21. Cycle performance of the rechargeable capacities and charge-discharge efficiencies for the In / LiCoO$_2$ cell with the 80Li$_2$S·20P$_2$S$_5$ glass-ceramics under the current density of 64 µAcm^{-2} at room temperature.

the cell at the 300th cycle is shown in Fig. 3.2.22. The excellent cycling performance with no capacity loss of the cells using Li$_2$S-P$_2$S$_5$ glass-ceramics is superior to that of the similar cells using Li$_2$S-SiS$_2$ based oxysulfide glasses mentioned above. During the formation of Li$_2$S-P$_2$S$_5$ based glass-ceramics, a softening of glass concurrently occurred [16]. The unique process with softening the glassy fine powders can be applied to form the close solid/solid contact between electrolyte and electrode powders, which would enhance the cell performance.

3.2.3.3 Electrode materials in all-solid-state cells

In this section, we reports electrochemical properties of all-solid-state cells with several positive and negative electrodes. At first, we demonstrated the cell performance using LiCoO$_2$, LiNi$_{1/2}$Mn$_{1/2}$O$_2$, Li$_{4/3}$Ti$_{5/3}$O$_4$, and amorphous V$_2$O$_5$ as active materials. Layered LiNi$_{1/2}$Mn$_{1/2}$O$_2$ is attractive from viewpoints of both economy and safety, and this active material exhibited high rechargeable capacity of 200 mAh g^{-1} in nonaqueous lithium cells [17,18]. Li$_{4/3}$Ti$_{5/3}$O$_4$ with a defect spinel-framework structure has a favorable feature of negligible volume change during charge-discharge processes [19], which suppress the loss of electrical contact between the active material and the other components of the solid electrolyte and the conductive additive. Amorphous V$_2$O$_5$ is a typical amorphous positive electrode and shows reversible cycling performance in liquid-type lithium cells [20,21].

Figure 3.2.22 shows charge-discharge curves at the 300th cycle for all-solid-state cells In or In-Li alloy / LiCoO$_2$ [10], LiNi$_{1/2}$Mn$_{1/2}$O$_2$ [22], Li$_{4/3}$Ti$_{5/3}$O$_4$, or amorphous V$_2$O$_5$. The 80Li$_2$S·20P$_2$S$_5$ glass-ceramic was used as a solid electrolyte. The electrochemical measurements were carried out at room temperature under a constant current density of 64 µA cm^{-2}. In spite of the 300th cycle data, high capacities and 100 % efficiency are observed in the charge-discharge performance of all the cells with various electrode materials. The cells show different voltage plateaux based on the combination of electrode materials. The glass-ceramic

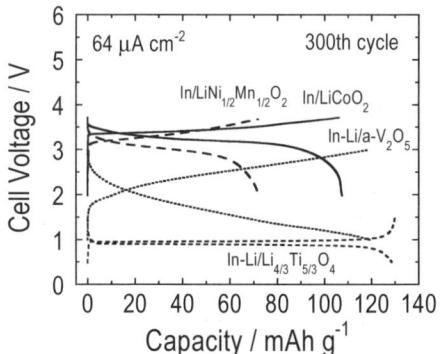

Fig. 3.2.22. Charge-discharge curves at the 300th cycle for all-solid-state cells In or In-Li alloy / LiCoO$_2$, LiNi$_{1/2}$Mn$_{1/2}$O$_2$, Li$_{4/3}$Ti$_{5/3}$O$_4$, or amorphous V$_2$O$_5$. The 80Li$_2$S·20P$_2$S$_5$ glass-ceramic was used as a solid electrolyte.

electrolytes are compatible with various positive electrode materials for all-solid-state cells with excellent cyclability.

Recently, LiCoPO$_4$ with olivine structure was reported to be a 5 V class positive electrode material [23], which is expected to develop energy density of current lithium ion cells. Solid electrolytes with a wide electrochemical window over 5 V is more suitable for the electrode than conventional liquid electrolytes, which would be decomposed during cycling. All-solid-state cells using LiCoPO$_4$ as a positive electrode were constructed and their electrochemical properties were examined [24]. The In / LiCoPO$_4$ cell with a Li$_2$S-based oxysufide glass was charged and then discharged at 4.1 V against In (4.7 V vs Li/Li$^+$), although the discharge capacity was smaller than the theoretical capacity. The increase in electronic conductivity of LiCoPO$_4$ is a key issue to enhance the cell performance.

Much attention has also been devoted to develop negative electrode materials instead of conventional graphite in order to realize lithium secondary batteries with higher energy density. Here, Sn-based glassy materials as an alternative negative electrode are picked up. Tin-oxide based glasses in the system SnO-B$_2$O$_3$-P$_2$O$_5$-Al$_2$O$_3$ [25,26] were reported to work as a negative electrode material with high capacity in lithium ion cells. The cell performance of SnO-based glasses exhibited larger reversible capacity than the corresponding crystalline material [27]. The cell capacity depends on the composition of glasses, which determines the local structure of glass matrix. The SnO-B$_2$O$_3$ glasses were prepared by melt-quenching in the composition range 0 ≤ SnO (mol %) ≤ 75 and the capacity of the cell with the SnO-B$_2$O$_3$ glasses was estimated to increase monotonically with increasing SnO content as active center of lithium ions. However, the highest capacity was obtained at the composition of 50 mol% SnO, where the glass matrix was mainly formed by four-coordinated boron anions (BO$_4^-$) [28]. The cell capacity was found to be closely related to local structure of the glass.

Very recently, the SnS-P_2S_5 glassy materials were synthesized by mechanical milling and applied as an electrode to all-solid-state cells with the Li_2S-P_2S_5 solid electrolyte [29]. Because of using a common P_2S_5 component, a continuous sulfide network between electrode and electrolyte was formed in the cells. The cell performance with the $80SnS \cdot 20P_2S_5$ glassy electrode was compared to that with the SnS electrode. The effects of the addition of a network former P_2S_5 to SnS on the cell capacity and cyclability are discussed.

Figure 3.2.23 shows the first charge-discharge curves of all-solid-state cells Li-In / $80SnS \cdot 20P_2S_5$ and Li-In / SnS. The $80Li_2S \cdot 20P_2S_5$ glass-ceramic solid electrolyte with high conductivity was used in these cells. The charge-discharge measurements were carried out at a current density of 64 µA cm^{-2} at 25°C. The ordinate on the left hand side denotes cell potential $vs.$ the Li-In counter electrode, and that on the right hand side denotes potential vs. the Li electrode, which was calculated from the basis of the difference on the potential between Li-In and Li [6]. A charge corresponds to an insertion of lithium ions to the working electrode, while a discharge corresponds to an extraction of lithium ions from the working electrode.

Two plateaux at around 1.3 V and 0.5 V ($vs.$ Li) are observed on the charge curve and one plateau at around 0.5 V is observed on the discharge curve in both cells. The SnS-P_2S_5 amorphous materials prove to work as electrode materials for all-solid-state rechargeable lithium batteries. The charge-discharge capacities of the cell with the $80SnS \cdot 20P_2S_5$ electrode are larger than those with the SnS electrode. The discharge capacity of the cell with the $80SnS \cdot 20P_2S_5$ electrode (590 mAh g^{-1}) is 1.5 times as large as the capacity of conventional graphite electrodes used in commercialized lithium ion secondary batteries. The charge-discharge efficiency for the cell with the $80SnS \cdot 20P_2S_5$ electrode is 56%, which is larger than that of the cell with SnS (43%). Although a large irreversible capacity is observed at the 1st cycle for the cell with the SnS-P_2S_5 electrode, the addition of P_2S_5 as a network former component to SnS enhances the reversible capacity of the cells.

The charge-discharge profiles with two-step of the cells with SnS-P_2S_5 electrodes are quite similar to those of the nonaqueous lithium cells with SnO-P_2O_5 electrodes [30]. The mechanism of electrochemical lithium insertion to SnO-based glassy materials has been widely investigated. The first plateau on the charge curve is due to the formation of metallic Sn nanoparticles ($Sn^{2+} \rightarrow Sn^0$) and Li_2O-based glassy matrix (irreversible process), and the second plateau is due to the formation of Li-Sn alloy domains ($Sn^0 \rightarrow Li$-Sn) (reversible process) [26-28].

Based on the charge-discharge process on SnO-based glasses, it is presumed that the first plateau of the cell with the $80SnS \cdot 20P_2S_5$ electrode as shown in Fig.3.2.23 is due to the formation of Sn nanoparticles and the Li_2S-P_2S_5 matrix. The formation of the highly Li$^+$ conductive Li_2S-P_2S_5 matrix around Sn active particles provides a close electrode-electrolyte interface, where smooth electrochemical reaction would occur during the charge-discharge cycle. On the other hand, the less conductive Li_2S matrix would be formed in the cell with SnS electrode. The sulfide glassy matrix played an important role in not only preventing the aggregation of Sn domains but also endowing the electrode with high conductivity. Hence, the $80SnS \cdot 20P_2S_5$ electrode exhibited better performance than the SnS electrode as shown in Fig.3.1.23.

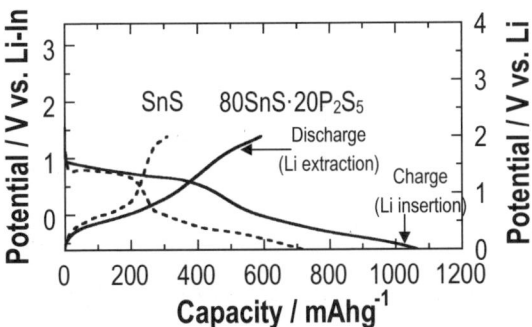

Fig. 3.2.23. Charge-discharge curves at the first cycle of all-solid-state cells Li-In / 80SnS·20P$_2$S$_5$, and Li-In / SnS. The 80Li$_2$S·20P$_2$S$_5$ glass-ceramic was used as a solid electrolyte. The charge-discharge measurements were carried out at a current density of 64 μAcm^{-2} at 25°C.

Figure 3.2.24 shows the charge-discharge curves at the 50th cycle of the all-solid-state 80SnS·20P$_2$S$_5$ / LiCoO$_2$ cell. The 80Li$_2$S·20P$_2$S$_5$ glass-ceramic electrolyte was used in the cells. The charge-discharge measurements were carried out at a current density of 64 μA cm^{-2} in the potential range from 2 to 4 V. The charge-discharge plateaux of the cell are about 3.4 V. The inset shows the cycling performance of the cell. The reversible capacity gradually decreases and settled in over 400 mAh g^{-1} at the 50 cycle. The electrochemical performance of the cells with tin-based electrodes is very sensitive to cut-off voltages, which affect the size of Sn domains in glass matrix [26]. The constant reversible capacity of about 400 mAh g^{-1} was obtained for 50 cycles under the measurements in the potential range from 2.5 to 4 V.

As a future development of all-solid-state cells, Souquet et al. proposed a glassy monolithic cell [31]: a common network former is used for the electrolyte and electrodes materials. The monolithic cell with a continuity for the lithium ion surrounding during its transfer from one electrode to the other would have a favorable feature of reducing the interfacial polarization during ion transfer. The all-solid-state cells with the combination of the SnS-P$_2$S$_5$ electrode and the Li$_2$S-P$_2$S$_5$ electrolyte is a first step for the realization of the monolithic cell.

3.2.3.4 Conductive additives in all-solid-state cells

In the cells with liquid electrolytes, favorable electrode-electrolyte interfaces are formed just by soaking electrodes in liquid electrolytes, while in the cells with solid electrolytes, electrode and electrolyte powders should be purposely contacted to form close solid-solid interfaces, where electrochemical reactions occur. In the solid-state cells, composite electrodes with a conductive additive are commonly used in order to form continuous electron conducting paths. This section focuses on conductive additives in composite positive electrodes in order to construct favorable solid-solid interfaces in all-solid-state cells. The effects of conductive additives in composite

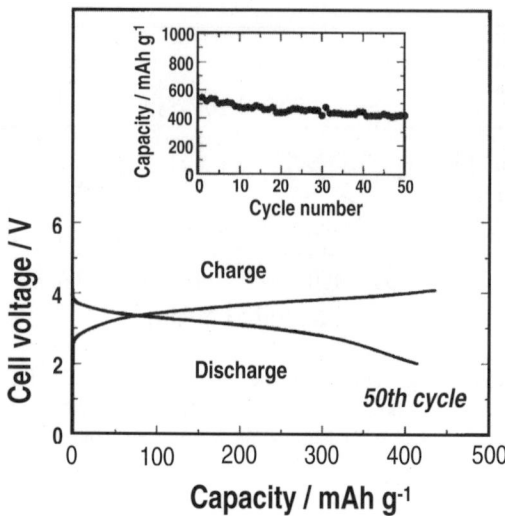

Fig. 3.2.24. Charge-discharge curves at the 50th cycle of the all-solid-state 80SnS·20P$_2$S$_5$ / LiCoO$_2$ cell. The 80Li$_2$S·20P$_2$S$_5$ glass-ceramic electrolyte was used in the cells. The inset shows the cycling performance of the cell.

positive electrodes on charge-discharge behaviors of all-solid-state In / LiCoO$_2$ cells with the 80Li$_2$S·20P$_2$S$_5$ glass-ceramic electrolyte were investigated. Acetylene black (AB), vapor grown carbon fiber (VGCF), titanium nitride (TiN) and nickel (Ni) were used as conductive additives.

The morphologies of the conductive additives were examined using a field emission scanning electron microscope (FE-SEM). Figure 3.2.25 shows FE-SEM photographs of AB, VGCF, TiN and Ni. AB is composed of aggregates with primary particles of about 15 nm in average size, whereas VGCF is composed of fibers of about 150 nm in average diameter and about 10 μm in average length. The average particle sizes of TiN and Ni are about 1.0 and 0.5 μm, respectively.

Using a composite positive electrode without a conductive additive (LiCoO$_2$: solid electrolyte = 40 : 60 (wt. %)), the solid-state In / LiCoO$_2$ cell was assembled. The cell voltage rapidly reached 6 V in charging and the cell was not able to be discharged at a current density of 64 μA cm^{-2} (the charge-discharge curves are shown in Fig.3.2.26). In this case, it is difficult to transport electrons in the composite positive electrode without a conductive additive. The addition of conductive additives is very important to achieve sufficient electronic conduction of composite positive electrodes.

The additional amounts to endow a composite positive electrode with sufficient electron conductivity were determined by ac impedance measurements for the model composite materials consisting of conductive additive and insulative sulfur [32]. Carbon materials AB and VGCF were more effective to decrease the resistivities of composite materials than TiN and Ni. Figure.3.2.26 shows the first charge and discharge curves of all-solid-state In / LiCoO$_2$ cells with various conductive

Fig. 3.2.25. FE-SEM photographs of acetylene black (AB), vapor grown carbon fiber (VGCF), titanium nitride (TiN) and nickel (Ni).

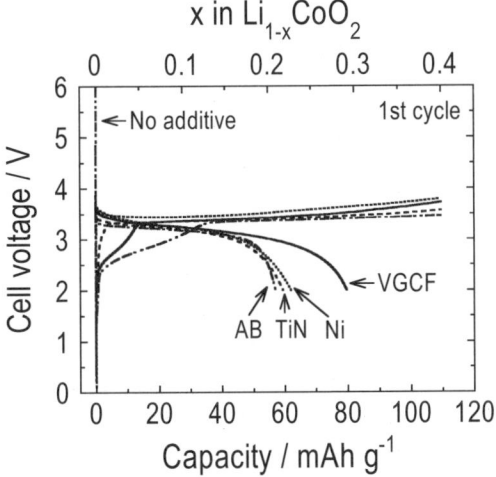

Fig. 3.2.26. Charge and discharge curves at the first cycle of all-solid-state In / LiCoO$_2$ cells with various conductive additives. The 80Li$_2$S·20P$_2$S$_5$ glass-ceramic was used as a solid electrolyte. The amounts of AB, VGCF, TiN and Ni added to the composite positive electrodes were respectively 6, 4, 60 and 40 wt%, and a current density was 64 μAcm^{-2}.

additives. The amounts of AB, VGCF, TiN and Ni added to the composite positive electrodes were respectively 6, 4, 60 and 40 wt%, and a current density was 64 μA cm^{-2}. All cells with conductive additives work as rechargeable batteries at room temperature. Under a cut off voltage of 2.0 V, the cells with AB, TiN and Ni exhibit discharge capacities of about 60 mAh g^{-1}, while the cell with VGCF exhibits larger discharge capacities of about 80 mAh g^{-1}. All cells were charged up to x=0.40 in Li$_{1-x}$CoO$_2$ (110 mAh g^{-1}) and discharged to 2.0 V until the 2nd cycle, and then charge-discharge cycles were repeated between the maximum charge voltage of the 2nd cycle and 2.0 V after the 3rd cycle. The cell with VGCF maintained the discharge

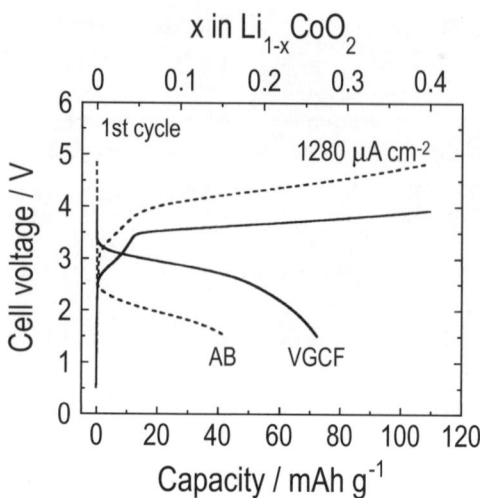

Fig. 3.2.27. Charge and discharge curves at the first cycle of all-solid-state In / LiCoO$_2$ cells with AB and VGCF. The amounts of AB and VGCF added to the composite positive electrodes were respectively 6 and 4 wt%, and a current density was 1280 μAcm^{-2}.

capacities of about 100 mAh g^{-1} and charge-discharge efficiencies of 100% up to the 300th cycle. On the other hand, discharge capacities of the cells with AB, TiN and Ni monotonically decreased with an increase in cycle number, although charge-discharge efficiencies of 100% were obtained. At the 30th cycle, discharge capacities of the cells with AB, TiN and Ni were respectively about 70, 50 and 60 mAh g^{-1}. The fact that the cells with AB and VGCF exhibited larger discharge capacities than those with TiN and Ni suggests that the carbon materials would be suitable to carry out reversible electrochemical reactions at solid-solid interfaces in the composite positive electrodes as compared with TiN and Ni.

Cell performances under high current densities over 1 mA cm^{-2} were examined for all-solid-state cells with carbon conductive additives. Figure 3.2.27 shows the first charge and discharge curves of all-solid-state In / LiCoO$_2$ cells with AB and VGCF. The amounts of AB and VGCF added to the composite positive electrodes were respectively 6 and 4 wt%, and a current density was 1280 μA cm^{-2}. The cells with AB and VGCF are charged and discharged even at a current density over 1 mA cm^{-2}, although the drops of the discharge potentials are observed. The cells with AB and VGCF exhibit discharge capacities of about 40 and 70 mAh g^{-1} under a cut off voltage of 1.5 V, respectively. The cell with AB shows a lower discharge plateau and a smaller discharge capacity than the cell with VGCF, suggesting that an overpotential observed in the cell with AB is larger than that in the cell with VGCF. Both cells were charged up to x=0.40 in Li$_{1-x}$CoO$_2$ (110 mAh g^{-1}) and discharged to 1.5 V until the 2nd cycle, and then charge-discharge cycles were repeated between the maximum charge voltage of the 2nd cycle and 1.5 V after the 3rd cycle. After the 3rd cycle, discharge capacities of both cells monotonically decreased with an increase in cycle number, although charge-discharge efficiencies of about 100% were obtained. The discharge capacities of the cells with AB and VGCF at the 50th cycle were about 10 and 40 mAh g^{-1}, respectively, and the cell with VGCF kept larger discharge capacities during 50 cycles than the cell with AB [33]. The difference in the cycling performance for the cells with AB and VGCF would be closely related to the morphologies of conductive additives in composite positive electrodes. AB with nano-ordered primary particles would make contact with LiCoO$_2$ at the nanodimensional level. However, it is difficult to homogeneously disperse AB in the composite electrodes by dry-mixing because AB is easy to aggregate by itself. On the other hand, VGCF would make contact with LiCoO$_2$ at multiple points along submicron-ordered fibers. Since it is easy for VGCF to form a continuous electron conducting path within an electrode, the overpotential in the cell with VGCF was smaller than that with AB. Therefore, the cell with VGCF showed larger discharge capacity during 50 cycles than the cell with AB. The design of continuous electron conducting paths from the point of view of morphology for conductive additives is important to enhance cell performances of all-solid-state lithium secondary batteries. Fibrous VGCF is a promising conductive additive suitable for all-solid-state batteries.

References

1. C. Julien and G.-A. Nazri, *Solid State Batteries: Materials Design and Optimization*, Kluwer Academic Publishers, Boston, p. 579 (1994).
2. J.-M. Tarascon and M. Armand, *Nature*, **414** (2001) 359.
3. J.B. Bates, *Electron. Eng.*, **69** (1997) 63.
4. B.J. Neudecker, N.J. Dudney and J.B. Bates, *J. Electrochem. Soc.*, **147** (2000) 517.
5. K. Iwamoto, N. Aotani, K. Takada and S. Kondo, *Solid State Ionics*, **70/71** (1994) 658.
6. K. Takada, N. Aotani, K. Iwamoto and S. Kondo, *Solid State Ionics*, **86–88** (1996) 877.
7. R. Komiya, A. Hayashi, H. Morimoto, M. Tatsumisago and T. Minami, *Solid State Ionics*, **140** (2001) 83.
8. A. Hayashi, R. Komiya, M. Tatsumisago and T. Minami, *Solid State Ionics*, **152–153** (2002) 285.
9. N. Machida, H. Maeda, H. Peng and T. Shigematsu, *J. Electrochem. Soc.*, **149** (2002) A688.
10. F. Mizuno, S. Hama, A. Hayashi, K. Tadanaga, T. Minami and M. Tatsumisago, *Chem. Lett.*, **2002**, 1244.
11. K. Takada, T. Inada, A. Kajiyama, H. Sasaki, S. Kondo, M. Watanabe, M. Murayama and R. Kanno, *Solid State Ionics*, **158** (2003) 269.
12. T. Minami, A. Hayashi and M. Tatsumisago, *Solid State Ionics*, **136–137** (2000) 1015.
13. A. Hayashi, H. Yamashita, M. Tatsumisago and T. Minami, *Solid State Ionics*, **148** (2002) 381.

14. F. Mizuno, A. Hayashi, K. Tadanaga, T. Minami and M. Tatsumisago, *Electrochemistry*, **71** (2003) 1196.
15. A. Hayashi, S. Hama, T. Minami and M. Tatsumisago, *Electrochem. Commun.*, **5** (2003) 111.
16. M. Tatsumisago, S. Hama, A. Hayashi, H. Morimoto and T. Minami, *Solid State Ionics*, **154–155** (2002) 635.
17. T. Ohzuku and Y. Makimura, *Chem. Lett.*, **2001**, 744.
18. Y. Makimura and T. Ohzuku, *J. Power Sources*, **119** (2003) 156.
19. T. Ohzuku, A. Ueda and N. Yamamoto, *J. Electrochem. Soc.*, **142** (1995) 1431.
20. M. Nabavi, C. Sanchez, F. Taulelle, J. Livage and A Guibert, *Solid State Ionics*, **28–30** (1988) 1183.
21. N. Machida, R. Fuchida and T. Minami, *J. Electrochem. Soc.*, **136** (1989) 2133.
22. F. Mizuno, A. Hayashi, K. Tadanaga, T. Minami and M. Tatsumisago, *J. Power Sources*, **124** (2003) 170.
23. K. Amine, H. Yasuda and M. Yamachi, *Electrochem. Solid-State Lett.*, **3** (2000) 178.
24. K. Tadanaga, F. Mizuno, A. Hayashi, T. Minami and M. Tatsumisago, *Electrochemistry*, **71** (2003) 1192.
25. Y. Idota, T. Kubota, A. Matsufuji, Y. Maekawa and T. Miyasaka, *Science*, **276** (1997) 1395.
26. I.A. Courtney and J.R. Dahn, *J. Electrochem. Soc.* **144** (1997) 2943.
27. Y.W. Xiao, J.Y. Lee, A.S. Yu and Z.L. Liu, *J. Electrochem. Soc.*, **146** (1999) 3623.
28. A. Hayashi, M. Nakai, M. Tatsumisago, T. Minami and M. Katada, *J. Electrochem. Soc.*, **150** (2003) A582.
29. A. Hayashi, T. Konishi, K. Tadanaga, T. Minami and M. Tatsumisago, *J. Power Sources*, in press (2005).
30. A. Hayashi, T. Konishi, M. Nakai, H. Morimoto, K. Tadanaga, T. Minami and M. Tatsumisago, *J. Ceram. Soc. Jpn.*, **112** (2004) S695.
31. J.L. Souquet and M. Duclot, *Solid State Ionics*, **148** (2002) 375.
32. F. Mizuno, A. Hayashi, K. Tadanaga and M. Tatsumisago, *J. Electrochem. Soc.*, in press.
33. F. Mizuno, A. Hayashi, K. Tadanaga and M. Tatsumisago, *J. Power Sources*, in press (2005).

3.2.4 Thin film batteries

3.2.4.1 Background

The concept of the thin film battery is very simple; just to construct solid films of anode, solid electrolyte and cathode sequentially on a substrate as shown in Fig.3.2.28. Thus, some naive cells were reported already 35 years ago, *eg.* Li/LiI/AgI [1]. By using polymer or gel electrolytes, a "thick" film battery of less than 0.3mm has been already available [2]. The "thin film battery" invoked here belongs to another category, which is constructed by physical vacuum vapor processes matching

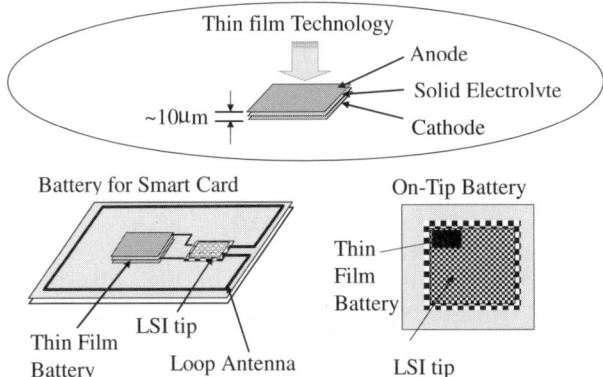

Fig. 3.2.28. Concept of thin-film battery and its possible applications.

to modern semiconductor technologies. The cell is usually less than 10μm in total thickness and is aimed to integrated with semiconductor devices.

The first epoch making report of practical thin film battery was announced in 1982 from Hitachi Co., Ltd. Japan. However it was too early to be used widely in electric devices, since the power of the battery was too small to operate contemporary electric devices those days [3]. During these 20 years, the advancement of semiconductor technology allows us to use low power devices such as CMOS, FE-RAM and liquid crystal displays etc. Now, it is possible to operate some small electric devices even by the thin film battery. Moreover, the recent developments of micro-electric devices such as wearable computers, RF-ID tags, micro-machines etc. strongly demand micro-power sources with high energy density. Fortunately, now we have much more choices of the materials for battery components owing to the developments in solid state ionics filed, thus we will prospect the practical use of the thin film batteries in near future [4–15].

In order to construct a thin film battery, it is necessary to fabricate all the battery components, as an anode, a solid electrolyte, a cathode and current leads into multi-layered thin films by suitable techniques. Usually, the lithium metal used for anode is prepared by vacuum thermal vapor deposition (VD). Solid electrolytes and cathode or sometimes anode materials of oxides are prepared by various sputtering techniques as RF sputtering (RFS), RF magnetron sputtering (RFMS). In some cases chemical vapor deposition (CVD) and electrostatic spray deposition (ESD) are used. Recently, pulsed laser deposition (PLD) is often used especially for cathode materials. Typical thin film batteries reported so far are listed in Table 3.2.1, where the preparation methods are also shown in brackets. Owing to the initial cost for the construction of the cluster vacuum chambers and expensive preparation devices, only some restricted groups, mainly in big companies of Japan, France and USA have succeeded in fabrication of thin film batteries. Recently, the group in Tohoku University Japan succeeded to fabricate thin-film batteries only by a sequential PLD technique, which uses only one vacuum chamber and a laser source. This technique

Table 3.2.1. Thin film lithium batteries reported in literatures.

Anode	Electrolyte	Cathode	Voltage [V]	Current [$\mu A/cm^2$]	Capacity	Ref.
Li (VD)	$Li_{3.6}Si_{0.6}P_{0.4}O_4$ (RFS)	TiS_2 (CVD)	2.5	16	45~150$\mu Ah/cm^2$	[3]
Li (VD)	$Li_{3.6}Si_{0.6}P_{0.4}O_4$ (RFS)	TiS_2 (CVD)	2.5	16~30	-	[38]
Li (VD)	$Li_{3.6}Si_{0.6}P_{0.4}O_4$ (RFS)	WO_3, WO_3-V_2O_5 (RFS)	1.8~2.2	16	60~92Ah/cm^2	[18]
Li (VD)	$LiBO_2$ (MBD)	In_2Se_3 (MBD)	1.2	0.1	-	[33]
Li (VD)	Li_2SO_4-Li_2O-B_2O_3 (RFS)	TiS_xO_y (MBD)	2.6	1~60	40~15$\mu Ah/cm^2$	[39]
Li (VD)	Li_2S-SiS_2-P_2S_5 (RFS)	V_2O_5-TeO_2	2.8~3.1	0.5~2	-	[34]
$Li_xV_2O_5$(RFMS)	LiPON (RFMS)	V_2O_5 (RFMS)	3.5~3.6	10	6$\mu Ah/cm^2$	[29]
V_2O_5 (RFMS)	LiPON (RFMS)	$LiMn_2O_4$ (RFMS)	3.5~1	>2	18$\mu Ah/cm^2$	[30]
Li/LiI (VD)	LiI-Li_2S-P_2S_5-P_2O_5 etc. (SP)	TiS_2 (SP)	1.8~2.8	300	70mAh/cm^3	[6]
Li (VD)	LiBP, LiPON (SP)	$LiMn_2O_4$ (SP)	3.5~4.5	70	100mAh/g	[8]
Li (VD)	$Li_{6.1}V_{0.61}Si_{0.39}O_{5.36}$ (RFS)	$MoO_{2.89}$ (RFS)	2.8	20	60$\mu Ah/cm^2$	[20]
Li (VD)	$Li_{6.1}V_{0.61}Si_{0.39}O_{5.36}$ (RFS)	$LiMn_2O_4$ (RFS)	3.5~5	10	33.3 $\mu Ah/cm^2$	[22]
Li (VD)	LiPON (RFS)	$LiMn_2O_4$ (RFS)	4.5~2.5	2~40	11~81 $\mu Ah/cm^2$	[40]
Cu	LiPON (RFS)	$LiCoO_2$ (RFS)	4.2~3.5	1~5	130$\mu Ah/cm^2$	[27]
Li (VD)	LiPON (RFS)	$LiCoO_2$ (RFS)	4.2~2.0	50~400	35$\mu Ah/cm^2$	[26]
Li (VD)	LiPON (RFS)	$Li_x(Mn_yNi_{1-y})_{2-x}O_2$ (RFS)	4~3.5	1~10	100mAh/g	[41,42]
Li (VD)	LiPON (RFS)	$LiMn_2O_4$ (RFS)	4~5.3	10	10~30$\mu Ah/cm^2$	[43]
Li (VD)	LiPON (RFS)	Li-V_2O_5 (RFS)	1.5~3	2~40	10~20$\mu Ah/cm^2$	[44]
$SiSn_{0.87}O_{1.20}N_{1.72}$	LiPON (RFS)	$LiCoO_2$ (RFS)	2.7~4.2	~5000	340~450mAh/g	[45]
Li (VD)	LiPON (RFS)	$LiMn_2O_4$ (RFMS)	4.3~3.7	~800	45$\mu Ah/cm^2$-μm	[28]
SnO (PLD)	$Li_{6.1}V_{0.61}Si_{0.39}O_{5.36}$ (PLD)	$LiCoO_2$ (PLD)	2.7~1.5	10~200	4~10$\mu Ah/cm^2$	[16,17]

Preparation technique is shown in bracket where, MBD: molecular beam deposition, CVD: chemical vapor deposition, VD: vacuum vapor evaporation, S: sputtering, RFS: RF-sputtering, RFMS: RF magnetron sputtering. Voltage and current are in normal charging/discharging condition. Capacity of a cell is expressed in $\mu Ahcm^{-2}$, although some are shown in original form due to the difficulty of conversion.

reduces the initial cost for developments and will be useful for further research of thin film batteries [16,17].

In this section, a short review of the historical efforts of developing thin film batteries is given followed by the approach by using the PLD technique.

3.2.4.2 History of thin film batteries

The history of the practical thin film battery started in1982, when an "all solid-state thin film battery" was first announced from Hitachi Co., Ltd. Japan. It comprised a TiS_2 cathode prepared by CVD, a $Li_{3.6}Si_{0.6}P_{0.4}O_4$ glass electrolyte by RF sputtering and a metallic lithium as anode deposited by a vacuum evaporation [3,38]. Also, a-WO_3 or $0.63WO_3 \cdot 0.37V_2O_5$ cathode prepared by reactive sputtering in an H_2-Ar plasma was combined with the above battery [18]. The first battery was 4x4 mm in dimension, total thickness is 6 ~ 8 μm comprising of 1 ~ 3.7 μm of TiS_2 cathode, 2 ~ 4 μm electrolyte and 4μm lithium anode, which generated 2.5 V and a maximum current density of 3 ~ 16 μA cm^{-2} and capacity was 45 ~ 150 μA cm^{-2}.

Also, NTT Co. Group in Japan [9,19–22] developed thin film batteries by using $Li_{3.4}V_{0.6}Si_{0.4}O_4$ glass for electrolyte and $LiCoO_2$ [20] and $LiMn_2O_4$ [22] for cathodes by using RF sputtering method. The battery size is about 1cm^2 and the thickness is 1 ~ 5 μm of cathode, 1 μm of electrolyte and 4 ~ 8 μm of lithium anode. The cell performance is shown in Table 3.2.1.

Thin film batteries were also developed by Ever-ready Battery Co., Ltd. and Bellcore Co., Ltd. USA in 1980s using sulfide glass of $Li_4P_2S_7$ or Li_3PO_4-P_2S_5 for electrolytes, TiS_2 cathode and Li and LiI for anode. [6,23,24]. The battery of 1.5 ~ 2.8 V, 10 ~ 135 μA cm^{-2} operated more than 1000 cycles [8]. Bellcore Co., Ltd. also announced the lithium cell consisting of $LiMn_2O_4$ cathode, a lithium borophosphate (LiBP) glass or lithium phosphorus oxynitride glass (LiPON) for electrolyte and metallic lithium as anode. The cell has 4.2V OCV and operated at 3.5 ~ 4.3V, 70 μA cm^{-2} for more than 150 cycles.

The group in Oak Ride National Laboratory (ORNL) in USA [7,13,25] have been energetically studying thin film batteries using lithium phosphonitroxide glass (LiPON). The LiPON was prepared by RF sputtering of Li_3PO_4 target in nitrogen gas, which is rather stable in comparison with other lithium oxide or sulfide based glasses in spite of moderate ionic conductivity of 2.3×10^{-6} Scm^{-1} at room temperature and activation energy of 0.55 eV [25]. The current potential curve indicates the stability range of LiPON is from 0 to 5.5V with respect to a Li electrode. Lithium metal anode was prepared by vacuum evaporation, and LiPON electrolyte and cathodes of $LiCoO_2$, $LiMn_2O_4$ etc. were by RF sputtering. They have reported some combinations of anodes and cathodes with LiPON electrolyte, which exhibited very good performance as voltage range from 2 to 5V, current density up to 10 mAcm^{-2} and cyclability more than 10000 [26,27]. It is noteworthy that Neudecker *et al.* reported a Li-free thin-film battery with an *in situ* plated Li anode on copper electrode [27]. This technique is useful to avoid the presence of low melting lithium (mp 178°C) during solder reflow conditions.

LiPON is now recognized as a standard solid electrolyte for thin film lithium batteries and have been used by many groups especially private companies in USA. It is also used in Park *et al.* in Korea [28] and Baba and Kumagai in Iwate University Japan. Baba *et al.* reported thin film batteries using LiPON electrolyte, using $Li_xV_2O_5$ as anode and V_2O_5 or $LiMn_2O_4$ as cathode by means of RF magnetron sputtering [29,30]. These locking-chair type batteries without using lithium metal anode have strong merit in preparation and safety, although they need initial charging processes before use. Their battery exhibited so called the "forming behavior" increasing the capacity with increasing the cycle number up to 20 cycles when the capacity reaches a maximum of 10 μAhcm^{-2} [30]. This behavior is attributed to the gradual decreasing of interfacial resistance. Baba *et al.* also proposed high-voltage and high-current batteries by stacking two cells on the same substrate, which comprises of 9 layers including current collectors. It works 3 to 6.5V in 2 μA cm^{-2} operation [31].

Another big group is in France, where the thin film batteries have been studied since 1980s using chalcogenide intercalation compound such as In_2Se_3, TeO_2, $GeSe_{5.5}$ and TiOS for the cathode and lithium ion conducting glasses as the

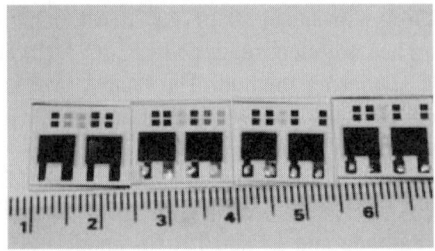

Fig. 3.2.29. Thin film batteries prepared by sequential pulsed laser depositions by Kuwata et al. [16,17].

electrolyte [5,12,14,15, 32–37]. Balkanski *et al.* reported in 1989 the thin film batteries composed of In_2Se_3 cathode, lithium borate glass and lithium metal anode. The cathode and electrolyte film was prepared by molecular beam deposition (MBD) and a flash evaporation [33]. Creus *et al.* reported thin film battery cells of Li / Li_2S-SiS_2 / $GeSe_{5.5}$ and Li / $0.7(0.4Li_2O \cdot 0.2B_2O_3 \cdot 0.4P_2O_5) \cdot 0.3LiCl / 0.7V_2O_5 \cdot 0.3TeO_2$ by RF sputtering and flash evaporation [5]. Levasseur et al. in Bordeaux University and French company Hydromecanique et Frottements (HEF) has collaborated to deriver thin film battery prototypes using TiOS cathode . The battery has 2.5V output, with 50 ~ 300 uAhcm^{-2} capacity and operates more than several 1000 cycles [14].

3.2.4.3 Thin-film battery by sequential PLD

Recently, Kuwata *et al.* prepared a thin film secondary lithium battery of $LiCoO_2$ cathode, $Li_{6.1}V_{0.61}Si_{0.39}O_{5.36}$ (LVSO) glassy electrolyte and SnO anode, by means of only a sequential PLD technique as shown in Fig.3.2.29 [16,17].

A Q-switched Nd:YAG laser (λ = 266 nm, repetition rate of 10 Hz) or XeCl Eximer laser (λ = 308 nm, 5 ~ 10 Hz) was used to ablate the different targets. The laser beam was focused on the target to obtain fluencies of 3.5 mJcm^{-2} at 45° incidence angle. The target was mounted on a rotating holder. Films were prepared in a flowing oxygen atmosphere at a pressure of 10^{-1} Pa. The substrate temperature was kept at room temperature. Target of $Li_{3.4}V_{0.6}Si_{0.4}O_4$, was prepared following the literature by Ohtsuka and Yamaki [9,19–22]. The fabrication process of thin-film battery includes: (1) Pt/Cr cathode current collector, (2) $LiCoO_2$ cathode, (3) amorphous LVSO solid electrolyte, (4) amorphous SnO anode, and (5) Pt anode current collector as shown in Fig.3.2.30; all the layers fabricated by the same laser chamber system changing the targets and hard masks.

The batteries typically have an area of about 0.23 cm^2 and are about 2 μm thick. The as-deposited LVSO and SnO films are X-ray amorphous, while the as-deposited $LiCoO_2$ films are partially crystallized. The $LiCoO_2$ films were post-annealed at 600°C for 1 h in air to form a single phase of the layered $LiCoO_2$ structure.

Good quality amorphous films were obtained for a lithium ion conductor $Li_{3.4}V_{0.6}Si_{0.4}O_4$ (LVSO), which was previously investigated by Ohtsuka and Yamaki as an electrolyte for thin film lithium battery [9,22]. They have prepared the thin

3.2 Lithium ion conductors

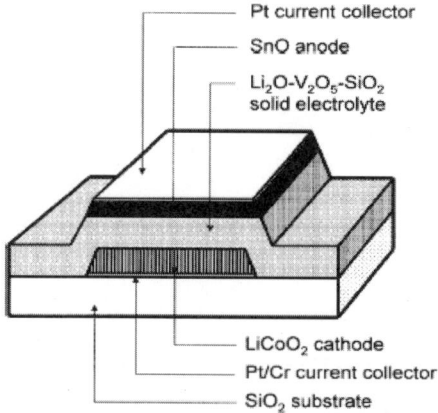

Fig. 3.2.30. Structure of thin-film battery prepared in Tohoku university by sequential PLD.

film of LVSO by using RF-sputtering method. Their film was partially crystallized showing X-ray diffraction peak of (002) plane of crystalline LVSO. On the other hand, the film prepared by PLD method is completely amorphous state. The composition of the film was in fairly good agreement with the target. The conductivity of the present LVSO was evaluated from impedance analysis to be $ca.10^{-7}$ S cm^{-1} at room temperature, which is one order of magnitude smaller than the value of 10^{-6} S cm^{-1} reported for the film prepared by RF-sputtering method [9,22]. The difference is probably due to the complete amorphous nature of the present film and the partially crystallized form of the RF-sputtered film. Actually, the crystalline sample of the target showed $ca.$ 10^{-5} S cm^{-1} at room temperature, which is two orders higher than the present amorphous thin film.

An example of charge-discharge profiles of the cell is shown in Fig.3.2.31. The thin-film battery was cycled at a current density of 44 µA cm^{-2} between 3.0 to 0.7 V. The discharge capacity in the second cycle is about 9.5 Ah cm^{-2}, which corresponds to 10 ~ 20% of LiCoO$_2$ utilization. The discharge voltage gradually decreases in the range from 2.7 to 1.5 V, probably because of the amorphous nature of the SnO film. A considerable capacity loss on the first discharge is observed on the first discharge. This initial capacity loss is explained as that the SnO reacts with inserted Li$^+$ ions to produce the metallic Sn-Li alloy and amorphous Li$_2$O, which causes an irreversible capacity [15]. This cell can be cycled for more than 100 cycles, maintaining good efficiency. However, after the first discharge, there is a continuous decrease in the charge and discharge capacity. The discharge capacity after 100 cycles is about 45% of that of the first cycle. The cell can discharge at a current density up to 200 µA cm^{-2}, although the potential drop is large owing to the resistance of the solid electrolyte or the interface (Fig. 3.2.32).

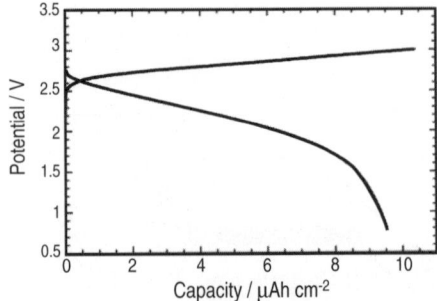

Fig. 3.2.31. Charge discharge character of SnO/LVSO/LiCoO$_2$ cell prepared by PLD [16].

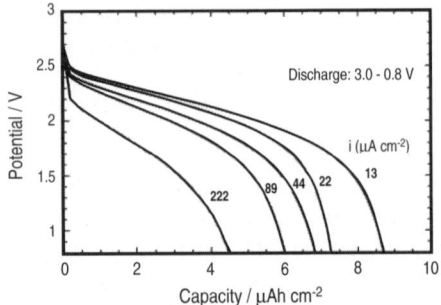

Fig. 3.2.32. Discharge character of SnO/LVSO/LiCoO$_2$ cell prepared by PLD[16].

3.2.4.4 General performance of thin film batteries and their futures

The performance of a thin film battery is evaluated in (i) open circuit voltage, (ii) maximum current, (iii) capacity, (iv) cyclability etc. The open circuit voltage is almost determined by the combination of cathode and anode materials as in the case of conventional lithium batteries [46], although the decomposition voltage of the solid electrolyte limits the maximum.

Other three parameters are depending with each other. The maximum available current of the battery is strongly depends on the interfacial condition between the cathode, electrolyte and anode as well as on the diffusivity of electrons and ions in the components. Since the thin film surface is very flat, interface area between the electrodes and electrolyte is almost equal to the apparent contact area. This is a merit to investigate the interfacial reaction phenomena in contrast to the conventional batteries, where the interface area is much larger than the apparent dimension due to the rough surface morphology of the electrodes. On the other hand, it is a drawback of the application of the thin film battery whose interfacial impedance is limited by the geometry. Generally, the current density of the thin film batteries is 10 to 100 µA cm^{-2} in normal charging and discharging processes and the maximum current for short circuit is a few mA. The small timer IC with liquid crystal display can be easily operated less than 10 µA by single thin film cell. Good cycle performance even more than 10000 has been reported, which is a strong advantage of thin film batteries [26,27].

The most important factor is the capacity of the battery, which depends primarily on the volume (thickness × area) of the cathode and anode as well as their qualities. In case of typical size of 1cm^2 and 1μm thickness, the capacity is about 10 ~ 50 μAh which can operate a timer IC for one hour by one charge. By optimizing the condition, one can realize more than 80% of the cathode utilization [8]. Increase of the apparent capacity can be achieved by stacking two or more cells on one substrate [31], which will be a promising approach for future.

Large volume change of electrode material during charge discharge process is another important factor of limiting the size and cycle performance. The shrink of the anode during discharging (also the cathode in charging) causes a clack or an exfoliation of the anode or cathode, which limits the maximum thickness and area of the electrodes. Good and strong contact between electrode and electrolyte is necessary to overcome this difficulty, which can be achieved sometimes (not always) by post-anneal treatments. The flexibility of the materials may be useful to absorb the stress at interface, which is partly achieved by using glassy electrolytes.

Relatively low ionic conductivity of lithium oxide glasses used for electrolyte (about 10^{-6} S cm^{-1}) limit the maximum current density to 1 mAcm^{-1}, where the thickness of the electrolyte is assumed 1μm and allowing voltage drop is 0.1 V. Actually, larger voltage drop appearing at the interface further limits the maximum current. If one can use lithium oxysulfide glasses whose conductivity is about 10^{-3} S cm^{-1} [47], the available current density will be up to 1 A cm^{-2}. This is a reason of high performance of Eveready battery [6], although a similar cell by Creus et al. reported only 10 μAcm^{-2} probably due to some unfavorable reaction between the electrolyte and electrode materials. Although their inherent superiority of oxysulfide glasses, they have not studied enough for thin film battery applications. It is remaining to be addressed in future.

Battery technology is now in transition stage from traditional electrochemistry to solid state physics concerning ionic transport in solids and their interface. Thin film battery is a good entrance to this new challenging field.

References

1. C. C. Liang, J. Epstein and G. H. Boyle, *J.Electrochem.Soc.*, **116** (1969) 1452.
2. K. Murata, S. Izuchi and Y. Yoshihisa, *Electrochimica Acta*, **45** (2000) 1501.
3. K. Kanehori, K. Matsumoto, K. Miyauchi and T. Kudo, *Solid State Ionics*, **9–10** (1983) 1445.
4. Y. Ito, K. Kanabori and K. Miyauchi, *Kotaibutsuri (Japanese)* ,**19** (1984) 233.
5. R. Creus, J. Sarradin and M. Ribes, *Solid State Ionics*, **53–56** (1992) 641.
6. S. D. Jones and J.R. Akridge, *Solid State Ionics*, **53–56** (1992) 628.
7. J. B. Bates, G.R. Gruzalski, N.J. Dudney, C.F. Luck and X.H. Yu, *Solid State Ionics*, **70** (1994) 619.
8. S. D. Jones, J. R. Akridge and F. K. Shokoohi, *Solid State Ionics*, **69** (1994) 357.
9. J. Yamaki, H. Ohtsuka and T. Shodai, *Solid State Ionics*, **86–88** (1996) 1279.

10. P. Birke, W. F. Chu and W. Weppner, *Solid State Ionics*, **93** (1996) 1.
11. C. Julien, *The CRC Handbook of Solid State Electrochemistry*, ed. P.J.Gellings, H.J.M.Bouwmeester, CRC Press, Roca Raton, 1997, 371.
12. A. Levasseur and P. Vanatier, in *Solid State Ionics: Science & Technology*, ed. B.V.R.Chowdari et al., World Scientific Publ.Co., (1998) p.421.
13. J. B. Bates, N. J. Dudney, B. Neudecker, A. Ueda and C. D. Evans, *Solid State Ionics*, **135** (2000) 33.
14. M. Duclot and J. L. Souquet, *J. Power Sources*, **97–98** (2001) 610.
15. J. L. Souquet and M. Duclot, *Solid State Ionics*, **148** (2002) 375.
16. N. Kuwata, J. Kawamura, K. Toribami and T. Hattori, in *Solid State Ionics: The Science and Technology of Ions in Motion (Proc. 9th Asian Conf. Solid State Ionics)*, edited by B.V.R.Chowdari et al. World Sci.Publ.Co., Singapor, 2004, p. 637.
17. N. Kuwata, J. Kawamura, K. Toribami, T. Hattori and N. Sata, *Electrochem.Commun.*, **6** (2004) 417.
18. F. Kirino, Y. Ito, K. Miyauchi and T. Kudo, *Nippon Kagakukaishi (Japanese)*, (1986) 445.
19. H. Ohtsuka and A. Yamaji, *Solid State Ionics*, **8** (1983) 43.
20. H. Ohtsuka and J. Yamaki, *Jpn. J. Appl. Phys.*, **28** (1989) 2264.
21. H. Ohtsuka and J. Yamaki, *Solid State Ionics*, **35** (1989) 201.
22. H. Ohtsuka, S. Okada and J. Yamaki, *Solid State Ionics*, **40-41** (1990) 964.
23. J. R. Akridge and H. Vourlis, *Solid State Ionics*, **18–19** (1986) 1082.
24. J. R. Akridge and H. Vourlis, *Solid State Ionics*, **28–30** (1988) 841.
25. X. Yu, J. B. Bates, G. E. Jellison and F. X. Hart, *J. Electrochem. Soc.*, **144** (1997) 524.
26. B. Wang, J. B. Bates, F. X. Hart, B. C. Sales, R. A. Zuhr and J. D. Robertson, *J. Electrochem. Soc.*, **143** (1996) 3203.
27. B. J. Neudecker, N. J. Dudney and J. B. Bates, *J. Electrochem. Soc.*, **147** (2000) 517.
28. Y. S. Park, S. H. Lee, B. I. Lee and S. K. Joo, *Electrochem. Solid-State Lett.*, **2** (1999) 58.
29. M. Baba, N. Kumagai, H. Kobayashi, O. Nakano and K. Nishidate, *Electrochem. Solid-State Lett.*, **2** (1999) 320.
30. M. Baba, N. Kumagai, N. Fujita, K. Ohta, K. Nishidate, S. Komaba, H. Groult, D. Devilliers and B. Kaplan, *J. Power Sources*, **97-98** (2001) 798.
31. M. Baba, N. Kumagai, H. Fujita, K. Ohta, K. Nishidate, S. Komaba, B. Kaplan, H. Groult and D. Devilliers, *J. Power Sources*, **119** (2003) 914.
32. M. Menetrier, A. Levasseur, C. Delmas, J. F. Audevert and P. Hagenmuller, *Solid State Ionics*, **14** (1984) 257.
33. M. Balkanski, C. Julien and J. Y. Emery, *J. Power Sources*, **26** (1989) 615.
34. R. Creus, J. Sarradin, R. Astier, A. Pradel and a. M. Ribes, *Mater. Sci. Eng.*, **B3** (1989) 109.
35. L. Jourdaine, J. L. Souquet, V. Delord and M. Ribes, *Solid State Ionics*, **28–30** (1988)1490.

36. M. Ribes, *Technology of thin films. fabrication and characterization* in "Proceedings" of the International seminar Solid Ionic Devices (ed. B. V. R. Chowdari and S. Radhakrishna), World Scientific Publ., Singapore (1988), p.85.
37. M. Ribes and V. Delord, *Thin films of conductive glasses: applications in microionics* in *"Proceedings" of the International seminar Solid Ionic Devices (ed. B. V. R. Chowdari and S. Radhakrishna)*, World Scientific Publ., Singapore (1988) p.147.
38. K. Kanehori, Y. Ito, F. Kirino, K. Miyauchi and T. Kudo, *Solid State Ionics*, **18–19** (1986) 818.
39. G. Meunier, R. Dormoy and A. Levasseur, *Mater. Sci. Eng.*, B3 (1989) 19.
40. N. J. Dudney, J. B. Bates, R. A. Zuhr, S. Young, J. D. Robertson, H. P. Jun and S. A. Hackney, *J. Electrochem. Soc.*, **146** (1999) 2455.
41. B. J. Neudecker, R. A. Zuhr, J. D. Robertson and J. B. Bates, *J. Electrochem. Soc.*, **145** (1998) 4148.
42. B. J. Neudecker, R. A. Zuhr, J. D. Robertson and J. B. Bates, *J. Electrochem. Soc.*, **145** (1998) 4160.
43. J. B. Bates, D. Lubben, N. J. Dudney and F. X. Hart, *J. Electrochem. Soc.*, **142** (1995) L149.
44. J. B. Bates, N. J. Dudney, D. C. Lubben, G. R. Gruzalski, B. S. Kwak, X. Yu and R. A. Zuhr, *J. Power Sources*, **54** (1995) 58.
45. B. J. Neudecker, R. A. Zuhr and J. B. Bates, *J. Power Sources*, **81–82** (1999) 27.
46. C. M. Julien, *Mater. Sci. Eng. R* 283 (2003) 1.
47. M. Tatsumisago, S. Hama, A. Hayashi, H. Morimoto and T. Minami, *Solid State Ionics*, **154** (2002) 635.

3.3 Proton conductors

3.3.1 Sol-gel ionics

3.3.1.1 Background

The application of the sol-gel method to solid state ionics, so-called "sol-gel ionics" [1,2] has been attracting much attention in recent years. In the sol-gel method, metal alkoxides are often used as a starting material and are hydrolyzed and polymerized in an alcohol solution with water containing acids, bases or salts. The wet gel obtained by the method is macroscopically solid, whereas it microscopically consists of interpenetrating inorganic solid and liquid-like phases. Thus the liquid-like phase is regarded as suitable paths for the fast ion transport and the interface between the inorganic solid and the liquid is expected to enhance the ionic conduction [3].

Sol-gel derived silica gels are an excellent host matrix to design proton conductors because they contain a large number of micropores surrounded by the silanol, Si-OH, groups which may desirable for proton transfer in the remaining liquid-like phase. Highly proton-conductive materials are promising electrolytes for the electrochemical devices such as gas sensors, electrochromic cells, capacitors, and fuel cells [4].

74   3. Recent development of amorphous solid electrolytes

Silica gel films containing dodecamolybdophosphoric acid, $H_3PMo_{12}O_{40}\cdot 29H_2O$ (MPA) have been prepared by the sol-gel method. The gel films exhibit considerably high proton conductivities in a highly humid atmosphere, and the conductivity increases with increasing MPA content. The gel films obtained can be applied in humidity sensors [3].

Remarkable coloration has been observed in the electrochromic cell using a $H_3PW_{12}O_{40} \cdot 29H_2O$ (WPA)-doped silica gel as a solid electrolyte. The reversibility and response of the coloration and bleaching are comparable to the case using a typical liquid electrolyte of H_2SO_4 [5,6].

It has been demonstrated that $HClO_4^-$, and WPA-doped silica gels exhibit a high conductivity of 10^{-2}–10^{-1} Scm^{-1} at room temperature. The conductivity increases with a shift to the higher values of IR absorption band at around 3500 cm^{-1}, indicating that the stronger O-H bond in silanol groups in the silica gels causes the higher proton conduction [7,8].

3.3.1.2 Doping of protonic acid to silica gels

The ionic conductivity of the silica gel doped with $HClO_4$, H_2SO_4, or H_3PO_4 increases with an increase in the amounts of the acid added, and attains 10^{-1} to 10^{-2} S cm^{-1} in an ambient atmosphere at room temperature although that of the silica gel doped with HCl or HNO_3 is about 10^{-5} S cm^{-1} independent of the amounts of the acid added. Protonic acids with hydrated water such as $HClO_4\cdot nH_2O$, $H_2SO_4\cdot nH_2O$ and H_3PO_4 ($P_2O_5\cdot nH_2O$) are, thus, considered to act as an effective proton donor and increase the proton conductivity of the resultant acid-doped silica gel [9].

Temperature dependence of the ionic conductivity in dry N_2 atmosphere for silica gels doped with $HClO_4$, H_2SO_4, or H_3PO_4 is shown in Fig.3.3.1. The conductivities of all the acid-doped silica gels exponentially increase with an increase in temperature. The temperature dependence of conductivity of the gels is not the Arrhenius type but the Vogel-Tamman-Fulcher (VTF) type [10], suggesting that proton is transferred through a liquid-like path formed in micropores of the gels. The conductivities of silica gels doped with $HClO_4$, H_2SO_4, and H_3PO_4 decrease in this order. The conductivities of acid-doped silica gel can be related to the property of the acid on the basis of acid dissociation constant, K_a. The values of K_a of $HClO_4$, H_2SO_4, and H_3PO_4 are respectively 1×10^{10}, 1×10^{2} and 8×10^{-3}; namely acidity drastically decreases in this order [11]. The acid with hydrated water and larger K_a provides the resultant acid-doped silica gel with higher proton conductivity.

All-solid-state electric double-layer capacitors have been successfully fabricated using the highly proton-conductive silica gels as an electrolyte and activated carbon powder (ACP) hybridized with the silica gels as a polarizable electrode [12,13]. The capacitor has a three layered structure of polarizable electrode / electrolyte / polarizable electrode. The electrolyte part consists of polyvinyl alcohol-containing silica gels doped with $HClO_4$, H_2SO_4, or H_3PO_4. Electrodes are composed of ACP hybridized with the acid-doped silica gels.

The capacitance at room temperature of the all-solid-state electric double-layer capacitors fabricated is shown in Fig.3.3.2 as a function of the acid / tetraethoxysilane

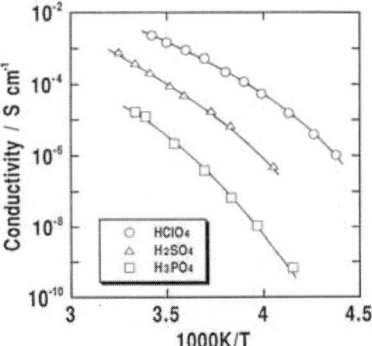

Fig. 3.3.1. Temperature dependence of the ionic conductivity in dry N_2 atmosphere for silica gels doped with $HClO_4$, H_2SO_4, or H_3PO_4.

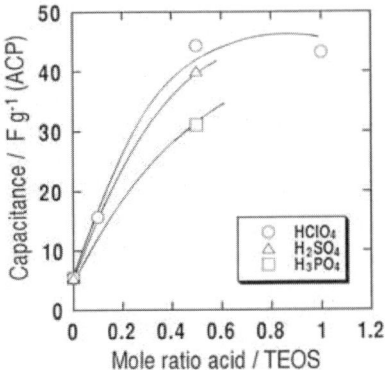

Fig. 3.3.2. The capacitance at room temperature of the all-solid-state electric double-layer capacitors fabricated as a function of the acid / tetraethoxysilane mole ratio.

mole ratio. The capacitance is enlarged with increasing the acid concentration. The values of capacitance of the capacitors doped with $HClO_4$, H_2SO_4, or H_3PO_4 with the ratio = 0.5 are 44, 40 and 31 F / (gram of total ACP), respectively. These values are comparable to those of the conventional capacitors with liquid electrolytes [14]. The large capacitance of the all-solid-state electric double layer capacitors fabricated is attributable to the electric double-layer sufficiently formed at the interface between the highly proton conductive silica gel and ACP.

3.3.1.3 Control of pore structure of the gels

The control of the pore structure such as pore diameter and pore volume of the gels is also very important to design the proton conducting paths. The pore diameter and pore volume of the silica gels can be designed using surfactants as a template [15–18]. Pores of the surfactant-templated mesoporous silica gels are

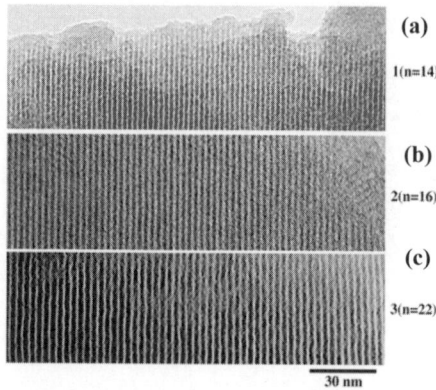

Fig. 3.3.3. TEM images viewed perpendicular to the pore axis of the mesoporous C_n-silica gels (n = 14, 16 and 22) for proton conductors. (a), (b) and (c) are for the C_{14}-, C_{16}- and C_{22}-silica gels, respectively.

cylindrical, regularly spaced and sharply distributed in size, compared with those of the conventional ones prepared by the usual sol-gel method. A new type of proton conductor which is the surfactant-templated mesoporous silica gel impregnated with sulfuric acid has been proposed [19]. The sulfuric acid-impregnated mesoporous silica gels show much higher proton conductivities than the conventional ones probably due to the diameter-controlled pore structure and the large pore volume.

Highly proton-conductive acid-impregnated mesoporous silica gels have been also prepared using alkyltrimethylammonium chlorides (C_nTAC) with different alkyl chain length (n = 14, 16 and 22) as a template and several kinds of protonic acids such as $HClO_4$, H_2SO_4 and H_3PO_4 as a proton donor [20,21].

Typical transmission electron microscopy (TEM) images viewed perpendicular to the pore axis of the mesoporous C_n-silica gels (n = 14, 16 and 22) are compared in Fig.3.3.3. Figures 3.3.3 (a), (b) and (c) are for the C_{14}-, C_{16}- and C_{22}-silica gels, respectively. Cylindrical channels are regularly spaced in all the mesoporous C_n-silica gels. The d spacings of mesoporous C_n-silica gels increase with increasing n, i.e. the number of carbon atoms in the alkyl groups of C_nTAC. The specific surface area and pore volume of the mesoporous silica gels increase with an increase in the number of carbon atoms in the alkyl chain. The electric conductivities of all the acid-impregnated mesoporous silica gels also tend to increase with increasing the number of the carbon atoms in the alkyl chain in the surfactant. This phenomenon can be ascribed to an increase in the amounts of impregnated acid due to the enlarged pore volume in the mesoporous silica gels. It is noteworthy that the conductivities of the $HClO_4$- and H_2SO_4-impregnated mesoporous silica gels obtained are as high as 10^{-1} S cm^{-1} under 60% relative humidity at 85°C.

3.3.1.4 Blending of the gels with elastic polymers

Improvement of the molding property of the proton-conductive silica gels is desired for the practical application of the gels to electrochemical devices. Highly proton-conductive elastic composites have been successfully prepared from H_3PO_4-doped silica gel and a styrene-ethylene-butylene-styrene (SEBS) block elastic copolymer. Ionic conductivities of the composites depend on the concentration of H_3PO_4 and the heat-treatment temperature of the H_3PO_4-doped silica gel [22]. The composite composed of H_3PO_4-doped silica gel with a mole ratio of $H_3PO_4/SiO_2 = 0.5$ heat-treated at temperatures below 200°C and SEBS elastomer in 5 wt% shows a high conductivity of 10^{-5} Scm^{-1} at 25°C in an dry N_2 atmosphere. The temperature dependence of conductivity of the composites is the VTF type, indicating that proton is transferred through a liquid-like phase formed in micropores of the H_3PO_4-doped silica gels. The temperature dependence of the elastic modulus of the composite is similar to that of the SEBS elastomer. The thermoplastically deforming temperature of the composite is around 100°C, which is higher by 30°C than that of the SEBS elastomer.

Sulfonated organic polymers have been designed to assist the proton conduction in the composite as well as good molding characteristics due to the thermoplasticity [23]. Polyisoprene (PI) and styrene-isoprene-styrene block copolymer (SIS) have been selected as the base polymers for sulfonation (**Scheme 3.3.1**). The average molecular weights of PI and SIS are 40,000 and 220,000, respectively. The styrene/isoprene mole ratio in SIS is 22/78. Sulfonation ratios are varied from 0 to 50% for PI and from 0 to 20% for SIS, where the ratio is defined as a percentage of the amount of SO_3H groups to the total amounts of hydrogens and SO_3H groups, and is represented as X in Scheme 3.3.1. The solutions containing polymers above are mixed with H_3PO_4-doped silica gel powders using an agate mortar in ambient air. Highly proton-conductive and thermoplastic composites from H_3PO_4-doped silica gel and organic polymers with sulfo groups have been fabricated. The composite composed of H_3PO_4-doped silica gel with a mole ratio of $H_3PO_4/SiO_2 = 0.5$ in 80 wt% and sulfonated styrene-isoprene-styrene (SIS-SO_3H) block copolymer in 20 wt% shows a high conductivity of 10^{-4} Scm^{-1} at 25°C in dry N_2 atmosphere and good molding characteristics.

3.3.1.5 Phosphosilicate gels

Solid state proton conductors which show high conductivities in the medium temperature range (100–200°C) with low humidity have been required as an electrolyte for fuel cells for electric vehicles and cogeneration systems [24–27]. The poisoning of Pt catalysts with CO is depressed at medium temperatures. In addition, the weight and volume of humidifiers, which are indispensable for the operation of the polymer electrolyte fuel cells, must be reduced for the practical application to the electric vehicles. These situations described above are the background for the requirement of the highly proton conductive solid state materials in the medium temperature range with low humidities.

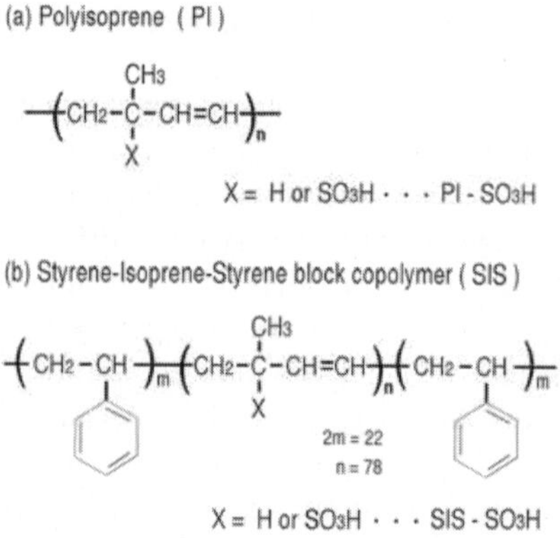

(a) Polyisoprene (PI)

$$-(CH_2-\underset{X}{\underset{|}{C}}(\overset{CH_3}{\overset{|}{}})-CH=CH)_n-$$

X = H or SO₃H · · · PI - SO₃H

(b) Styrene-Isoprene-Styrene block copolymer (SIS)

$$-(CH_2-CH(\phi))_m-(CH_2-\underset{X}{\underset{|}{C}}(\overset{CH_3}{\overset{|}{}})-CH=CH)_n-(CH_2-CH(\phi))_m-$$

2m = 22
n = 78

X = H or SO₃H · · · SIS - SO₃H

Scheme 3.3.1.

Proton conductive organic polymers with functional groups or containing proton donors like acids and salts tend to degrade at temperatures higher than 100°C [24]. Some proton donors like heteropolyacids and protonic acids are dehydrated or thermally decomposed at temperatures in the medium range with relatively low humidity [28]. Among the protonic acids examined, phosphoric acid, H_3PO_4, is expected to form chemical bondings with silica matrix, Si-O-P, leading to the formation of phosphosilicate, *i.e.* P_2O_5-SiO_2, gels, which are promising solid state proton conductors in the low and medium temperature ranges [29–32].

Variations in conductivities with holding time under 0 (open circles), 0.7 (closed circles) and 1.4 (closed diamonds) % R.H. at 130°C for the phosphosilicate gels with the P/Si mole ratio of 1.0 are shown in Fig.3.3.4. The broken line shows the temperature profile for the gel samples. The conductivity of the gel monotonically decreases from 10^{-1} to 8×10^{-4} Scm^{-1} while the gel was kept in the atmosphere of 0% R.H. The conductivity of the gel decreases in 20 min and tends to level off at around 10^{-2} Scm^{-1} during holding under 0.7% R.H. It is noteworthy that the decrease in conductivity of the gel is so small during holding under 1.4% R.H.; the conductivity is as high as 8×10^{-2} Scm^{-1} even after holding for about 400 min. These results indicate that a very small amount of water vapor is sufficient for the gels to retain a high conductivity of about 10^{-1} Scm^{-1} at temperatures in the medium range.

In phosphosilicate gels, the presence of condensed phosphoric acid with a bridging oxygen (Q^1 unit) as well as isolated orthophosphoric acid (Q^0 unit) has been proved from ^{31}P MAS-NMR spectra. Phosphorus Q^1 units which have Si-O-P-OH groups are expected to enhance the retention of the adsorbed water in the phosphosilicate gels and thus improve the proton conductivity of the gels even in a low humidity atmosphere in the medium temperature range.

In conclusion, the technical background and the research progress in the sol-gel ionics with respect to solid state proton conductors have been described. In the progress, (1) doping of protonic acids such as $HClO_4$, H_2SO_4, or H_3PO_4 to silica gels, (2) control of pore structures of the silica gels using a surfactant as a template, (3) blending of the gels with elastic polymers to improve the molding properties,

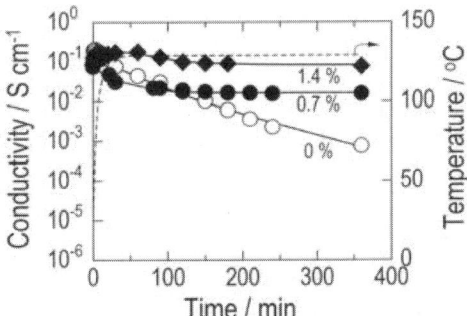

Fig. 3.3.4. Variations in conductivities with holding time under 0 (open circles), 0.7 (closed circles) and 1.4 (closed diamonds) % R.H. at 130°C for the phosphosilicate gels with a P/Si mole ratio of 1.0. The broken line shows the temperature profile for the gel samples.

and (4) medium temperature characterization of phosphosilicate gels for the fuel cells have been illustrated. Sol-gel derived solid state proton conductors offer the advantages of total solidification of electrochemical devices. Phosphosilicate gels are promising as an electrolyte especially for the fuel cells in medium temperature operations.

References

1. L. C. Klein, *Solid State Ionics*, **32–33** (1989) 639.
2. J. Livage, *Solid State Ionics*, **50** (1992) 307.
3. M. Tatsumisago and T. Minami, *J. Am. Ceram. Soc.*, **72** (1989) 484.
4. P. Colomban, *Ann. Chim. Sci. Mater.*, **24** (1999) 1.
5. M. Tatsumisago, K. Kishida and T. Minami, *Solid State Ionics*, **59** (1993) 171.
6. H. Honjo, M. Tatsumisago and T. Minami, *J. Mater. Sci. Lett.*, **14** (1995) 783.
7. M. Tatsumisago, H. Honjo, Y. Sakai and T. Minami, *Solid State Ionics*, **74** (1994) 105.
8. M. Tatsumisago, Y. Sakai, H. Honjo and T. Minami, *J. Ceram. Soc. Jpn*, **103** (1995) 189.
9. A. Matsuda, H. Honjo, M. Tatsumisago and T. Minami, *Chem. Lett.*, (1998) 1189.
10. F. M. Gray, J. R. MacCallum, C. A. Vincent and J. R. M. Giles, *Macromolecules*, **21** (1988) 392.
11. D. F. Shriver, P. W. Atkins, and C. H. Langford, *Inorganic Chemistry*, Oxford University Press, Oxford (1991), Chap. 5, p. 150.
12. A. Matsuda, H. Honjo, M. Tatsumisago and T. Minami, *Solid State Ionics*, **113–115** (1998) 97.
13. A. Matsuda, M. Tatsumisago and T. Minami, *J. Sol-Gel. Sci. Tech.*, **19** (2000) 581.
14. I. Tanahashi, A. Yoshida and A. Nishino, *Bull. Chem. Soc. Jpn.*, **63** (1990) 3611.
15. J. S. Beck, J. C. Vartuli, W. J. Roth, M. E. Leonowicz, C. T. Kresge, K. D. Schmitt, C. T.-W. Chu, D. H. Olson, E. W. Sheppard, S. B. McCullen and J. B. Higgens, *J. Am. Chem. Soc.*, **114** (1992) 10834.

16. M. Ogawa, *J. Am. Chem. Soc.*, **116** (1994) 7941.
17. M. Kruk, M. Jaroniec and A. Sayari, *J. Phys. Chem. B*, **101** (1997) 583.
18. G. Alberti, M. Casciola, S. Cavalaglio and R. Vivani, *Solid State Ionics*, **125** (1999) 91.
19. S. Nishiwaki, K. Tadanaga, M. Tatsumisago and T. Minami, *J. Am. Ceram. Soc.*, **83** (2000) 3004.
20. A. Matsuda, Y. Nono, T. Kanzaki, K. Tadanaga, M. Tatsumisago and T. Minami, *Solid State Ionics*, **145** (2001) 135.
21. A. Matsuda, T. Kanzaki, K. Tadanaga, T. Kogure, M. Tatsumisago and T. Minami, *J. Electrochem. Soc.*, **145** (2001) 135.
22. K. Hirata, A. Matsuda, T. Hirata, M. Tatsumisago and T. Minami, *J.Sol-Gel Sci. Tech.*, **17** (2000) 61.
23. A. Matsuda, K. Hirata, M. Tatsumisago and T. Minami, *J. Ceram. Soc. Jpn.*, **108** (2000) 45.
24. P. L. Antonucci, A. S. Arico, P. Creti, E. Ramunni and V. Antonucci, *Solid State Ionics*, **125** (1999) 431.
25. M. Watanabe, H. Uchida and M. Emori, *J. Phys. Chem. B*, **102** (1998) 3129.
26. I. Honma, S. Hirakawa, K. Yamada and J. M. Bae, *Solid State Ionics*, **118** (1999) 29.
27. G. Alberti and M. Casciola, *Solid State Ionics*, **97** (1997) 177.
28. U. B. Moič, S. K. Milonjić, D. Malović, V. Samanković, P. Colomban, M. M. Mitrović and R. Dimitrijević, *Solid State Ionics*, **97** (1997) 239.
29. A. Matsuda, T. Kanzaki, Y. Kotani, M. Tatsumisago and T. Minami, *Solid State Ionics*, **139** (2001) 113.
30. A. Matsuda, T. Kanzaki, K. Tadanaga, M. Tatsumisago and T. Minami, *Electrochim. Acta*, **47** (2001) 939.
31. A. Matsuda, T. Kanzaki, K. Tadanaga, M. Tatsumisago and T. Minami, *Solid State Ionics*, **154–155** (2002) 687.
32. A. Matsuda, T. Kanzaki, K. Tadanaga, M. Tatsumisago and T. Minami, *J. Ceram. Soc. Jpn.*, **110** (2002) 131.

3.3.2 Proton-conducting composite materials

3.3.2.1 Background

Polymer electrolyte fuel cells (PEFCs) are promising power sources for the application to electric vehicles, cogeneration systems and mobile electronic devices [1–5]. Solid state proton conductors with high conductivity in the medium temperature range (100–200°C) with low humidity are required as the electrolytes for PEFCs [6–14]. The operation of PEFCs in the medium temperature range improves the efficiency of electric generation in the cells and depresses the poisoning of Pt catalysts with CO in the fuel gases. In addition, working of PEFCs under the conditions of low humidity permits the reduction of the weight and volume of humidifiers. As described in the preceding section 3.3.1, the sol-gel method using metal alkoxides as starting materials is a versatile way to prepare solid proton conductors [15–17]. It has been demonstrated that the sol-gel derived phosphosilicate (P_2O_5-SiO_2) gels containing large amounts of phosphorus keep a high conductivity of 1×10^{-2} Scm^{-1} even at 150°C and 0.4% RH. [18–20]. For the practical application of the phosphosilicate gels as an electrolyte for PEFCs, sheet formation of the gel is indispensable.

Thermally stable proton-conducting composite sheets have been successfully fabricated from phosphosilicate gel (P/Si=1 mole ratio) powders and polyimide precursor [21]. Polyimide is selected as an organic polymer matrix because of its excellent thermal stability and good sheet forming property. High proton conductivity and excellent mechanical properties of the composite sheets in a wide temperature range have been shown. Moreover, medium temperature range operation of the fuel cell using the composite sheet has been demonstrated [22].

3.3.2.2 Fabrication of composite sheets and MEAs

Proton-conductive composite sheets of about 100 μm in thickness were prepared from phosphosilicate gel powders and commercially available polyimide precursor. Phosphosilicate gel powders were obtained from tetraethoxysilane and H_3PO_4 by the sol-gel method. The mole ratio of P/Si was 1.0. The dry gel powders heat-treated at 150°C were ground into fine powders of a few micrometers in diameter using a planetary type of ball mill. Furthermore, the phosphosilicate gel powders were sieved with a 145 mesh to control the particle size of the powder less than about 5 μm. The sieved gel powder is then mixed with the N,N-dimethylacetoamide solution of a polyimide precursor with stirring for 30 min and the mixture was stirred under ultrasonic wave irradiation for additional 30 min. The gel powder : polyimide weight ratio was varied from 20:80 to 80:20. Yellowish composite slurries thus obtained were cast on a glass plate and developed using a squeegee and a spacer to control their thickness. The developed composite sheets on glass plates were heat-treated at 150°C for 3 h and then 180°C for 3 h.

The single fuel cell tests were carried out for the membrane/electrode assemblies (MEAs) which consisted of the composite sheet as an electrolyte and commercially available Pt-loaded (1 mg cm^{-2}) carbon paper sheets as electrodes. The MEAs were obtained by hot pressing the composite sheet between the Pt-loaded carbon paper sheets at 130°C under about 11 MPa *in vacuo* for 10 min. When current-voltage profiles of the fuel cell were measured, humidified H_2 and air by passing through the bubbler were provided to the cells. Flow rates of pure hydrogen for anode and of air for cathode were 20 and 100 ml min^{-1}, respectively.

3.3.2.3 Characterization of the composite sheets

Sieving of the pulverized phosphosilicate gel powder was effective to prepare homogeneous and flexible composite sheet containing larger amounts of gel powder. Flexible composite sheets were obtained up to 75 wt% of gel content using the sieved gel powders of less than 5 μm in size. The elimination of agglomerate gel

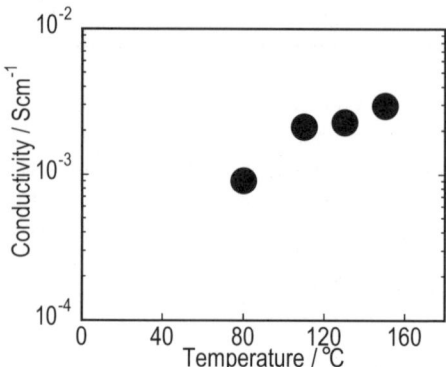

Fig. 3.3.5. Temperature dependence of conductivity of the composite sheet composed of 75 wt% of phosphosilicate gel and polyimide under constant water vapor pressure of 150 mmHg.

particles with large size should be an important factor to improve the flexibility and homogeneity of the composites.

Important characteristics of the composite sheet composed of phosphosilicate gel and polyimide should have water molecules with a good affinity for the gel matrices at temperatures up to around 230°C and have an excellent thermal stability even at around 300°C.

Conductivity of the composite sheets increases with an increase in the amount of phosphosilicate gel. The sheet containing sieved gel powder of 75 wt% showed high conductivities of 2.2 x 10^{-3} Scm^{-1} at 30°C under 60% R.H. and maintained almost the same value even after the consecutive keeping at 150°C and 0.4% R.H. for 6 h.

The temperature dependence of conductivity for the composite sheet containing 75 wt% phosphosilicate gel powders under a constant water vapor pressure of 150 mmHg is shown in Fig.3.3.5. At 80, 110, 130 and 150°C, the water vapor pressure corresponds to 42, 14, 7 and 4% R.H., respectively. The temperature was raised stepwise and the measurements were carried out after keeping at each temperature for 2 h. Conductivities of the composite sheet increase with increasing temperature from 80 to 150°C and the value is as high as 3.0 x 10^{-3} Scm^{-1} at 150°C. At 160°C and 150 mmHg of water vapor pressure (3% R.H.), conductivities of composite sheet gradually decrease with keeping. This is probably caused by the evaporation of strongly adsorbed water in the phosphosilicate gel [20,21]. When conductivities of the composite sheet were measured during cooling steps from 150 to 80°C, the sheet showed almost the same conductivities at each temperature as those in the preceding heating steps. Therefore, water vapor pressure of 150 mmHg is sufficient to keep constant proton conductivity in a wide temperature range up to 150°C.

The relative humidity dependence of conductivity for the composite sheet containing 75 wt% phosphosilicate gel powders at 150°C is shown in Fig.3.3.6.

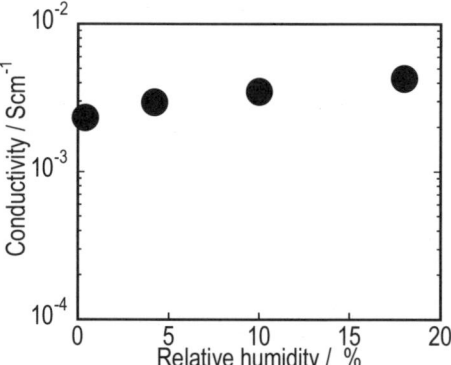

Fig. 3.3.6. Relative humidity dependence of conductivity of the composite sheet composed of 75 wt% of phosphosilicate gel and polyimide at 150°C.

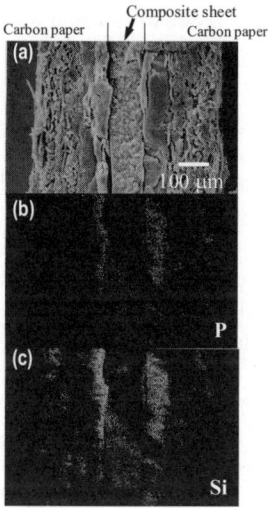

Fig. 3.3.7. SEM photograph (a) and P (b) and Si (c) distribution map determined by EDX of the cross-section of an MEA with composite sheet containing 75 wt% of phosphosilicate gel and Pt-loaded carbon electrodes immersed in phosphosilicate sol.

Conductivities of the composite sheet increase with an increase in the relative humidity. It is noteworthy that very slight humidification of 0.4% R.H. is sufficient to keep conductivities higher than 2×10^{-3} Scm^{-1} at 150°C, and the fluctuation in conductivity with humidity change is relatively small.

3.3.2.4 Characteristics of MEAs and performance of the fuel cells

Figure 3.3.7 shows (a) cross-sectional SEM image and corresponding EDX analyses of (b) P and (c) Si for the MEA which consists of the composite sheet and Pt-loaded carbon paper sheets. The composite sheet has good contact with the carbon paper electrodes. Phosphorus and silicon elements are widely distributed in the MEA, indicating that phosphosilicate sol has permeated into carbon paper to act as a paste at the interface between the composite electrolyte and the carbon paper electrodes. Cell voltage *vs.* current density with the resulting power density plots for a fuel cell using a MEA which consists of the composite sheet containing 75 wt% phosphosilicate gel as an electrolyte and Pt-loaded carbon paper sheets as electrodes are shown in Fig.3.3.8. The cell operates under 150 mmHg water vapor pressure at various temperatures, which corresponds to 42, 20, 10 and 7% R.H. at 80, 100, 120 and 130°C, respectively. Closed diamonds, triangles, squares and circles represent the current-voltage characteristics of the fuel cell at 80, 100, 120 and 130°C, respectively. Open diamonds, triangles, squares and circles represent the calculated current-power density plots of the fuel cells at 80, 100, 120 and 130°C, respectively. The open circuit voltage of the cell at 80°C is about 0.85 V, whereas the voltage of the cell decreases to be about 0.70 V at temperatures higher than 100°C. Open circuit voltage remains almost constant, *i.e.* 0.7 V, when the temperature increases from 110 to 130°C. At temperatures lower than 100°C, the pores in the composite sheets are closed due to absorbed water, whereas at temperatures higher than 100°C, the physically absorbed water evaporates to lead to H_2 cross-over. The slopes of current-voltage curves decrease with increasing the operation temperature, which is probably due to an increase in proton conductivity of the composite sheet as shown in Fig.3.3.5. The maximum power density increases with an increase in the operation temperature, which demonstrates that the fuel cells using MEA consisting of the composite sheet as an electrolyte have high potentiality to operate in the medium temperature range.

Relative humidity dependence of the cell voltage *vs.* current density with the resulting power density plots for this fuel cell at 150°C is shown in Fig.3.3.9. The cell construction is the same as in Fig.3.3.8. Closed circles, triangles and squares represent the current-voltage characteristics of the fuel cells under 4, 10 and 18% R.H., respectively. Open circles, triangles and squares represent the calculated current-power density plots of the fuel cells under 4, 10 and 18% R.H., respectively. Open circuit voltage is about 0.70 V under each relative humidity. The slopes of current-voltage curves decrease with an increase in the relative humidity. This corresponds to a decrease in the total resistance of the cell which is mainly due to an increase in the proton conductivity of the composite sheet as an electrolyte.

Continuous power generation under a current density of 50 mAcm^{-2} at 150°C and 4% R.H. has been confirmed so far. At the beginning of the operation the voltage was 0.33 V and increased to more than 0.4 V with operating time for 10 h. The increase in the cell voltage can be ascribed to a decrease in the total resistance of the cell, which is probably due to the decreases in charge transfer resistance of oxidation reaction and in the resistance of composite sheet. The performance of the cell is still

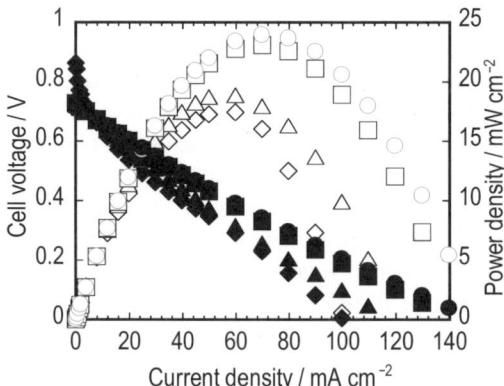

Fig. 3.3.8. Cell voltage versus current density with the resulting power density plots for a fuel cell using an MEA consisting of the composite sheet containing 75 wt% of phosphosilicate gel. The cell operated at 80°C (◆◇), 100°C (▲△), 120°C (■□) and 130°C (●○) under 150 mmHg water vapor pressure.

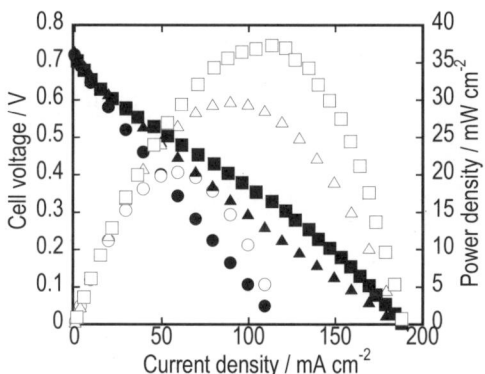

Fig. 3.3.9. Cell voltage versus current density with the resulting power density plots for a fuel cell with the same constructions as in Fig.3.3.8. The cell operated at 150°C under 4%R.H. (●○), 10%R.H. (▲△) and 18%R.H. (■□).

low for the practical application. Reliability testing for longer operation time is still required. However, these results demonstrate that the fuel cell using the composite sheet as an electrolyte can continuously operate at medium temperatures even under low humidity.

In conclusion, flexible composite sheets containing phosphosilicate gel up to 75 wt% are obtained using mechanically milled and then sieved gel powders of less than 5 μm in size. The conductivity of the composite sheets containing 75 wt% of gel powder is 2×10^{-3} Scm^{-1} at 150°C under 0.4% R.H. and increased to be 4×10^{-3} Scm^{-1} under 18% R.H. at the same temperature. These results indicate that the composite sheets obtained maintain high conductivities at medium temperatures even under very low humidity and the fluctuation in conductivity with humidity change is small. A fuel cell using an MEA which consists of the composite sheet as an electrolyte operates up to 150°C under water vapor pressure of 150 mmHg which corresponds to 4% R.H. Power density of the fuel cell increases with an increase in the relative humidity at 150°C. No degradation is observed for the fuel cell during continuous operation at 150°C under 4% R.H. with 50 mAcm^{-2}. Reliability testing and further performance improving of the fuel cells are still required. Successful fabrication of thermally stable proton conductive composite sheets with high conductivity in medium temperature range with low humidity should contribute the practical application of PEFCs to electric vehicles, cogeneration systems and mobile electronic devices.

References

1. G. Alberti and M. Casciola, *Solid State Ionics*, **145** (2001) 3.
2. P. Jannasch, *Curr. Opin. Collo. Interface Sci.*, **8** (2003) 96.
3. A.S. Arico, V. Baglio, A. Di Blasi, P. Creti, P.L. Antonucci and V. Antonucci, *Solid State Ionics*, **161** (2003) 251.
4. K. Miyake, N. Asano and M. Watanabe, *J. Polymer Sci. A Polymer Chem.*, **41** (2003) 3901.
5. I. Honma, H. Nakajima, O. Nishikawa, T. Sugimoto and S. Nomura, *J. Electrochem. Soc.*, **149** (2002) A1389.
6. T. Norby, *Nature*, **410** (2001) 877.
7. P. L. Antonucci, A. S. Arico, P. Creti, E. Ramunni and V. Antonucci, *Solid State Ionics*, **125** (1999) 431.
8. M. Watanabe, H. Uchida and M. Emori, *J. Phys. Chem. B*, **102** (1998) 3129.
9. I. Honma, S. Hirakawa, K. Yamada and J. M. Bae, *Solid State Ionics*, **118** (1999) 29.
10. Y. Park and M. Nagai, *J. Electrochem.Soc.*, **148** (2001) 149.
11. K.T. Adjemian, S.J. Lee, S. Srinivasan, J. Benziger and A.B. Bocarsly, *J. Electrochem. Soc.*, **149** (2002) A256.
12. S.Malhotra and R. Datta, *J. Electrochem. Soc.*, **144** (1997) L23.
13. K.D. Kreuer, *Solid State Ionics*, **97** (1997) 1.
14. P.Colomban, *Annales Chim.*, **24** (1999) 1.
15. M. Tatsumisago and T. Minami, *J. Am. Ceram. Soc.*, **72** (1989) 484.
16. M. Tatsumisago, H. Honjo, Y. Sakai and T. Minami, *Solid State Ionics*, **74** (1994) 105.
17. A. Matsuda, H. Honjo M. Tatsumisago and T. Minami, *Chem. Lett.*, (1998) 1189.
18. A. Matsuda, T. Kanzaki, Y. Kotani, M. Tatsumisago and T. Minami, *Solid State Ionics*, **139** (2001) 113.
19. A. Matsuda, T. Kanzaki, K. Tadanaga, M. Tatsumisago and T. Minami, *Electrochim. Acta*, **47** (2001) 939.
20. A. Matsuda, T. Kanzaki, K. Tadanaga, M. Tatsumisago and T. Minami, *J. Ceram. Soc. Jpn.*, **110** (2002) 131.

21. A. Matsuda, N. Nakamoto, K. Tadanaga, T. Minami and M. Tatsumisago, *Solid State Ionics*, **162–163** (2001) 113.
22. N. Nakamoto, A. Matsuda, K. Tadanaga, T. Minami and M. Tatsumisago, *J. Power Sources*, **138** (2004) 51.

3.3.3 Proton-conducting hybrid materials

3.3.3.1 Background

Inorganic-organic hybrid materials, where inorganic and organic components are chemically bonded each other, have extensively been studied because blending of organic and inorganic components allows the development of materials with novel properties [1]. The sol-gel process is known to be a practical method for preparing the hybrid materials. Alkoxysilanes with organic groups are often used as starting materials since inorganic network is formed by hydrolysis and condensation reaction of the alkoxy group in alkoxysilane, and organic network can also be introduced by selecting the organic groups.

Recently, proton conductive inorganic-organic hybrid membranes have been proposed as an alternative to the perfluorinated polymers in polymer electrolyte fuel cells (PEFCs). As described in the previous section, solid state proton conductors with high conductivity in the medium temperature range (100–200°C) with low humidity are required as the electrolytes for PEFCs, since the operation of PEFCs in the medium temperature range improves the efficiency of electric generation in the cells and depresses the poisoning of Pt catalysts with CO in the fuel gases. Since the incorporation of inorganic network into organic network will give good thermal, and mechanical and chemical stability, proton conductive inorganic-organic hybrid membranes are attractive for the membrane for PEFCs in the medium temperature range. For example, Poinsignon *et al.*[2] reported an inorganic-organic proton conductive hybrid using grafted arylsulfonic anions. Depre *et al.* reported the preparation of alkylsulfonatealkoxysilanes [3] or slufon/sulfonaimide functionalized hybrids [4]. Bonnet *et al.* prepared hybrid membranes using sulfonated polyetheretherketone as the organic component and metal phosphate and sulfophenylphosphonates and silica as the inorganic component [5]. They also reported that the membrane can be used as the electrolyte of fuel cells with operation up to 120°C. Stangar *et al.* reported the preparation of organic-inorganic gel matrix from 3-isocyanatopropyltriethoxysilane and ethoxysilane-functionalized polypropylene glycol, and proton conductive hybrids were obtained by dispersing phosphotungstic acid or silicotungstic acid in the gel matrix [6]. Park and Nagai reported the preparation of silicotungsitc acid or α-zirconium phosphate doped organic-inorganic hybrid from 3-glycidoxypropyltrimethoxysilane and tetraethoxysilane [7,8]. Honma *et al.* reported the preparation of hybrid polymer membrane from alkoxysilylated polyethylene oxide and acidic tungsten oxide cluster [9]. They also prepared the hybrid using bridged polysilsesquioxanes [10], or zirconium oxide and polydimethylsiloxane [11], and the zirconia bridged hydrocarbon phosphotungstic acids [12]. Our group

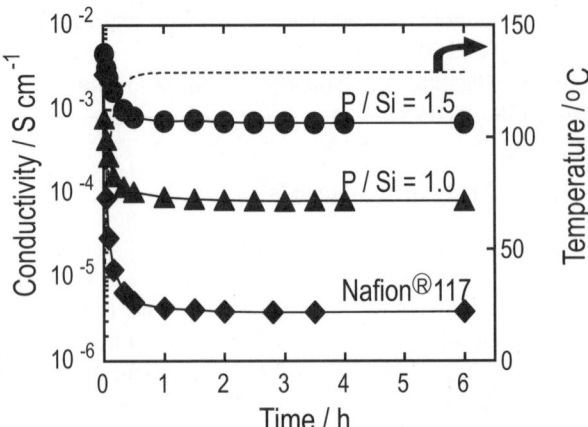

Fig. 3.3.10. Variations in conductivities with time for the hybrid films with P/Si of 1.5, 1.0 and Nafion® membrane, kept at 130°C under 0.7 % R.H.

has been studying the preparation of proton conductive inorganic-organic hybrid from 3-glycidoxypropyltrimethoxysilane and orthophosphoric acid [13–17]. It was found that test cells using the hybrid films as an electrolyte membrane in PEFC can be operated at 130°C [15].

3.3.3.2 Inorganic-organic hybrid membrane prepared from 3-glycidoxypropyltrimethoxysilane and orthophosphoric acid

Proton conductive inorganic-organic hybrid films, which showed high proton conductivities at temperatures higher than 100°C with low humidity, have been successfully prepared from 3-glycidoxypropyltrimethoxysilane (GPTMS), tetramethoxysilane (TMOS) and orthophosphoric acid by the sol-gel method [13–15]. GPTMS and TMOS were dissolved in ethanol, and aqueous orthophosphoric acid was added to the solution to hydrolyze the silanes. After being stirred for 2 h, precursor sol was obtained. The sol was poured into the Petri dishes and gelled at room temperature. Gel films obtained were then dried at 50°C for 1 day, and consecutively at 100°C for 5 h and at 150°C for 5 h. Self-supporting, flexible, transparent and brownish films with a thickness of about 200 to 300 μm were obtained in the composition of P/Si ranging from 0 to 1.5 (molar ratio). From the nitrogen adsorption measurements, the hybrid films were found to have almost no pores and BET surface areas were estimated to be less than 0.5 m^2g^{-1}, indicating that the films were dense. Differential thermal analysis and thermogravimetric measurements revealed that the films were stable up to about 200°C. The ^{31}P MAS NMR spectra of the hybrid films showed that phosphorus was mainly present as isolated phosphoric acid in the hybrid and only a small portion of phosphorus made P-O-P or Si-O-P bonds, which were observed as the so-called Q^1 units [14].

Figure 3.3.10 shows the variations in conductivities of with time for the films with P/Si of 1.5, 1.0 and Nafion® 117 membrane, with being kept at 130°C under 0.7% R.H. The conductivities of the hybrid films become constant after 1 h, and increase with an increase in the content of phosphoric acid in the films. The film with P/Si of 1.5 keeps a high conductivity of about 7×10^{-4} S cm^{-1} even after holding for 6 h. This value is higher by nearly three orders of magnitude than that of Nafion® under the same conditions. Nafion® is known to show high proton conductivities by forming water channels inside of the membranes, and the membrane loses water at low relative humidity, causing a decrease in conductivity. In contrast, the hybrid films keep high proton conductivities because phosphoric acid in the films can retain water by the hydrogen bond with P-OH in the medium temperature range over 100°C. The conductivities of the hybrid films were lower by one or two orders of magnitude than those of phosphosilicate gels at the same conditions [18]. Phosphosilicate gels, which are hydrophilic and porous, can easily adsorb much water, and exhibit high conductivities even at temperatures higher than 100°C. In the hybrid films, an organic component, which should be hydrophobic, was introduced. In addition, the hybrid films had a dense structure. Thus, the hybrid films are expected to show much lower conductivities. However, the hybrid films show high proton conductivities at temperatures higher than 100°C even at low R.H., suggesting that the epoxy groups in GPTMS play a very important role in the hybrid films: the epoxy groups in GPTMS were cleaved under the acidic condition to form hydroxy groups, and hydroxy groups formed must have strong interaction with phosphoric acid and also water.

Figure 3.3.11 shows the relative humidity dependence of conductivity for the films with P/Si of 1.5 and 1.0 at 130°C [15]. The ionic conductivity of the films increases with an increase in the relative humidity, and is about 5×10^{-2} S cm^{-1} under 20% R.H. at 130°C for the films with P/Si ratio of 1.5. These results indicate that the amount of adsorbed water in the gels must increase with an increase in the relative humidity or the content of phosphoric acid, and the adsorbed water contributes to the increase in conductivity. Although nitrogen adsorption measurements showed that the hybrid films had dense structure, paths suitable for fast proton conduction must be formed in the hybrid films by the adsorption of water. At 30°C, the conductivity of the films with P/Si of 1.5 was about 2×10^{-4} S cm^{-1} under 20% R.H. [14], indicating that the conductivity was increased by two orders of magnitude by raising temperature from 30 to 130°C with the same relative humidity.

For the fabrication of a single cell, the composite electrodes were prepared from the precursor sol of the hybrid films and Pt-loaded carbon sheets [15, 16]. Nafion® solution is usually used for making proton paths in the electrode. However, in the cell operation at temperatures higher than 100°C, Nafion® loses the proton conductivity without high hydration, and thus a proton conductor which keeps high conductivities even under temperatures higher than 100°C and low humidity should be used. Since the viscosity of the precursor sol of the hybrid is low enough and the precursor sol becomes gel with heating, the precursor sol was used for the formation of proton paths in the electrodes and for making good contact between the electrodes and hybrid films. The composite electrodes prepared from the precursor sol of the hybrid

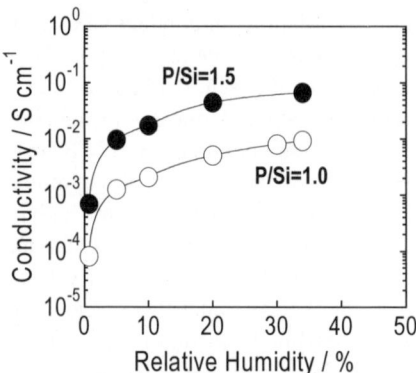

Fig. 3.3.11. Humidity dependence of ionic conductivity for GPTMS-TMOS-H_3PO_4 films at 130°C. Mole ratio of P / Si in films is 1.0 (○) and 1.5 (●) and mole ratio of GPTMS:TMOS in the films is 0.75:0.25.

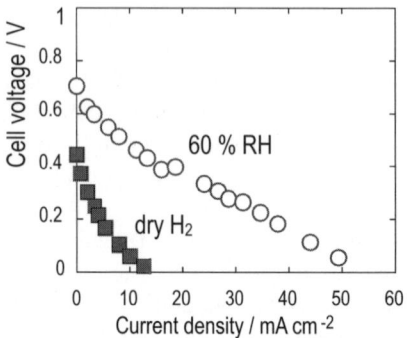

Fig. 3.3.12. Cell voltage versus current density for a single cell with a hybrid film (P/Si = 1) at room temperature with dry hydrogen and the ambient air, and gas of about 60 % RH.

films and Pt-loaded carbon sheets (1 mg cm^{-2}) (ElectroChem Inc, 20 wt% Pt/Vulcan XC-72). Precursor sols were separately prepared, and the Pt-loaded carbon sheets were immersed in the sols. The hybrid films were sandwiched with the two composite electrodes, and hot-pressed at 130°C with 20 MPa [15, 16].

Figure 3.3.12 shows a single cell performance with a hybrid film (P/Si = 1) at room temperature with dried gases of hydrogen and the ambient air, and with the humidified gases of about 60% R.H. [16]. Although relatively small current density is only observed, the cell performance is obtained under dried gases. By using humidified gases, the cell performance is largely improved. Open circuit voltage of the cells is 0.44 V with the dried gases and 0.77 V with humidified gases, which suggests that hydrogen gas permeation through the hybrid films is large under dried conditions, and the gas permeation is suppressed by the absorbed water in the hybrid films. Figure 3.3.13 shows a performance of a single cell using hybrid film (P/Si = 1) under relative humidity of 7% and 17% at 130°C [16]. The test cells work

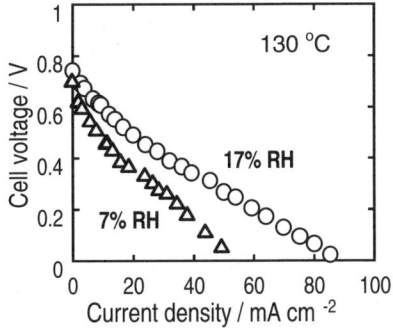

Fig. 3.3.13. Cell voltage versus current density for a test cell at 130°C byusing hydrogen and air of 7% and 17% RH. The hybrid film with P/Si = 1.0 was used as an electrolyte in this experiment.

as a fuel cell at 130°C, and the larger current density is observed under higher relative humidity. This must be due to the increase in proton conductivity with an increase in relative humidity as shown in Fig. 3.3.11. A maximum power density of 15 mWcm^{-2} is obtained for the cell with hybrid of P/Si=1 and 130°C, 17% R.H. Figure 3.3.14 shows cell voltage - current density characteristics for a test cell with the hybrid film of P/Si = 1.0 and 1.5 as an electrolyte under H$_2$ and air at 130°C, 17% R.H. [16]. The open circuit voltage of both cells is about 0.88 V. The larger current density is observed under the cell with P/Si =1.5, and a power of about 50 mW cm^{-2} is obtained with a current density of 150 mA cm^{-2}. The larger power density is mainly due to the increase in proton conductivity with an increase in the P/Si ratio. The cell performances show the high potentiality of the inorganic-organic hybrids as an electrolyte and also a component of composite electrodes for fuel cells at temperatures higher than 100°C.

To obtain hybrid films with more flexible and larger mechanical strength, trialkoxysilane with a longer organic group and an epoxy group should be used as substitute for TMOS. Proton conductive inorganic-organic hybrid films have also been prepared from epoxycyclohexylethyltrimethoxysilane (EHTMS), which has a long organic chain with an epoxy group, GPTMS and orthophosphoric acid (H$_3$PO$_4$) by the sol-gel method [17]. The addition of EHTMS to GPTMS-H$_3$PO$_4$ hybrid firms led to the increase in the tensile strength of the film.

In conclusion, proton conductive inorganic-organic hybrid membranes are very attractive as an alternative to the perfluorinated polymers since the incorporation of inorganic network into organic network gives good thermal, and mechanical and chemical stability, and production cost must be much lower compared with perfluorosulfonic acids. However, inorganic-organic hybrid membranes which have high proton conductivity, high mechanical strength, and high chemical stability do not seem to have been prepared, although many inorganic-organic hybrid systems are proposed up to now, as described in the introduction section. For example, inorganic-organic hybrid membranes based on GPTMS and H$_3$PO$_4$ described above showed

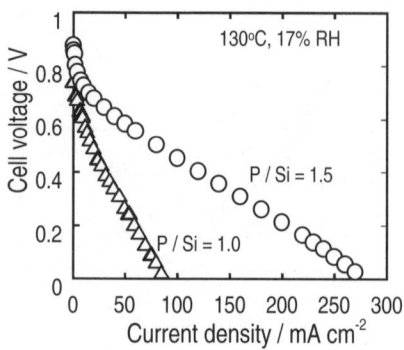

Fig. 3.3.14. Cell voltage - current density characteristics for a test cell with the hybrid film of P/Si = 1.0 and 1.5 as an electrolyte under H_2 and air at 130 °C, 17% RH.

high proton conductivity at temperatures higher than 100°C with low humidity and test cells using the hybrid films as an electrolyte membrane in PEFC were operated at 130°C. Improvement of chemical stability and mechanical strength is indispensable for practical applications. Nevertheless, the inorganic-organic hybrids have high potentiality as an electrolyte and also a component of composite electrodes for fuel cells or other electrochemical devices at temperatures higher than 100°C.

References

1. U. Schubert, N. Husing and A. Lorenz, *Chem. Mater.*, **7** (1995) 2010.
2. I. Gautier-Luneau, A. Deonyelle, J.Y. Sanchez and C. Poinsignon, *Electrochim. Acta*, **37** (1992) 1615.
3. L. Depre, J. Kappel and M. Popall, *Electrochim. Acta*, **43** (1998) 1301.
4. L. Depre, M. Ingram, C. Poinsignon and M. Popall., *Electrochim. Acta*, **45** (2000) 1377.
5. B. Bonnet, D. J. Jones, J. Roziere, L. Tchicaya, G. Alberti, J.M. Casciola, L. Massinelli, B. Bauer, A. Peraio and E. Ramunni , *J. New Mater. Electrochem. Systems*, **3** (2000) 87.
6. U.L. Stranger, N. Groselj, B. Orel and P. Colomban, *Chem. Mater.*, **12** (2000) 2745.
7. Y. Park and M. Nagai, *J. Electrochem. Soc.*, **148** (2001) A616.
8. Y. Park and M. Nagai, *Solid State Ionics*, **145** (2001) 149.
9. H. Nakajima and I. Honma, *Solid State Ionics*, **148** (2002) 607.
10. I. Honma, H. Nakajima, O. Nishikawa, T. Sugimoto and S. Nomura, *J. Electrochem. Soc.*, **150** (2003)A616.
11. J.D. Kim and I. Honma, *Electrochim. Acta*, **48** (2003) 3633.
12. J.D. Kim and I. Honma, *Electrochim. Acta*, **49** (2004) 3179.
13. K. Tadanaga, H. Yoshida, A. Matsuda, T. Minami and M. Tatsumisago, *Electrochemistry*, **70** (2002) 998.

14. K. Tadanaga, H. Yoshida, A. Matsuda, T. Minami and M. Tatsumisago, *Chem. Mater.*, **15** (2003) 1910.
15. K. Tadanaga, H. Yoshida, A. Matsuda, T. Minami and M. Tatsumisago, *Electrochem. Commun.*, **5** (2003) 644.
16. K. Tadanaga, H. Yoshida, A. Matsuda, T. Minami and M. Tatsumisago, *Electrochim. Acta.*, **50** (2004) 705
17. K. Tadanaga, H. Yoshida, A. Matsuda, T. Minami and M. Tatsumisago, *J. Sol-Gel Sci. Tech.*, **31** (2004) 365.
18. A. Matsuda, T. Kanzaki, K. Tadanaga, M. Tatsumisago and T. Minami, *J. Ceram. Soc. Jpn.*, **110** (2002) 131.

3.4 Conclusions

Recent development of amorphous-based lithium ion conductors and proton conductors has been described. In the case of lithium ion conducting glasses, high concentration of carrier ions is most important for high lithium ion conductivity, so that the rapid melt quenching and mechanical milling techniques are very useful to increase in the content of Li^+ ions in the sulfide and oxysulfide glassy systems. Superionic metastable crystalline phases with conductivities higher than 10^{-3} Scm^{-1} at room temperature, which could not be obtained by the usual solid state reaction, were precipitated during the heat treatment of the mechanochemically prepared sulfide glasses to form highly conductive glass-ceramics. The formation of superionic metastable phase only by the crystallization of glass is a remarkable advantage of glassy materials in fabrication of highly conductive solid electrolytes. The mechanical milling technique is also useful in preparation of various kinds of cathode and anode materials for all-solid-state lithium secondary batteries. All-solid-state lithium secondary batteries using sulfide glass-based solid electrolytes exhibited excellent cycling performance under low current densities. However, the high rate performance of the batteries should be improved for the application to high power energy devices. In order to improve the rate performance, characterization and improvement of the interface between electrode active materials and solid electrolytes are indispensable.

Inorganic gel-based materials with high proton conductivity were prepared using the sol-gel process. Sol-gel derived inorganic gels are excellent host to design proton conductors because they contain a large number of micropores surrounded by hydrophilic surfaces. Protonic acid-doped silica gels and their composites with elastic polymers exhibited high proton conductivities at room temperature, which can be used as solid electrolytes for humidity sensors, electrochromic display, and electric double-layer capacitors. Phosphosilicate gels, which keep high conductivities in the temperature range above 100°C under very low relative humidity, are promising as an electrolyte for the fuel cells of medium temperature operations. Inorganic-organic composite and hybrid films based on the phosphosilicate gels were successfully prepared and showed high proton conductivity of 10^{-3}–10^{-2} Scm^{-1} at 130–150°C under 1–20% R.H. Successful fabrication of thermally stable proton conductive composite and hybrid films contributes the practical application of PEFCs operating in the medium temperature range.

4

Recent development of electrode materials in lithium ion batteries

4.1 Candidate cathode material with high voltage for lithium ion batteries: Structural study of 5V class $LiNi_{0.5}Mn_{1.5}O_4$ and $LiNi_{0.5}Mn_{1.5}O_{4-\delta}$

4.1.1 Introduction

In mid-1970s, Whittingham [1] first proposed the feasibility of using an inorganic compound, TiS_2 for non-aqueous secondary batteries of high specific and power density. This compound has good metallic character and undergoes lithium intercalation reversibly. Despite its constant discharge voltage in excess of 2V with current densities of 1-10 mAcm^{-2}, the difficulties with practical non-aqueous batteries and reversible deposition over lithium negative electrode has restricted commercial Li_xTiS_2/Li cells to button-size units. In 1980, Mizushima et al. [2] proposed possibility of using layered $LiCoO_2$ with α-$NaFeO_2$ structure as an intercalation cathode. This oxide is structurally more stable than TiS_2. The open circuit voltage of the cell, $LiCoO_2$/Li was approximately twice (3.6V) that of $LiTiS_2$ giving a theoretical energy density of 1.1 kWhkg^{-1}. Nevertheless, it took almost 10 years to materialize $LiCoO_2$ in commercial batteries.

The technological advancement in the areas of microelectronics and the onset of pure and hybrid-type electric vehicles (HEV) demand in the near future necessitate low cost, environment friendly and thermally stable lithium ion batteries with high power and high energy density. As an important part of these requirements, lithium manganese spinel $LiMn_2O_4$(Fd3m) is considered to be a more attractive cathode material than cobalt and nickel oxides. Although the $LiMn_2O_4$ can be cycled at 4V range, gradual capacity degradation is caused, especially at elevated temperatures [3,4], i.e. above 50°C [5]. The capacity fading was ascribed by Thackeray's group. [6], and Tarascon et al. [7], to the dissolution of Mn into the electrolyte by a disproportionation reaction $2Mn^{3+}$(solid)$\rightarrow Mn^{4+}$ (solid) + Mn^{2+}(liquid). In order to improve the structural stability of $LiMn_2O_4$, Bittihn et al. [8] investigated partial substitution of Co for Mn and improved cycleability was reported using $LiCo_yMn_{2-y}O_4$. Our research group [9] systematically studied charge-discharge properties of $LiM_yMn_{2-y}O_4$(M=Co,Cr,Ni) in the 4V range and observed better cycleability than the parent $LiMn_2O_4$, even though 1/12 of octahedral MO_6 is substituted by other transition metals. Gummow et al. [6] and we [9] have pointed out that the substitution enhances the stability of spinels, in other words, M-O bondings rather than Mn-O bondings in the MO_6 should be strong. We presented the thermodynamic stability of some MO_6 including MnO_6 in spinels by using part of Born-Haber cycle assuming a simple ionic model [10]. Recently our research group extended the study of the structural stability of spinels based on the calculation of molecular dynamics (MD) considering a partially ionic model [11,12]. The results of the calculation supported the experimental data relating to the electrochemical performances.

Furthermore, the extensive charge-discharge properties in the 5V class have been studied using partially substituted spinel oxides. When we consider the development of large scale batteries for electric vechcles(EV) etc., the operation voltage becomes 300V–400V. Therefore, the number of single cell stack becomes important. If the discharge voltage of single cell itself is high, the cell stack number, which relates to higher quality control, can be reduced.

The discovery of highly oxidation resistant electrolyte [13,14] has enabled the study of intercalation compounds up to charging voltages higher than 5V. As an electrolyte for the study, 1M solution of either $LiPF_6$ or $LiClO_4$ in EC and DMC or EC and DEC mixtures are used. Several research groups have reported transition-metal substituted spinel materials, $LiM_yMn_{2-y}O_4$, (M=Cr^{3+} [15–17], Fe^{3+} [16,18,19], Ni^{2+} [16,20,21], Co^{3+} [22]) with high voltage plateaus at around 5V. Usually these oxide spinels exhibit two plateaus at around 4V and 5V. The plateau at around 4V corresponds to reduction of Mn^{4+} to Mn^{3+} whereas reduction of Cr^{4+} to Cr^{3+} or Co^{4+} to Co^{3+} is observed at about 4.9V. In $LiNi_yMn_{2-y}O_4$, two plateaus are present at around 4.3V and 4.7V, respectively. The plateau at 4.3V arises from the Mn^{3+}/Mn^{4+} redox couple, whereas the plateau at 4.7V is due to the Ni^{2+}/Ni^{4+} redox. For the ideal $LiNi_{0.5}Mn_{1.5}O_4$ composition [23,24], it has special interest for its high discharge capacity and dominant plateau at around 4.7V because the oxidation state of Mn should be fixed at +4, resulting in only the Ni^{2+}/Ni^{4+} redox couple during the charge / discharge process. However, some difficulties are encountered with the synthesis of the $LiNi_{0.5}Mn_{1.5}O_4$. Because high temperature calcinations above 800°C lead to a partial reduction of Mn from Mn^{4+} to Mn^{3+}. Impurities such as NiO and Li_xNi_yO can be incorporated during the solid-state reaction. Myung et al. [23] synthesized pure $LiNi_{0.5}Mn_{1.5}O_4$ by an emulsion drying method. Kim et al. [24], synthesized $LiNi_{0.5}Mn_{1.5}O_{4-\delta}$(Fd$\bar{3}$m) by a molten salt method at 900 °C. After annealing it at 700 °C in air, they obtained a single cubic phase of $LiNi_{0.5}Mn_{1.5}O_4$(P$4_3$32).

When these two types of compositions are concerned as cathodes for lithium ion batteries, distinct differences are observed especially in the charge-discharge properties. Such differences seem to come from structural order-disorder which relates to the diffusion of lithium in the structure.

In this chapter, we discuss such electrochemical property with experimental results of XRD, ND(neutron diffraction), and Coulomb potential calculation.

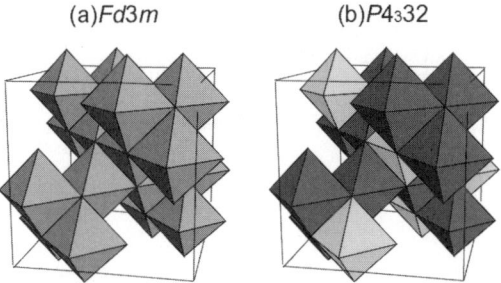

Fig. 4.1.1. MO_6 octahedral network in (a)$Fd\bar{3}m$ and (b)$P4_332$ spinel structure (a) all octahedral metal sites are crystallographycally equal (16d), while (b) two kinds of metal sites exist (4a site in the brighter octahedron, 12c site in the darker octahedron)

4.1.2 Crystal structure of spinel type phase

The structure of stoichiometric $LiNi_{0.5}Mn_{1.5}O_4$ has been first reported by Blasse [25]. Gryffroy et al. [26,27] have proposed that the structure is ordered-type ($P4_332$) by the measurement of ND [26] and IR [27] and supported the proposal of Blasse. Similar results were obtained later from the measurement of ^6Li MAS-NMR [28]. Kim et al. [24] recently determined the structure of $LiNi_{0.5}Mn_{1.5}O_4$ and nonstoichiometric $LiNi_{0.5}Mn_{1.5}O_{4-\delta}$ by careful measurements of XRD and electron diffraction. Their results on the $LiNi_{0.5}Mn_{1.5}O_4$ are consistent with others while $LiNi_{0.5}Mn_{1.5}O_{4-\delta}$ belongs to cubic spinel($Fd\bar{3}m$). Figure 4.1.1 shows MO_6 octahedral network in $Fd\bar{3}m$(a) and $P4_332$(b). In our group, stoichiometric $LiNi_{0.5}Mn_{1.5}O_4$ was synthesized by mixing Li_2CO_3, NiO and Mn_2O_3 with moler ratio, Li : Ni : Mn = 1.0 : 1.5 : 1.5 in ethanol. NiO was preliminarily obtained by decomposing $NiC_2O_4 \cdot 2H_2O$ in air at 700 °C for 24h. Mn_2O_3 was also obtained by heating $MnCO_3$ at 600 °C for 48h because of prevention of deviation of composition by hygroscopic property of $MnCO_3$. After evaporating ethanol, condensed materials were heated at 820 °C in O_2 atmosphere for 4 days and then slowly cooled to room temperature with 0.5°Cmin^{-1}.

Figure 4.1.2 shows XRD pattern for $LiNi_{0.5}Mn_{1.5}O_{4-\alpha-\delta}(\alpha \approx 0)$ synthesized in our group. Nonstoichiometoric $LiNi_{0.5}Mn_{1.5}O_{4-\delta}$ was synthesized by heating the stoichiometric $LiNi_{0.5}Mn_{1.5}O_4$ at 820 °C under controlled partial pressure of oxygen by changing the mixing ratio of O_2 and N_2 using O_2 and $N_2 - O_2$ balanced gas. Actually δ values were evaluated from the wait change using TG (Seiko Denshi, EXSTAR6000TG/DTA). In order to get the data of electrochemical and structural properties for various compositions of δ, the nonstoichiometric samples were quenched from 820 °C to 0 °C by controlling the partial pressures of oxygen in the same manner using separate electric furnace. Slight amount of impurity $LiNi_{1-y}O$ coexists. In the range of δ values between 0 and 0.31, disordered-type cubic spinel ($Fd\bar{3}m$) is observed. The lattice parameter changes by the intercalation reaction of lithium as reported by Kim et al. [12]. Figure 4.1.3 shows their results [12]. The ordered $LiNi_{0.5}Mn_{1.5}O_4$ has three cubic phases, while disordered $LiNi_{0.5}Mn_{1.5}O_{4-\delta}$ forms only two phases. Figure 4.1.4 shows ND patterns for

98 4 Recent development of electrode materials in lithium ion batteries

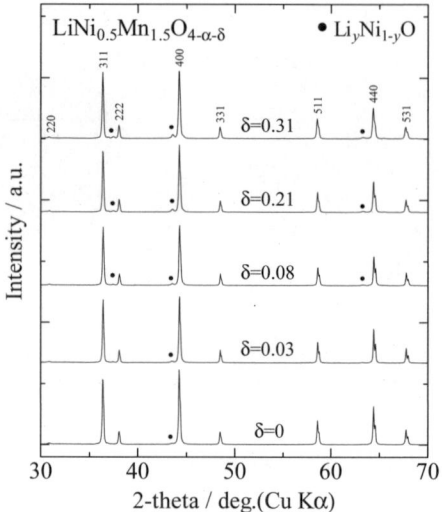

Fig. 4.1.2. Powder XRD patterns of $LiNi_{0.5}Mn_{1.5}O_{4-\alpha-\delta}$

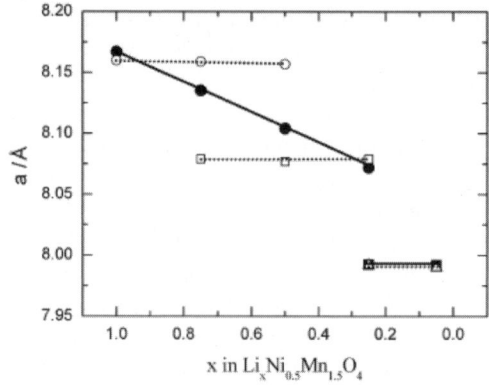

Fig. 4.1.3. Variation in lattice parameter, a, for $Li_xNi_{0.5}Mn_{1.5}O_{4-\delta}$ (closed symbols) and $Li_xNi_{0.5}Mn_{1.5}O_4$ (open symbols) electrodes on charge.

$LiNi_{0.5}Mn_{1.5}O_{4-\alpha-\delta}$ (δ=0.03 and 0.21) obtained in our group. As references, ND patterns of $LiNi_{0.5}Mn_{1.5}O_4$ (P4$_3$32) and $LiMn_2O_4$ (Fd$\bar{3}$m) estimated by diffraction simulation are included in the bottom part. Neutron diffraction was carried out at KEK, Neutron Scattering in Tsukuba, Japan. The diffraction measurement was done by TOF method using ND apparatus, Vega. Simulation of the diffraction data was performed. When δ values increase, the peaks belonging to the ordered $LiNi_{0.5}Mn_{1.5}O_{4-\alpha-\delta}$ ($\alpha \approx 0$) gradually diminished (solid circles). This supports the data of the electron diffraction presented by Kim *et al.* [12]. Accordingly, this would be attributed to the increase of disordering by increasing of defects.

4.1 Candidate cathode material with high voltage for lithium ion batteries

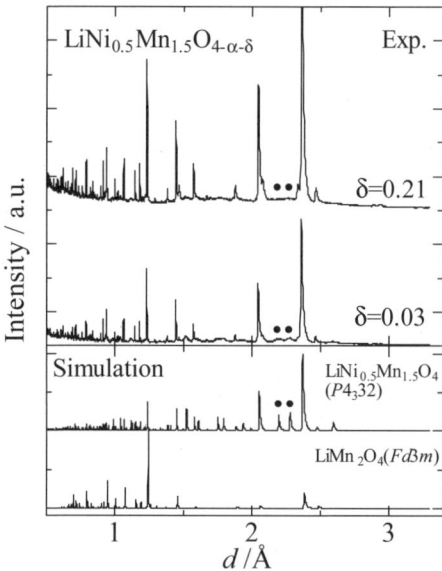

Fig. 4.1.4. Top: ND patterns of $LiNi_{0.5}Mn_{1.5}O_{4-\alpha-\delta}$ observed by ND measurement and bottom: ND patterns of $LiNi_{0.5}Mn_{1.5}O_4$ ($P4_332$) and $LiMn_2O_4$ ($Fd3m$) obtained by diffraction simulation. Solid circles (●) represent typical reflections due to super lattice structure.

Defect in the spinel oxides has been discussed by many researchers [29,30]. From careful density measurements of $LiMn_2O_{4-\delta}$[30] in our group, we have proposed that the defective structures of the spinels synthesized under suitable partial pressure of oxygen are not oxygen defects but metal excess structure. Probably metals in the octahedral $16d$ sites migrate to $16c$ octahedral sites which are usually vacant sites in the stoichiometric spinels. In $LiNi_{0.5}Mn_{1.5}O_{4-\delta}$, the same things would happen. Recently Koyama et al. [31] evaluated the formation energy for $LiMn_2O_{4-\delta}$ including oxygen-vacancy type and metal excess type by first principles plane-wave pseudopotential calculation. According to the calculation, metal excess defects showed the lowest formation energy which suggests the consistency with experimental results of the density measurements.

4.1.3 Electrochemical properties of $LiNi_{0.5}Mn_{1.5}O_4$ and $LiNi_{0.5}Mn_{1.5}O_{4-\delta}$

Figure 4.1.5 shows the first charge-discharge curves for $LiNi_{0.5}Mn_{1.5}O_{4-\delta}$($\delta$=0-0.31) at a low current density of $40\mu Acm^{-2}$ obtained recently. Even at δ=0, a very small shoulder at around 4.0V is observed. The shoulder part increases gradually with increasing δ values. Since the shoulder is due to the Mn^{4+}/Mn^{3+} redox reaction, small amounts of Mn^{3+} would exist even in stoichiometric $LiNi_{0.5}Mn_{1.5}O_4$. The plateau at around 4.7V corresponds to the $Ni^{2+/4+}$ redox reaction. The same charge-discharge profiles have been obtained in literature [20,23,24]. Compared with a 4.0V range cathode, the discharge voltage at 4.7V is 15% higher. If proper stable electrolyte

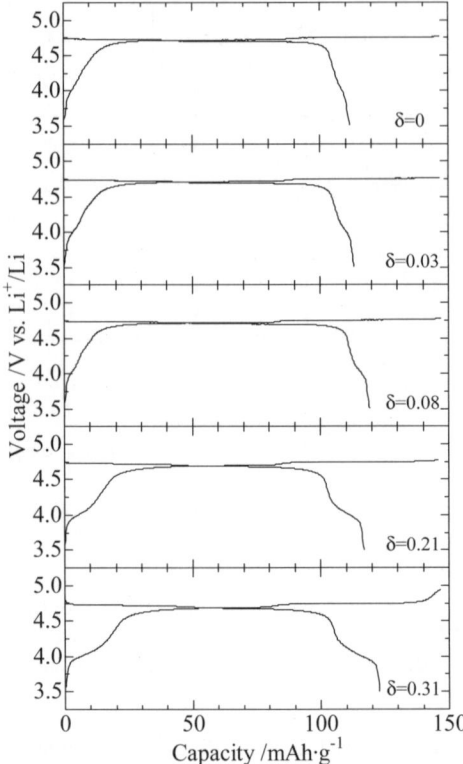

Fig. 4.1.5. First charge-discharge curves of LiNi$_{0.5}$Mn$_{1.5}$O$_{4-\alpha-\delta}$.

enduring the oxidation at around 5V is developed, LiNi$_{0.5}$Mn$_{1.5}$O$_{4-\delta}$ will become a good candidate of cathode which works at high voltage.

Figure 4.1.6 shows the rate performance of LiNi$_{0.5}$Mn$_{1.5}$O$_{4-\delta}$ (δ=0 and 0.31). Solid symbols indicate the data for δ=0, while open symbols show those for δ=0.31. The nonstoichiometric samples show excellent high-rate property. In other words, lithium diffusion in disordered spinels is easier than in ordered ones. This kind of phenomena may depend on the diffusion path of lithium in the structure. In the following section, we will discuss the site-potential energy for lithium migration in both ordered and disordered LiNi$_{0.5}$Mn$_{1.5}$O$_4$ using a simple Coulombic potential calculation.

4.1.4 Coulombic potential calculation of diffusion path for ordered (P4$_3$32) and disordered (Fd$\bar{3}$m) spinels

Figure 4.1.7 shows the diffusion path for disordered (Fd$\bar{3}$m)(a) and ordered (P4$_3$32)(b) spinels. In the disordered spinel, lithium at tetrahedral 8a site moves to vacant octahedral 16c site. Accordingly, diffusion path of 8a-16c arises. On the other

4.1 Candidate cathode material with high voltage for lithium ion batteries

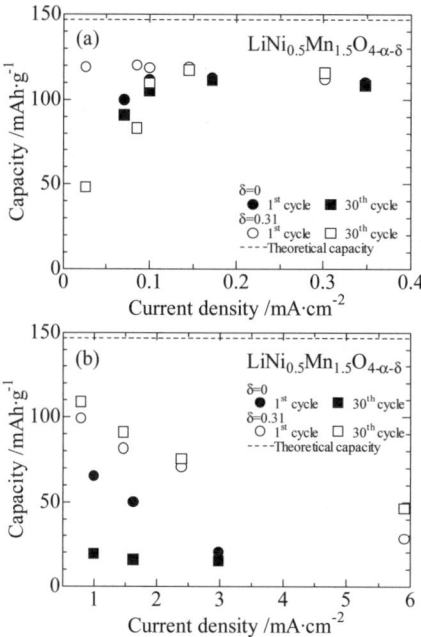

Fig. 4.1.6. Rate performance of LiNi$_{0.5}$Mn$_{1.5}$O$_{4-\alpha-\delta}$ (=0, 0.31). (a) Low rate range and (b) high rate range.

hand, in the ordered spinel, the octahedral vacant 16c sites are splitted into ordered 4a and 12d sites with a 1:3 ratio, forming diffusion paths, 8c-4a and 8c-12d.

The electrostatic potential calculation can be done by using the equation of Coulombic potential U_{ij} as follows:

$$U_{ij} = \frac{Z_i Z_j e^2}{r_{ij}} \tag{1}$$

where Z_i and Z_j are the charges of the ions i and j, respectively. Z values for each ion are set as follows: O^{2-}(-2), Mn^{4+}(+4), Ni^{2+}(+2), Li$^+$(+1). Figure 4.1.8 shows the calculated Coulomb potentials of the diffusion path between two lithium sites for the ordered and the disordered spinels (Fig. 4.1.7). In order to compare the diffusivity in the ordered spinel with that in the disordered spinel, the potential was normalized at lithium sites. The estimated results are shown in Fig. 4.1.8. The obtained potential curves for each diffusion path are quite similar and show a maximum at the vacant site position. The vacant position should be saddle point. The order of the potential at each bottle neck is 8c-4a (order) < 8a-16c (disorder) < 8c-12d (order). For the ordered spinel, the easiest diffusion path 8c-4a and the most difficult diffusion path 8c-12d are coexisting. As a result, the disordered-type containing 8a-16c bears easier diffusion than the ordered-type. Considering the data of ND, the diffusivity of lithium may increase with increasing δ values. These calculations support the experimental

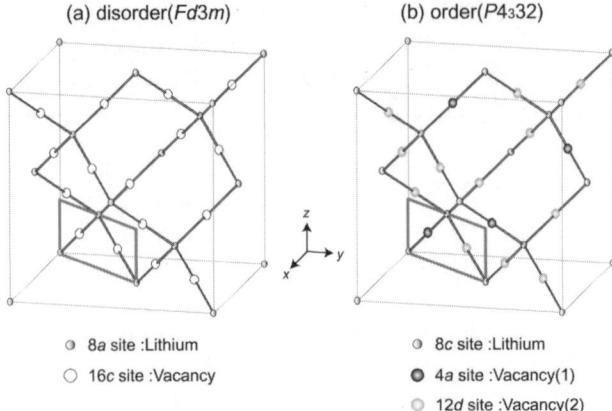

Fig. 4.1.7. Diffusion path of Li ion in LiNi$_{0.5}$Mn$_{1.5}$O$_4$ (a)Fd$\bar{3}$m and (b)P4$_3$32 spinel structure.

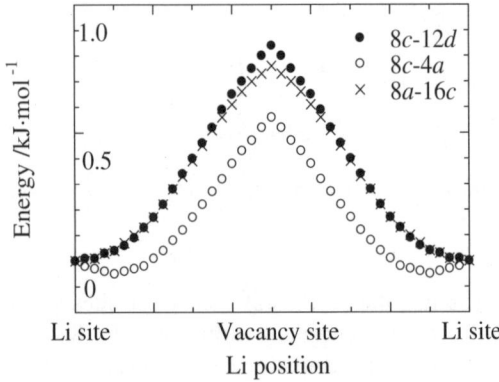

Fig. 4.1.8. Variation of the calculated Coulombic potential along with Li diffusion path.

charge-discharge property, *i.e.* the disordered spinels keep higher capacity than the ordered spinel (see Fig. 4.1.6). Introduction of defects to LiNi$_{0.5}$Mn$_{1.5}$O$_{4-\delta}$ seems to improve the rate property of the charge-discharge.

4.1.5 Conclusions

LiNi$_{0.5}$Mn$_{1.5}$O$_4$ and LiNi$_{0.5}$Mn$_{1.5}$O$_{4-\delta}$ were synthesized by solid-state reactions at 820 °C. It became clear from the ND data that the former stoichiometric spinel oxide had an ordered structure (P4$_3$32), while the latter nonstoichiometric oxides formed a disordered structure (Fd$\bar{3}$m). The disordered spinel showed better rate property than the ordered spinel. From calculated Coulombic potential for diffusion paths in both ordered and disordered spinels, it is suggested that the diffusivity in the disordered spinel is higher in comparison with the ordered spinel. The results of the calculation agreed well with the experimental electrochemical property.

References

1. M. S. Whittinghum, Prog. *Solid State Chem.*, **12** (1978) 4.
2. K. Mizushima, P. C. Jones, P. T. Weiseman and J.B. Goodenough, *Mater. Res. Bull.*, **15** (1980) 258.
3. Y. Xia and M. Yoshio, *J. Electrochem. Soc.*, 143 (1996) 825.
4. D. H. Jang, Y. J. Shin and S. M. Oh, *J. Electrochem. Soc.*, **143** (1996) 2204.
5. D. Song, H. Ikuta and M. Wakihara, *Electrochemistry*, **68** (2000) 460.
6. R. J. Gummow, A. de Kock and M. M. Thackery, *Solid State Ionics*, 69 (1994) 59.
7. J. M. Tarascon, E. Wang, F. K. Shokoohi, W. R. Mckinnon and S. Colson, *J. Electrochem. Soc.*, **138** (1991) 2859.
8. R. Bittihn, R. Herr and D. Hoge, *J. Power Sources*, **43–44** (1993) 223.
9. G. Li, H. Ikuta, T. Uchida and M. Wakihara, *J. Electrochem. Soc.*, **143** (1996) 178.
10. M. Wakihara, G. Li and H. Ikuta, *Lithium Ion Batteries*, Kodansha, Wiley-VCH, Tokyo, New York, Chapter2 (1998).
11. M. Kaneko, S. Matsuno, T. Miki, M. Nakayama, H. Ikuta, Y. Uchimoto, M. Wakihara and K. Kawamura, *J. Phys. Chem. B*, **107** (2003) 1727.
12. M. Nakayama, M. Kaneko, Y. Uchimoto, M. Wakihara and K. Kawamura, *J. Phys. Chem. B*, **108** (2004) 3754.
13. J. M. Tarascon and D. Guyomard, *Soled State Ionics*, **69** (1994) 293.
14. D. Guyomard and J. M. Tarascon, *J. Power Sources*, **54** (1995) 921.
15. C. Sigala, D. Guyomard, A. Verbaere, Y. Piffard and M. Tournoux, *Solid State Ionics*, **81** (1995) 167.
16. T. Ohzuku, S. Takeda and M. Iwanaga, *J. Power Sources*, **81–82** (1999) 90.
17. C. Sigala, A. Le Gal La Salle, Y. Piffaed and D. Guyomard, *J. Electrochem. Soc.*, **148** (2001) A812.
18. H. Kawai, M. Nagata, M. Tabuchi, H. Tsukamoto and A. W. West, *Chem. Mater.*, **10** (1998) 3266.
19. H. Kawai, M. Nagata, H. Tsukamoto and A. W. West, *Electrochem. Solid-State Lett.*, **1** (1998) 212.
20. Q. Zhong, A. Bonakdarpour, M. Zhang, Y. Gao and J. R. Dahn, *J. Electrochem. Soc.*, **144** (1994) 205.
21. T. Zheng and J. R. Dahn, *Phys. Rev. B*, **56** (1997) 3800.
22. H. Kawai, M. Nagata, H. Kageyamam, H. Tsukamoto and A. W. West, *Electrochem. Acta*, **45** (1999) 315.
23. S. T. Myung, S. Komada, N. Kumagai, H. Yashiro, H. T. Chung and T-H. Cho, *Electrochem. Acta*, **47** (2002) 2543.
24. J. H. Kim, S. T. Myung, C. S. Yoon, S. G. Kang and Y. K. Sun, *Chem. Mater.*, **16** (2004) 906.
25. G. Blasse, *Phylips Res. Rep. Suppl.*, **3** (1964) 1.
26. D. Gryffroy, R. E. Vandenberghe and E. Legrand, *Mater. Sci. Forum*, **79–82** (1991) 785.
27. D. Gryffroy, and R. E. Vandenberghe, *J. Phys. Chem. Solids*, **53** (1992) 77.

28. Y. J. Lee, C. Eng and C. P. Grey, *J. Electrochem. Soc.*, **148** (2001) A249.
29. J. Sugiyama, T. Tatsumi, T. Hioki, S. Noda and N. Kamegashira, *J. Alloy Compd.*, **235** (1996) 163.
30. M. Hosoya, H. Ikuta and M. Wakihara, *J. Electrochem. Soc.*, **144** (1997) L52.
31. Y. Koyama, I. Tanaka, H. Adachi, Y. Uchimoto and M. Wakihara, *J. Electrochem. Soc.*, **150** (2003) A63.

4.2 Electrode/electrolyte interfaces in all-solid-state lithium ion batteries

4.2.1 Introduction

In the past three decades, much attention has been paid to solid electrolytes because of their general potential, and particularly the possibility for all-solid-state lithium ion secondary batteries [1–3]. The great advantages of a solid electrolyte are safety (compared with conventional liquid electrolytes containing flammable organic solvents) and high energy density, which is an important issue both for portable small batteries and large-scale batteries [3,4]. In this chapter, we focus on two types of solid electrolytes; inorganic glass electrolytes and polymer electrolytes. Among inorganic glass electrolytes, the SiS_2-Li_2S-Li_4SiO_4 glass electrolyte which was developed by Minami & Tatsumisago's group is known to exhibit a high lithium ion conductivity of up to 10^{-3} S cm^{-1}, excellent ionic conductivity (transport number of lithium ion = 1.0) and an outstandingly wide electrochemical window [3,4]. In contrast, in a solid polymer electrolyte based on polyethers complexed with lithium salts, the low ionic conductivity and transport number of lithium ion are serious disadvantages for polymer electrolytes, caused by the low degree of dissociation of lithium salts, *i.e.* low concentration of carrier ions, and strong interactions between lithium ions and electron donor atoms of the polymer chains, for example ether oxygen atoms of polyether, leading to low ionic mobility [1,2]. However, the polymer electrolytes are flexible and therefore easily make good contact between electrode and electrolyte.

Many published studies have attempted to improve the bulk properties of these two types of electrolytes, although only a few investigations have been focused on the charge transfer reaction at the electrode/solid electrolyte interface [5,6]. However, enhancement of the interfacial reaction rate is important to fabricate high power density batteries, because battery performance is directly related to the rates of the charge-transfer reactions at the electrode/electrolyte interface as well as the bulk conductivity of the electrolyte, which is likely to be a more important problem with solid electrolytes than with conventional liquid electrolytes.

4.2.2 Control and charge transfer process at electrode/electrolyte interface

We will treat three topics in this section; (i) the important factors determining the charge transfer reaction rate at the polyether based electrolyte/electrode interfaces [7], (ii) the enhancement of ionic conductivity and transport number of lithium ions of the polymer electrolytes by the addition of PEG-borate ester as a Lewis acid in the electrolytes [8–15], and (iii) the enhancement of the reaction rate by the addition of PEG-borate ester [16].

(i) Study of factors influencing the charge transfer reaction rate at the polyether-based electrolyte/electrode interface

Clarifying the reaction rate at the interfaces as well as the ionic conduction mechanism of the bulk is necessary for the molecular design of advanced polymer electrolytes for batteries with higher power density. We investigated the electrokinetics of the Li^+/Li couple reaction on lithium metal electrodes in poly (ethylene glycol) dimethyl ether (PEGDME)-based electrolytes. PEGDME electrolyte solutions with dissolved $LiCF_3SO_3$ are used as a model for polymer electrolytes, similar to the amorphous conducting phase in high-molecular-weight PEO. In order to clarify the factors that influence electrokinetics in electrolyte solutions based on PEGDME, we measured the charge transfer reaction rate at polymer electrolyte/electrode interfaces by using PEGDME, whose average molecular weight was 500 (abbreviated as PEGDME500) by adding various amounts of a higher molecular weight polyether, PEGDME1000. This sub-section reports that the viscosity of the polymer electrolytes is one of the important factors determining the charge transfer reaction rate at the interfaces and the phenomenon is interpreted by the Marcus microscopic theories of the charge transfer reaction [17–19].

The electrokinetics of lithium deposition and dissolution on the working microelectrode in the electrolyte solution was studied using a double potential step technique, chronoamperometry. All measurements of the coulombic efficiency of lithium deposition and dissolution were over 90%. This is due to the fresh deposition and immediate dissolution of lithium on the microelectrode. For the estimation of the exchange current densities of the Li^+/Li couple in the electrolyte solutions, the steady-state currents (i) obtained at various overpotentials (E) by chronoamperometry were analyzed with the following Allen-Hickling equation,

$$\ln\left[\frac{i}{1 - \exp(FE/RT)}\right] = \ln i_0 - \frac{\alpha F}{RT}E \tag{1}$$

where F is the Faraday constant, R is the gas constant, T is the absolute temperature, α is the transfer coefficient and i_0 is the exchange current density [20]. Figure 4.2.1 shows Allen-Hickling plots for the steady-state currents at various overpotentials in 0.5 M $LiCF_3SO_3$/PEGDME500 at 333 K. In the Allen-Hickling plots, a linear relationship was found in the range of overpotentials around -150 to 170 mV vs. Li^+/Li, while at numerically higher potentials, some deviation from the relationship was found, which was due to the limiting effect of mass transport. The values of the exchange current density and the transfer coefficient were estimated from the intercept and slope of the Allen-Hickling plots. From the same procedure, the exchange current densities in electrolyte solutions with various amounts of PEGDME1000 added were obtained, and are summarized in Fig. 4.2.1. Figure 4.2.2

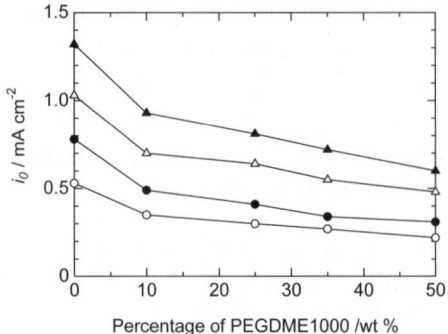

Fig. 4.2.1. i_0 vs. the fraction of PEGDME1000 in PEGDM500/PEGDME1000 mixed solution containing 0.5M LiCF$_3$SO$_3$ at various temperatures: ○ at 313 K, ● at 323 K, △ at 333 K, ▲ at 343 K.

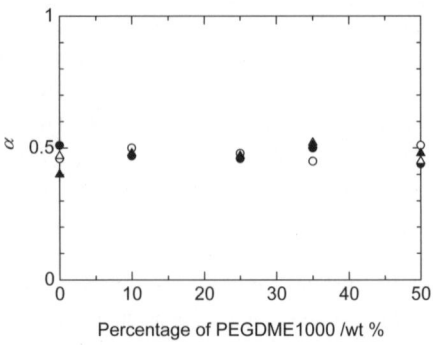

Fig. 4.2.2. α vs. the fraction of PEGDME1000 in PEGDM500/PEGDME1000 mixed solution containing 0.5M LiCF$_3$SO$_3$ at various temperatures: ○ at 313 K, ● at 323 K, △ at 333 K, ▲ at 343 K.

summarizes the estimated values of the transfer coefficient. All the values are close to 0.5, which implies a symmetric reaction of the Li$^+$/Li couple. As shown in the result, the exchange current densities decrease with increasing amounts of PEGDME1000.

The exchange current density i_0 can be expressed thus: [20]

$$i_0 = K_r a_{Li^+}^{1-\alpha} a_{Li}^{\alpha} \tag{2}$$

where a_{Li^+} is the activity of lithium ion, a_{Li} is the activity of lithium metal, and K_r is the standard rate constant, which can be expressed as the following equation (3) according to a recent advanced theory revealing dynamic solvent effects on electron transfer rates [17–19], [21–23],

$$K_r = \frac{A}{\tau_L} \exp(-\Delta G^*/RT) \tag{3}$$

where A is the pre-exponential factor, ΔG^* is the Gibbs activation energy of the reaction and τ_L is the longitudinal relaxation time of the solvent. The fact that all the obtained values of the transfer coefficient are close to 0.5 and a_{Li} is unity leads to the following equation (4) for the exchange current density,

$$i_0 = \frac{A}{\tau_L} \sqrt{a_{Li^+}} \exp(-\Delta G^*/RT) \tag{4}$$

As a consequence, equation (4) indicates that the exchange current density depends on the activity of lithium ions, the Gibbs activation energy, and the longitudinal relaxation time of the solvent.

In order to investigate the activity of lithium ions in the electrolyte solutions, Raman spectroscopic studies were carried out. Raman spectra are very useful to investigate the activity of lithium ions in electrolytes containing lithium salts, due to the internal vibration modes of the counter anions [24–26]. The spectral band of the SO_3 symmetrical stretching mode of $CF_3SO_3^-$ ($v_s(SO_3^-)$), which is observed around 1040 cm^{-1}, was chosen for the present electrolyte solutions. Furthermore, this band consists of three components corresponding to the state of the anion as follows: the lower frequency component at ~1034 cm^{-1} is attributed to free ions, the component at ~1042 cm^{-1} to ion pairs, and the highest frequency component at ~1052 cm^{-1} to multiply aggregated ions, respectively [24–26]. Therefore, the first component, which corresponds to a fraction of free lithium ions, in other words, the activity of lithium ions, is very useful in the present study. The fraction of free ions corresponds to the ratio of the area due to free ions (dissociated ions) to the areas of all the components (aggregated and ion-paired as well as free ions). Figure 4.2.3 shows Raman spectra, in the region of 1000–1080 cm^{-1}, of 0.5 M LiCF$_3$SO$_3$/PEGDME500 with various amounts of PEGDME1000. Each spectrum was deconvoluted to three components by fitting three distinct Lorentzian curves. Figure 4.2.4 summarizes the variations for the fractions of these three kinds of ions in the electrolyte solutions at 313 K and 343 K. As shown in the results, the fractions of the component corresponding to free ions are almost the same, at around 20%, even with increasing amounts of PEGDME1000, in the temperature range of 313 K to 343 K, indicating that activity of lithium ions is almost constant in electrolyte solutions with various amounts of PEGDME1000. Generally, the dissociation constant of a lithium salt is determined by the lattice energy of the salt itself and the solvation state of ions, which has to do with ion-solvent molecule interaction [27,28]. When a lithium salt is added to the PEO-based matrix, the lithium ions are incorporated into the helix structures of EO chains [27,28]. In the case of PEGDME500 and PEGDME1000, since ether oxygens act as donor polar groups, the geometry, i.e. the solvation state, of lithium ions in the PEGDME should be similar. Therefore, the ratio of free ions, i.e. the activity of the lithium ions in the electrolyte solutions was almost constant.

The values of the Gibbs activation energies were calculated from the slope of the linear relationship plots. The calculated Gibbs activation energies are summarized in Table 4.2.1. The electrokinetics of the Li$^+$/Li couple reaction involves a desolvation/solvation process. Therefore, the Gibbs activation energy may depend on the desolvation/solvation energy. In other words, the Gibbs activation energy depends on the solvation state of lithium ions in the solvent. As explained in the Raman spectroscopic studies, the solvation state of lithium ions in the PEGDME500

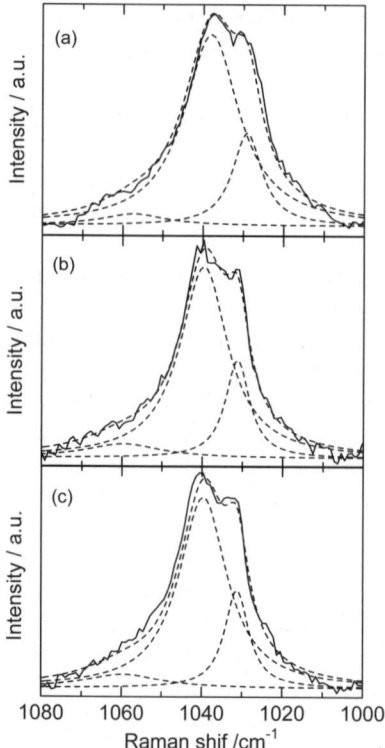

Fig. 4.2.3. Raman spectra of 0.5M LiCF$_3$SO$_3$ in the solutions at 313K: (a) PEGDME500 (100%), (b) PEGDME500 75wt% + PEGDME1000 25wt%, (c) PEGDME500 50wt% + PGDME1000 50wt%. Solid lines: experimental data; Dotted lines: fitting results with three distinct Lorentzian curves.

and PEGDME1000 should be similar, since lithium ions are incorporated into the helix structures of EO chains [27,28]. Therefore, the activation energy is independent of the amount of PEGDME1000.

These results clarify the fact that the activity of lithium ions and the Gibbs activation energy do not depend on the ratio of PEGDME500/PEGDME1000 in the electrolyte solutions. On the other hand, Fig. 4.2.1 clearly shows that the exchange current densities decreased with increasing amounts of PEGDME1000. Equation (4) suggests that the exchange current density depends on the relaxation time of the solvents. Studies of the solvent effect on charge transfer kinetics have elucidated that the reaction rate involves the frequency of solvent reorientation in the desolvation/solvation process of solvent dynamics [17–19,21–23]. Furthermore, this frequency is inversely proportional to the longitudinal relaxation time. As a consequence, the relaxation time can be seen to be related to the frequency of solvent reorientation in the desolvation/solvation process. In addition, the longitudinal relaxation time of the solvent, τ_L, is related to the Debye relaxation time τ_D through

4.2 Electrode/electrolyte interfaces in all-solid-state lithium ion batteries

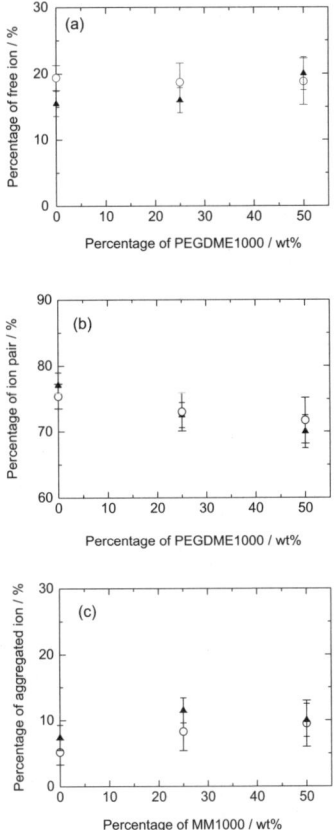

Fig. 4.2.4. Fraction of (a) free ions, (b) ion-pair, (c) aggregated ions in 0.5M LiCF$_3$SO$_3$/PEGDME500 with various amounts of PEGDME1000 at various temperatures: ○ at 313 K, ▲ at 343 K.

the following equation [29,30],

$$\tau_L = \tau_D \varepsilon_\infty / \varepsilon_s \qquad (5)$$

where ε_s is the static dielectric constant and ε_∞ is the infinite frequency dielectric constant. Furthermore, the Debye relaxation time, τ_D, is proportional to the viscosity, η [30,31]. Therefore, the longitudinal relaxation time of the solvent is proportional to the viscosity, η. Accordingly, the exchange current density, i_0, is inversely proportional to the viscosity from the above relationships and equation (4).

Figure 4.2.5 shows the relationship between the viscosities and the fraction of PEGDME1000 in the PEGDME500 based electrolyte solutions. It was found that the viscosity of the solutions based on PEGDME500 increased with increasing amounts of the higher molecular weight PEGDME1000. The relationships between the exchange current densities and the viscosities of the solutions are shown in Fig.

Table 4.2.1. Activation energies of Li$^+$/Li couple reaction in 0.5M LiCF$_3$SO$_3$/PEGDME500 with various amounts of PEGDME1000.

Amount of PEGDME1000 /wt %	G^* /kJ mol^{-1}
0	28.9
10	28.7
25	29.8
35	28.4
50	30.3

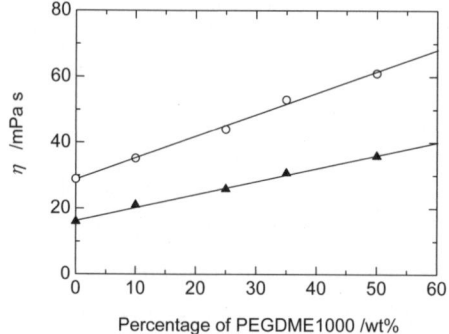

Fig. 4.2.5. Viscosity of electrolytic solutions, 0.5 M LiCF$_3$SO$_3$/PEGDME500 with various amounts of PEGDME1000, at various temperatures: ○ at 313 K, ▲ at 343 K.

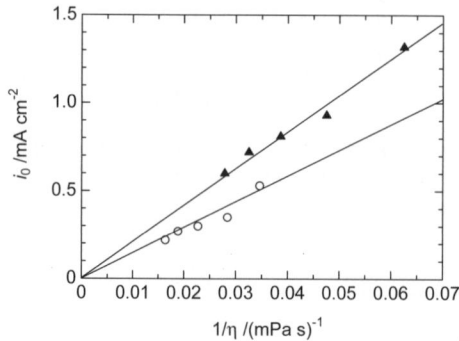

Fig. 4.2.6. Exchange current densities vs. inverse of viscosities of 0.5M LiCF$_3$SO$_3$/PEGDME500 with various amounts of PEGDME1000 at various temperatures: ○ at 313 K, at 343 K.

4.2.6. The exchange current density was inversely proportional to the viscosity. Therefore, the viscosity of the solvent is an important factor in determining the exchange current density of the polyether based electrolytes.

The studies in PEGDME-based electrolytes of the activity of lithium ion by Raman spectroscopy, the Gibbs activation energy and viscosity have revealed that only the last factor, viscosity, of the electrolyte solution has an effect on the charge transfer reaction rate at the electrode/electrolyte interfaces. These results imply that it will be difficult to get high rates of charge transfer reactions at the interfaces when using high-viscosity electrolytes, and especially polymer electrolytes.

In conclusion, it was found that the exchange current densities in the PEGDME500-based electrolytes decreased with increasing amounts of PEGDME1000. Raman spectroscopic studies of the electrolytes revealed that the activity of lithium ions was almost the same even with the addition of PEGDME1000. The results for the Gibbs activation energies for the reaction suggested that the solvation states of lithium ions were similar in the electrolytes to those in PEGDME whose molecular weight range was 500 to 1000. Meanwhile, the viscosity of the electrolytes increased with increasing amounts of PEGDME1000. Furthermore, inversely proportional relationships were found between the exchange current densities and the viscosities of the electrolytes. Therefore, the viscosity of the electrolyte is an important factor in determining the charge transfer reaction rate at the polyether-based electrolyte/electrode interfaces, which implies that it will be difficult to get high rates of charge transfer at the interfaces when using polymer electrolytes, or other highly viscous electrolytes.

(ii) Enhancement of ionic conductivity and transport number of lithium ions in polymer electrolytes by the addition of PEG-borate ester as a Lewis acid into the electrolytes

In order to utilize solid polymer electrolytes in practical electrochemical devices, it is necessary to enhance their ionic conductivity and the transport number of lithium ions. These goals can be achieved by enhancement of the dissociation of supporting salts and by decreasing the mobility of the counter anions. Addition of Lewis acid compounds, which are expected to interact with Lewis base anions, should enhance the transport number of lithium ions because the Lewis acid compounds trap the counter anions. Boron compounds have been reported to act as anion receptors because of the Lewis acid character of boron atoms. In addition, incorporation of boroxol into polymer structures, which achieve a high transport number for lithium ions in polymer electrolytes, has been proposed [32,33]. We have focused on PEG-borate ester, which acts as a Lewis acid. In the present study, we have investigated the effect of the Lewis acid character of borate ester groups, which are fixed to the chains of the matrix polymer, on ionic conductivity and transport number of lithium ions in the polymer electrolytes. We report the relationships between the concentration of borate ester groups and ionic conductivity or transport number of lithium ions in the polymer electrolytes. Furthermore, the interactions between the Lewis acidity of the borate ester groups and Lewis basicity of anions based on the Hard and Soft Acids and Bases (HSAB) theory have been investigated with *ab initio* calculations.

The backbone polymer was prepared from polyethylene glycol and boric acid anhydride as shown in Fig. 4.2.7. Polyethylene glycols with various molecular weights (PEG150, PEG200, PEG400, PEG600, whose average molecular weight was 150, 200, 400, 600, respectively) were supplied from NOF Co., Ltd. $LiN(CF_3SO_2)_2$, $LiCF_3SO_3$ and $LiClO_4$ were used for supporting salts without further

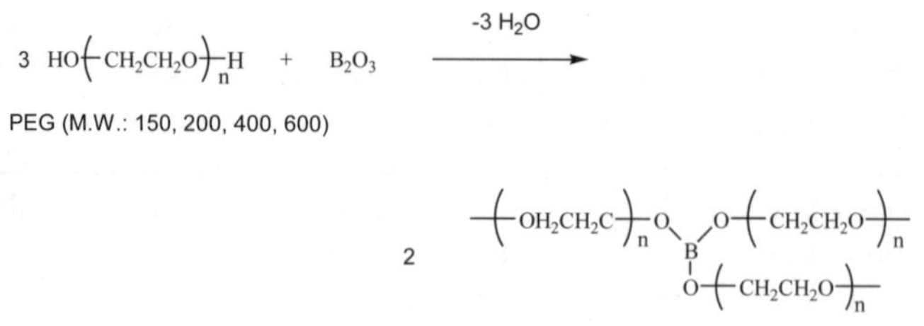

Fig. 4.2.7. Reaction scheme for preparation of backbone polymer from polyethylene glycol and boric acid anhydride.

purification. The composition of polymer electrolyte, for example, prepared from PEG200, boric acid anhydride and $LiN(CF_3SO_2)_2$, is represented as PEG200-B_2O_3 + $LiN(CF_3SO_2)_2$.

Ab initio Hartree-Fock (HF) self-consistent field molecular orbital calculations and density functional theory (DFT) calculations were performed with Gaussian98 [34]. Calculations for geometry optimizations were carried out at the HF level of the theory using the standard 3-21G basis set. Subsequently, single point calculation to investigate the energies of the optimized geometries was performed by using DFT methods with the B3LYP [35,36] form for the exchange-correlation function and the 6-311G** basis set.

T_g of the polymer electrolytes decreased with increasing molecular weight of PEG, in other words, with decreasing the concentration of the borate ester groups which play the role of crosslinking points on the polymer electrolytes. These results indicate that the mobility of the polymer chains increases with decreasing concentration of crosslinking points, *i.e.* borate ester groups, in the polymer electrolytes, because T_g is correlated with segmental motion of the polymer chains.

The temperature dependence of the ionic conductivity of the polymer electrolytes, PEGx–B_2O_3 (x: 150, 200, 400, 600) + $LiN(CF_3SO_2)_2$ (Li/EO = 1/24) is shown in Fig. 4.2.8. An increase in ionic conductivity of the polymer electrolytes is observed with increasing molecular weight of PEG. The highest ionic conductivity was found for the polymer electrolyte prepared from PEG600, to be 5.9×10^{-5} S cm^{-1} at 30 °C and 3.0×10^{-4} S cm^{-1} at 60 °C. The temperature dependence exhibits a convex profile in Arrhenius-type plots as shown in Fig. 4.2.8. This phenomenon indicates that the ionic conduction mechanism of the polymer electrolyte samples does not obey the hopping model of carrier ions. This temperature dependence of the ionic conductivity is typically observed in polymer electrolytes and is expected to obey the free volume theory of polymers [37–40], which leads to the temperature dependence for segmental motion of polymer chains expressed by the Williams-Landel-Ferry (WLF) relationship [41]. The WLF equation is represented as follows:

$$\log \frac{\sigma(T)}{\sigma(T_g)} = \frac{C_1(T - T_g)}{C_2 + (T - T_g)} \qquad (6)$$

4.2 Electrode/electrolyte interfaces in all-solid-state lithium ion batteries

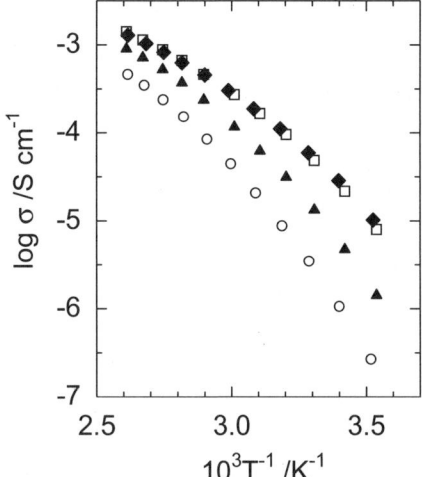

Fig. 4.2.8. Arrhenius plots of ionic conductivity for PEGx B$_2$O$_3$ + LiN(CF$_3$SO$_2$)$_2$ (Li/EO = 1/24). ○: PEG150, ▲: PEG200, □: PEG400, ◆: PEG600.

where $\sigma(T)$ and $\sigma(T_g)$ are the conductivity at temperatures T and T_g, respectively, and C_1, C_2 are the WLF parameters for the temperature dependence of ionic conductivity. However, the ionic conductivity at T_g, $\sigma(T_g)$, is difficult to measure in the present experiments because $\sigma(T_g)$ is too low to measure with a complex impedance measurement, and therefore, 10 °C was selected as the reference temperature, T_0. Equation (6) is rewritten as follows:

$$\log \frac{\sigma(T)}{\sigma(T_0)} = \frac{C_1'(T-T_0)}{C_2' + (T-T_0)} \qquad (7)$$

The parameters in equation (1) are calculated as follows:

$$C_1 = C_1'C_2'/[C_2' - (T_0 - T_g)] \qquad (8)$$

$$C_2 = C_2' - (T_0 - T_g) \qquad (9)$$

Figure 4.2.9 shows WLF plots for the ionic conductivity of the polymer electrolytes, PEGx–B$_2$O$_3$ (x: 150, 200, 400, 600) + LiN(CF$_3$SO$_2$)$_2$ (Li/EO = 1/24). From the inverse of equation (7), the temperature dependence of the ionic conductivity is plotted as $[\log(\sigma(T)/\sigma(T_0))]^{-1}$ vs. $1/(T - T_0)$. It is observed as shown in Fig. 4.2.9 that $[\log(\sigma(T)/\sigma(T_0))]^{-1}$ varies linearly with $1/(T - T_0)$, which indicates that the temperature dependence of the ionic conductivity for the polymer electrolytes follows a WLF-type equation.

The WLF parameters, C_1, C_2 and $\sigma(T_g)$, were estimated from Fig. 4.2.9 and equations (7–9), and are summarized in Table 4.2.2. The estimated parameters were found to be close to the universal values of WLF parameters, $C_1 = 17.4$ and $C_2 = 51.6/K$[47], which indicates that the temperature dependence of the ionic conductivity

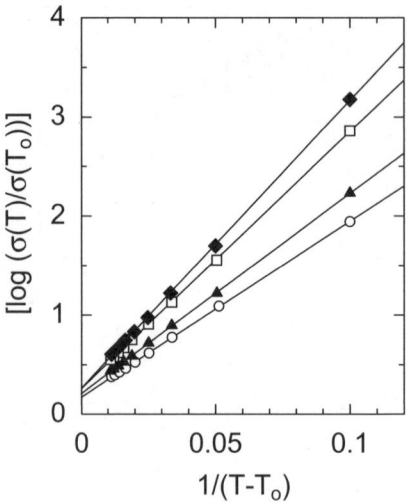

Fig. 4.2.9. WLF plots of ionic conductivity for PEGx B$_2$O$_3$ + LiN(CF$_3$SO$_2$)$_2$ (Li/EO = 1/24). ○: PEG150, ▲: PEG200, □: PEG400, ◆: PEG600. T_0=10°C.

Table 4.2.2. WLF parameters and ($\sigma(T_g)$) of PEGx B$_2$O$_3$+LiN(CF$_3$SO$_2$)$_2$ (Li/EO=1/24).

Molecular weight of PEG	T_g/°C	C_1/-	C_2/K	$\sigma(T_g)$ /S cm^{-1}
PEG150	28	10.6	59.1	1.00×10^{-11}
PEG200	39	12.1	37.7	1.87×10^{-13}
PEG400	48	14.7	22.1	1.58×10^{-16}
PEG600	52	16.9	18.0	7.94×10^{-19}

for the polymer electrolytes was dominated by that of the segmental motion of polymer chains. Therefore from the WLF plots and T_g values, the increase in ionic conductivity with increasing molecular weight of PEG is due to the increase in ionic mobility.

As shown in Table 4.2.2, $\sigma(T_g)$ increases with increasing concentration of borate ester groups. Since the segmental motion of polymer chains is frozen at the glass-transition temperature, it is considered that mobility of the polymer chains at that temperature may be the same. Accordingly, the difference in the ionic conductivity of the polymer electrolytes at T_g is ascribed to the following three reasons: 1) Concentration of carrier ions at T_g is different; 2) The dissociation constant of the lithium salt is different because the polymer electrolyte from different PEGs has different T_g; 3) Mobility of carrier ions is different at different T_g. The reason for the increase in $\sigma(T_g)$ is not clear in this case.

In order to investigate the effect of the borate ester groups on lithium salt, the transport number of lithium ions in the polymer electrolytes was estimated according to the equation presented by Abraham et al. [42] as follows:

4.2 Electrode/electrolyte interfaces in all-solid-state lithium ion batteries

Table 4.2.3. Transport number of PEGx – B$_2$O$_3$+LiN(CF$_3$SO$_2$)$_2$ (Li/EO=1/24) at 60°C.

Molecular weight of PEG	t_{Li}^+
PEG150	0.64
PEG200	0.44
PEG400	0.17
PEG600	0.16

$$t_{Li^+} = \frac{I_{(\infty)}R_{b(\infty)}(\Delta V - I_{(0)}R_{e(0)})}{I_{(0)}R_{b(0)}(\Delta V - I_{(\infty)}R_{e(\infty)})} \tag{10}$$

where I is the current, ΔV is the applied potential, R_b is the bulk resistance, R_e is the interface resistance, and 0 and ∞ refer to the initial and steady states, respectively. The estimated transport numbers of lithium ions in the polymer electrolytes, PEGx–B$_2$O$_3$ (x: 150, 200, 400, 600) + LiN(CF$_3$SO$_2$)$_2$ (Li/EO = 1/24), at 60 °C are summarized in Table 4.2.3. The transport number of lithium ions increases with increasing concentration of the borate ester groups in polymer electrolytes. Therefore, the borate ester groups interact with anions, and accordingly, the movement of anions is interrupted, which enhances the transport number of lithium ions in polymer electrolytes.

In order to clarify the interactions between borate ester groups and anions more precisely, other lithium salts, LiCF$_3$SO$_3$ and LiClO$_4$ as well as LiN(CF$_3$SO$_2$)$_2$ were used as supporting electrolytes for the polymer electrolytes. Accordingly, to investigate the effects of the Lewis acidity of the borate ester groups in polymer electrolytes on the anions, ionic conductivity and T_g of the polymer electrolytes containing LiCF$_3$SO$_3$ or LiClO$_4$ were examined.

Figure 4.2.10 shows the temperature dependence of the ionic conductivity of the polymer electrolytes, PEGx–B$_2$O$_3$ (x: 200, 600) + LiN(CF$_3$SO$_2$)$_2$, LiCF$_3$SO$_3$ or LiClO$_4$ (Li/EO = 1/24). Arrhenius plots for the ionic conductivity of all samples show a convex profile, and accordingly, the temperature dependence of their ionic conductivity is expected to obey the WLF relationship.

WLF plots for the ionic conductivity of the polymer electrolytes are shown in Fig. 4.2.9, and T_g and WLF parameters of the polymer electrolytes are summarized in Table 4.2.4. The temperature dependence of the ionic conductivity of the polymer electrolytes with LiN(CF$_3$SO$_2$)$_2$, LiCF$_3$SO$_3$ or LiClO$_4$, was found to be dominated by that of ionic mobility, which was correlated with the segmental motion of polymer chains. The differences in T_g of the polymer electrolytes with various lithium salts were small. In contrast, large differences in $\sigma(T_g)$ of the samples are found in Table 4.2.4. In this case, the difference in ionic conductivity at each T_g of polymer electrolytes reflects the difference in the degree of lithium salt dissociation. Accordingly, it is expected that the degree of dissociation of LiCF$_3$SO$_3$ or LiClO$_4$ in the polymer electrolytes is higher than that of LiN(CF$_3$SO$_2$)$_2$. It is generally known that the degree of dissociation of LiN(CF$_3$SO$_2$)$_2$ is higher than that of LiCF$_3$SO$_3$ or LiClO$_4$ in conventional matrices, PEO *etc.*, for polymer electrolytes [43–45].

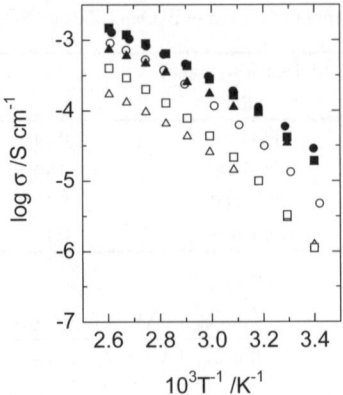

Fig. 4.2.10. Arrhenius plots of ionic conductivity for PEGx − B$_2$O$_3$ + Li-salt (Li/EO = 1/24). ○: PEG200 + LiN(CF$_3$SO$_2$)$_2$, ●: PEG600 + LiN(CF$_3$SO$_2$)$_2$, △: PEG200 + LiCF$_3$SO$_3$, ▲: PEG600 + LiCF$_3$SO$_3$, □: PEG200 + LiClO$_4$, ■: PEG600 + LiClO$_4$.

Table 4.2.4. WLF parameters and ($\sigma(T_g)$) of PEGx − B$_2$O$_3$+Li-salt(Li/EO=1/24).

Molecular weight of PEG	Supporting electrolyte	T_g/°C	C_1 /-	C_2 /K	$\sigma(T_g)$ /S cm^{-1}
PEG200	LiN(CF$_3$SO$_2$)$_2$	39	12.1	37.7	1.87×10^{-13}
	LiCF$_3$SO$_3$	38	9.9	66.0	2.51×10^{-11}
	LiClO$_4$	34	12.1	53.0	1.26×10^{-11}
PEG600	LiN(CF$_3$SO$_2$)$_2$	52	16.9	18.0	7.94×10^{-19}
	LiCF$_3$SO$_3$	48	18.6	13.7	1.29×10^{-16}
	LiClO$_4$	42	12.5	25.6	1.05×10^{-18}

Table 4.2.5. Transport number of PEG600 − B$_2$O$_3$+Li-salt(Li/EO=1/24) at 60°C.

Supporting electrolyte	t_{Li^+}
LiN(CF$_3$SO$_2$)$_2$	0.16
LiCF$_3$SO$_3$	0.40
LiClO$_4$	0.45

However, these results for the present polymer electrolytes indicate that the higher degree of dissociation of LiCF$_3$SO$_3$ or LiClO$_4$ compared to that of LiN(CF$_3$SO$_2$)$_2$ is induced by the borate ester groups.

Transport numbers of lithium ions in the polymer electrolytes, PEGx−B$_2$O$_3$ (x: 200, 600) + LiN(CF$_3$SO$_2$)$_2$, LiCF$_3$SO$_3$ or LiClO$_4$ (Li/EO = 1/24), are summarized in Table 4.2.5. It was found that the transport number of the polymer electrolytes containing LiCF$_3$SO$_3$ or LiClO$_4$ was higher than that of the electrolyte with LiN(CF$_3$SO$_2$)$_2$. These results also indicate that CF$_3$SO$_3^-$ and ClO$_4^-$ are attracted by the borate ester groups more effectively than N(CF$_3$SO$_2$)$_2^-$ is.

In order to clarify the qualitative relationships between the Lewis acidity of the borate ester groups and the Lewis basicity of the anions, *ab initio* calculations were

4.2 Electrode/electrolyte interfaces in all-solid-state lithium ion batteries

carried out. Initially, the highest energy occupied molecular orbital (HOMO) and the lowest energy unoccupied molecular orbital (LUMO) of the backbone polymer in the polymer electrolytes were investigated. To simplify the calculations, a PEG-borate ester whose EO chain length is n = 1 was chosen as a model for the matrix polymer. The most stable geometry of the PEG-borate ester obtained by the calculations for six starting geometries is shown in Fig. 4.2.11. Furthermore, the HOMO and LUMO for the most stable geometry of the PEG-borate ester were mapped, and the resulting molecular orbitals are shown in Figs. 4.2.12 and 4.2.13, respectively. It is found in Fig. 4.2.12 that the HOMO exists around the oxygen atoms, and electron density around the oxygen atoms apart from the boron atoms is higher than that around the oxygen atoms next to the boron atom. Therefore, lithium ions interact more strongly with the oxygen atoms that are apart from the boron atom than to other oxygen atoms. On the other hand, Fig. 4.2.13 shows that the LUMO exists perpendicular to the BO_3 plane, which indicates a location for interaction between the boron atom and a Lewis basic anion.

Furthermore, the interactions between the borate ester and the anions of the lithium salts, $LiN(CF_3SO_2)_2$, $LiCF_3SO_3$ and $LiClO_4$, were investigated. For the investigation, the electronegativity, χ, and the hardness, η, of the anions were estimated by following calculations. The total electron energies of each optimized anion whose charge valence is +1, 0 or −1 were calculated with DFT, and subsequently the ionization potentials, I, and electron affinities, A, were evaluated with the following equations,

$$I = E(X^{+1}) - E(X^0) \tag{11}$$

$$A = E(X^0) - E(X^{-1}) \tag{12}$$

where X denotes the anion molecule and E is the total electron energy of the anions. Then χ and η of anions or molecules are defined as follows [46,47],

$$\chi = \frac{I+A}{2}, \quad \eta = \frac{I-A}{2} \tag{13}$$

The obtained χ and η of the anions are summarized in Table 4.2.6. According to Hard and Soft Acids and Bases (HSAB) theory [48,49], "hard" Lewis acids prefer to interact strongly with "hard" Lewis bases, and on the other hand, "soft" Lewis acids prefer to interact strongly with "soft" Lewis bases. In this case, PEG-borate esters are hard Lewis acids [50]. It is found in Table 4.2.6 that the order of both χ and η of the anions is $CF_3SO_3^- > ClO_4^- > N(CF_3SO_2)_2^-$. Therefore, $CF_3SO_3^-$ or ClO_4^- should interact more strongly with the PEG-borate ester than $N(CF_3SO_2)_2^-$ does. These tendencies are in good agreement with the comparisons of $\sigma(T_g)$ (Table 4.2.4) and transport number of lithium ions (Table 4.2.5). Therefore, it is concluded that the boron atoms of the backbone polymer act as a "hard" Lewis acid center and prefer to interact strongly with "hard" Lewis bases such as $CF_3SO_3^-$ and ClO_4^- in the polymer electrolytes, which enhances the dissociation of lithium salts and the transport number of lithium ions.

118 4 Recent development of electrode materials in lithium ion batteries

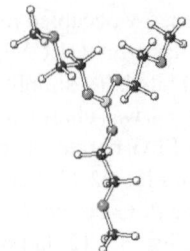

Fig. 4.2.11. Most stable geometry of the PEG-borate ester obtained with calculation (B3LYP/6-311G**//HF/3-21G*).

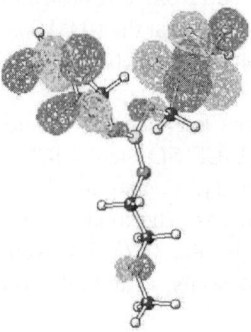

Fig. 4.2.12. Highest energy occupied molecular orbital (HOMO) of the PEG-borate ester.

Fig. 4.2.13. Lowest energy unoccupied molecular orbital (LUMO) of the PEG-borate ester.

In conclusion, we have studied ionic conductivity, thermal properties and transport number of lithium ions in polymer electrolytes containing borate ester groups that are fixed to chains of the matrix polymer. It was found that the ionic conductivity of the polymer electrolytes increased with decreasing concentration of the borate ester groups, which played the role of crosslinking points. The increase

4.2 Electrode/electrolyte interfaces in all-solid-state lithium ion batteries

Table 4.2.6. Ionization potential I, electron affinity A, electron negativity χ and hardness η of anions.

	I /eV	A /eV	χ	η
ClO_4	12.50	5.37	8.94	3.57
CF_3SO_3	13.37	5.24	9.31	4.07
$N(CF_3SO_2)_2$	10.82	5.51	8.17	2.66

in the ionic conductivity is due to the increase in the mobility of carrier ions, which was indicated by DSC studies. On the other hand, the transport number of lithium ions increased with increasing concentration of borate ester groups. Furthermore, the transport number in the polymer electrolyte containing $LiCF_3SO_3$ or $LiClO_4$ was found to be higher than that in the electrolyte with $LiN(CF_3SO_2)_2$. Comparison of $\sigma(T_g)$ of the polymer electrolytes with various lithium salts, $LiCF_3SO_3$, $LiClO_4$ and $LiN(CF_3SO_2)_2$, indicates that the degree of dissociation of $LiCF_3SO_3$ or $LiClO_4$ is higher than that of $LiN(CF_3SO_2)_2$.

In order to clarify the qualitative relationships between the Lewis acidity of the borate ester groups and the Lewis basicity of the anions, interactions between the borate ester group and anions were investigated by *ab initio* calculations, the Hartree-Fock level and the Density functional theory. It was found that the boron atoms of the PEG borate ester acted as a Lewis acid center and preferred to interact more strongly with the "hard" Lewis basic anions, $CF_3SO_3^-$ and ClO_4^-, than with $N(CF_3SO_2)_2^-$ in the polymer electrolytes. These tendencies are in good agreement with the obtained experimental results mentioned above. Therefore, it was confirmed by the *ab initio* calculations that the borate ester groups enhanced the dissociation of lithium salts and the transport number of lithium ions in the polymer electrolytes with "hard" Lewis basic anions.

(iii) Enhancement of the reaction rate by the addition of PEG-borate ester

In this sub-section, we will focus on factors affecting the charge transfer reaction at the electrode/electrolyte interfaces in detail, by examining the enhancement of the reaction rate by the addition of PEG-borate ester as a Lewis acid into the electrolytes. For this investigation, the exchange current densities of the Li^+/Li couple reaction in PEGDME-based electrolytes containing $LiCF_3SO_3$ were evaluated by chronoamperometry with a microelectrode. Electrolyte solutions based on PEGDME (average molecular weight 500) with dissolved $LiCF_3SO_3$ are used as a model for polymer electrolytes, similar to the amorphous conducting phase in high-molecular-weight poly(ethylene oxide) (PEO).

The electrokinetics of lithium deposition and dissolution at the working microelectrode in electrolyte solutions was studied using a double potential step technique, chronoamperometry. Figure 4.2.14 shows variations of the exchange current densities in the electrolyte solutions with various amounts of the PEG-borate ester, obtained using the same procedure as above with Allen-Hickling plots. Drastic increases in the exchange current densities of the electrolytes were found when the

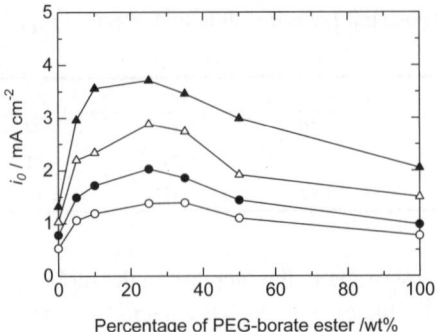

Fig. 4.2.14. i_0 vs. fraction of PEG-borate ester in 0.5M LiCF$_3$SO$_3$/PEGDME500 at various temperatures: ○ at 313 K, ● at 323 K, △ at 333 K, ▲ at 343 K.

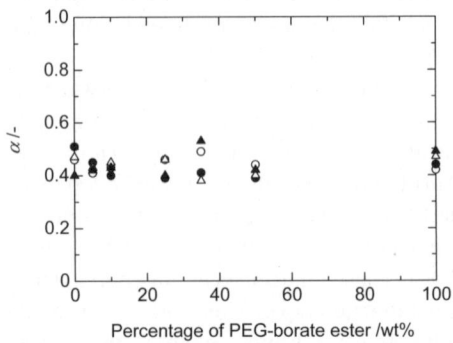

Fig. 4.2.15. α vs. fraction of PEG-borate ester in 0.5M LiCF$_3$SO$_3$/PEGDME500 at various temperatures: ○ at 313 K, ● at 323 K, △ at 333 K, ▲ at 343 K.

PEG-borate ester was added in. Furthermore, the values of the exchange current densities show a maximum at each temperature with the PEG-borate ester at 25 wt% of the standard solvent, PEGDME500, which corresponds to the molar ratio of the PEG-borate ester to the anion being almost 1:1 in the electrolyte. The maximum value at 333 K was 2.88 mA cm^{-2}, which was 2.8 times larger than that without the PEG-borate ester, and also, a maximum value at each temperature was around three times as much as that of the electrolyte without additive. The estimated values of the transfer coefficient α are summarized in Fig. 4.2.15. The transfer coefficient was found to be independent of the amount of PEG-borate ester and of temperature, and all the values are close to 0.5, which implies a symmetric reaction of the Li$^+$/Li couple.

All the Arrhenius plots exhibit a linear relationship between ln(i_0) and 1/T, as shown in Fig. 4.2.16; the values of the Gibbs activation energies were calculated from the slope of the linear relationship plots. The calculated Gibbs activation energies are summarized in Table 4.2.7. The electrokinetics of the Li$^+$/Li couple reaction involves the desolvation/solvation process. Therefore, the Gibbs activation energy

Table 4.2.7. Activation energies of Li$^+$/Li couple reaction in 0.5M LiCF$_3$SO$_3$/PEGDME with various amounts of PEG-borate ester.

Amount of PEG-borate ester /wt %	G^* /kJ mol^{-1}
0 (only PEGDME)	28.9
5	30.7
10	30.3
25	29.7
35	29.4
50	29.2
100	29.9

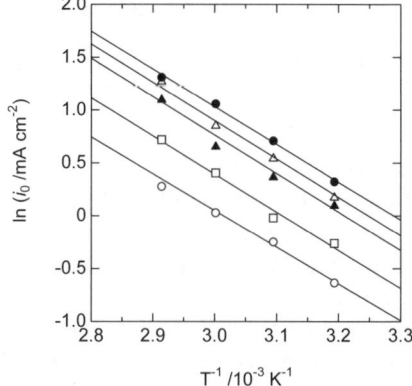

Fig. 4.2.16. Arrhenius plots for the exchange current densities in 0.5M LiCF$_3$SO$_3$/PEGDME500 with various amounts of PEG-borate ester: ○ PEGDME500 100%, △ PEGDME500 90wt% + PEG-borate ester 10wt%, ● PEGDME500 75wt% + PEG-borate ester 25wt%, ▲ PEGDME500 50wt% + PEG-borate ester 50wt%, □ PEG-borate ester 100%.

may depend on the desolvation/solvation energy. In other words, the Gibbs activation energy depends on the solvation state of lithium ions in the solvent. The Gibbs activation energies were found to remain almost constant, around 29 kJ mol^{-1}, even in the electrolyte solution with the added PEG-borate ester. This result indicates that the Gibbs activation energy does not depend on the amount of the PEG-borate ester in the electrolyte solution, *i.e.* the solvation states of lithium ions in the electrolyte solutions are quite similar, even in the electrolyte solution with added PEG-borate ester.

As described before, Raman spectroscopy is very useful for the qualitative investigation of the activity of lithium ions in electrolytes. The Raman spectrum was deconvoluted to three components by fitting three distinct Lorentzian curves, and the ratio of the free ions was estimated. As the results show in Fig. 4.2.17, the ratio of free ions was found to increase when PEG-borate ester was added to the electrolyte solution. Especially, the ratio of the free ions shows the maximum values at PEG-borate ester contents of 10 and 25 wt%, which agrees well with the obtained tendency for the exchange current density as shown in Fig. 4.2.14.

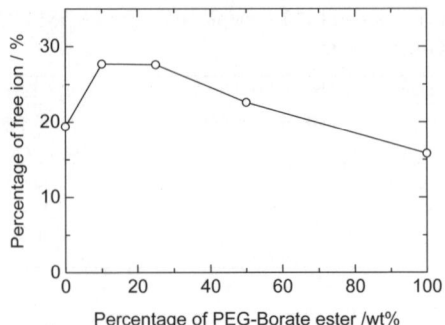

Fig. 4.2.17. Fraction of free ions in 0.5M LiCF$_3$SO$_3$/PEGDME500 with various amounts of PEG-borate ester at 313K.

Therefore, the Raman studies elucidated the increase in the activity of lithium ions in the electrolyte solutions by the addition of the PEG-borate ester, which should be due to the enhancement of dissociation of lithium salt induced by the interaction between the PEG-borate ester acting as a Lewis acid and the anions, leading to the increases in exchange current densities.

However, the reason why the ratio of the free ion, *i.e.* the activity of lithium ions, shows the maximum values, as shown in Fig. 4.2.17, in other words, the ratio decreases at weight ratios of over 50wt% of the PEG-borate ester, cannot be explained solely by the interaction between the Lewis acid and anions. Therefore, *ab initio* calculations were carried out to investigate the solvation state of the lithium ion and CF$_3$SO$_3$ anion in PEGDME and PEG-borate ester in more detail. The obtained stable geometries of a lithium ion with PEGDME and PEG-borate ester are shown in Fig. 4.2.18. As shown in Fig. 4.2.18(a), a lithium ion coordinates with five oxygen atoms in PEGDME. On the other hand, a lithium ion coordinates with four oxygen atoms in PEG-borate ester, as shown in Fig. 4.2.18(b). Additionally, the binding energies between a lithium ion and PEGDME or PEG-borate ester were calculated with the following equation,

$$\Delta E_{(bind)} = E_{(complex)} - \{E_{(solvent)} + E_{(lithium)}\} \quad (14)$$

where $E_{(complex)}$ is the total electron energy of the lithium ion state in PEGDME or PEG-borate ester. $E_{(solvent)}$ and $E_{(lithium)}$ are also the total electron energy of the solvent molecule, PEGDME or PEG-borate ester, and lithium ion, respectively. $\Delta E_{(bind)}$ for lithium ion with PEGDME was found to be higher than that with PEG-borate ester, as summarized in Table 4.2.8. This result suggests that a lithium ion coordinates preferentially with PEGDME rather than with PEG-borate ester. Meanwhile, interactions between CF$_3$SO$_3^-$ and the solvent molecule, PEGDME or PEG-borate ester, were also investigated with *ab initio* calculations. To simplify the calculations, a PEG-borate ester with an EO chain length of 1, B{(OC$_2$H$_4$)OCH$_3$}$_3$, was chosen as the model solvent molecule, and accordingly, the PEGDME with the corresponding EO chain length, CH$_3$O(CH$_2$CH$_2$O)$_5$CH$_3$, was also chosen. The

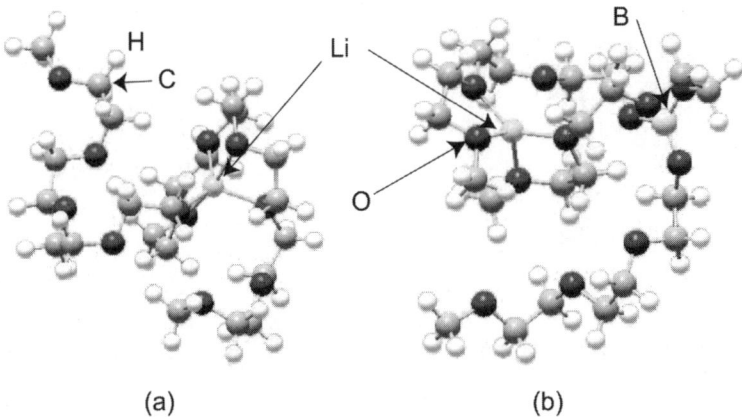

Fig. 4.2.18. Stable geometries of the complexes: (a) a lithium ion with PEGDME500, (b) a lithium ion with PEG-borate ester.

Table 4.2.8. Binding energies between lithium ion or CF_3SO_3 and PEGDME or PEG-borate ester.

Interaction		E /kJ mol^{-1}	Coordination number*
Li$^+$	PEGDME	464.5	5
Li$^+$	PEG-borate ester	415.4	4
CF_3SO_3	PEGDME	37.0	
CF_3SO_3	PEG-borate ester	44.6	

*Coordination number: The number of ether oxygen atoms coordinating a lithium ion

obtained stable geometries of $CF_3SO_3^-$ with PEGDME or PEG-borate ester are shown in Fig. 4.2.19. The binding energies between $CF_3SO_3^-$ and PEGDME or PEG-borate ester are also summarized in Table 4.2.8. This result means that PEG-borate ester prefers to interact with $CF_3SO_3^-$ as a Lewis acid, which leads to enhanced dissociation of the lithium salt. From these calculations for the solvation states of lithium ion and $CF_3SO_3^-$ in PEGDME or PEG-borate ester, lithium ions are preferentially solvated by PEGDME. On the other hand, PEG-borate ester interacts preferentially with $CF_3SO_3^-$. Therefore, we conclude that when PEG-borate ester is added to the electrolyte, the activity of lithium ions is increased by the interaction between the PEG-borate ester and $CF_3SO_3^-$, which is induced by the Lewis acidity of the PEG-borate ester. Furthermore, with excess amounts of PEG-borate ester in the electrolyte, the solvation state of lithium ions is not favored, when compared with solvation in PEGDME, and accordingly, the activity of lithium ions decreases. This consideration coincides with the trend of the activity of lithium ions obtained from Raman spectroscopic studies, as shown in Fig. 4.2.17.

As stated in equation (4), the exchange current density involves the longitudinal relaxation time of the solvent as well as the activity of lithium ions and the Gibbs activation energy. The longitudinal relaxation time of the solvent, τ_L, is related to the

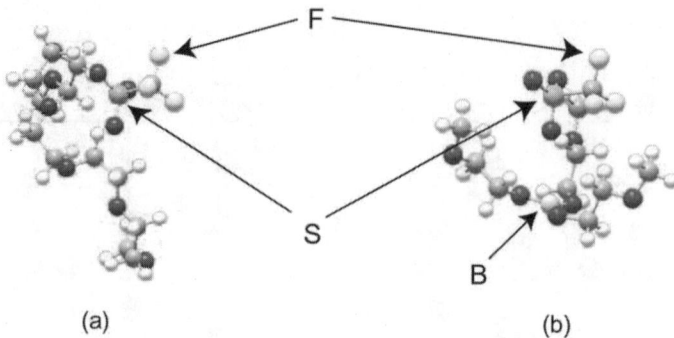

Fig. 4.2.19. Stable geometries of the complexes: (a) CF_3SO_3 with PEGDME, (b) CF_3SO_3 with PEG-borate ester.

Debye relaxation time τ_D through the following equation [29,30],

$$\tau_L = \tau_D \varepsilon_\infty / \varepsilon_s \tag{5}$$

where ε_s is the static dielectric constant and ε_∞ is the infinite frequency dielectric constant. Furthermore, the Debye relaxation time, τ_D, is proportional to the viscosity, η [29–31]. Therefore, the longitudinal relaxation time of the solvent is proportional to the viscosity, η. As a consequence, the exchange current density, i_0, is inversely proportional to the viscosity from the above relationships and equation (4). In the previous section, we showed that the exchange current density was inversely proportional to the viscosity in electrolyte solutions based on PEGDME500 by changing the viscosity with the higher molecular weight PEGDME1000 (see Fig. 4.2.6). More specifically, the activity of lithium ions and the Gibbs activation energy of the charge transfer reaction do not change in the case of polyether based electrolytes with PEGDME500 and PEGDME1000. Only the viscosity is an important factor for the interfacial charge transfer reaction. This phenomenon can be interpreted by the Marcus microscopic theories of charge transfer [17–19]. In the case of addition of the PEG-borate ester to the electrolyte solution based on PEGDME500, although the viscosity of the electrolyte increases slightly with increasing amounts of PEG-borate ester, the inverse relationship between the viscosity and the exchange current densities was not found, as shown in Fig. 4.2.20, which is explainable by the increase in the activity of lithium ions due to the PEG-borate ester acting as a Lewis acid. This result indicates that in spite of the slight increase in viscosity when PEG-borate ester is added to the electrolytes, which is not promising for the exchange current density as indicated by equation (4), the increase in activity of lithium ions induced by the Lewis acidity of the PEG-borate ester is dominant for the exchange current density.

The present studies have shown the usefulness of the PEG-borate ester acting as a Lewis acid on the charge transfer reaction rate at PEGDME-based electrolyte/electrode interfaces. This also means that this approach to increasing the activity of lithium ions is promising for getting a high interfacial reaction rate.

4.2 Electrode/electrolyte interfaces in all-solid-state lithium ion batteries

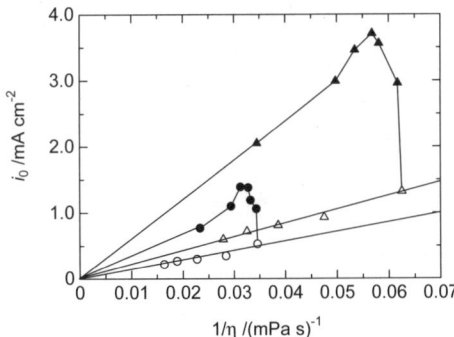

Fig. 4.2.20. Exchange current densities vs. inverse of viscosities of the electrolytic solutions containing 0.5M LiCF$_3$SO$_3$: ●▲ PEGDME500 with various amounts of PEG-borate ester, ○△ PEGDME500 with various amounts of PEGDME1000 (from the previous study), at various temperatures: ○● at 313K, △▲ at 343 K.

With addition of PEG-borate ester to the electrolyte solution, drastic increases in the exchange current density of he Li$^+$/Li couple were found despite a slight increase in the viscosity of the electrolyte, which was not promising for charge transfer according to the Marcus microscopic theories [17–19].

In conclusion, the influence of the PEG-borate ester on the charge transfer reaction rate at the PEGDME500 based electrolyte/electrode interfaces was investigated using a microelectrode technique. Drastic increases in the exchange current density of the Li$^+$/Li couple reaction in the electrolyte were observed upon the addition of PEG-borate ester. When the quantity of PEG-borate ester reached 25 wt% relative to PEGDME500, the values of the exchange current densities showed a maximum at each temperature, which corresponded to a molar ratio of almost 1:1 in for the PEG-borate ester and the anion in the electrolyte. The Gibbs activation energies for the charge transfer reaction were almost constant, even with added PEG-borate ester, indicating that the solvation states of lithium ions were quite similar. Raman spectroscopic studies elucidated the increases in the activity of lithium ions in the electrolyte solutions by the addition of the PEG-borate ester, which were due to the interaction between the PEG-borate ester acting as a Lewis acid and anions. This is likely to be an important factor underlying increases in the exchange current density obtained by the addition of PEG-borate ester to the electrolyte, which was also supported by the non-promising phenomenon for the charge transfer reaction rate that the viscosity of the solution increased with increasing amounts of PEG-borate ester. Therefore, the addition of PEG-borate ester with Lewis acidity is a very useful approach for getting a high charge transfer reaction rate in the polyether based electrolyte, implying the possibility of developing high power density lithium-ion batteries.

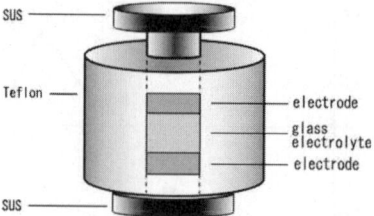

Fig. 4.2.21. All-solid-state cell for electrochemical measurement.

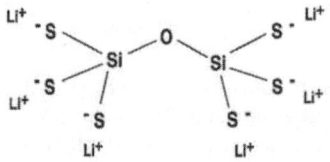

Fig. 4.2.22. Structural unit of glass electrolyte.

4.2.3 Characterization of electrode/solid electrolyte interface in solid-state batteries

The use of solid electrolytes in lithium ion batteries is desirable because of the batteries' excellent properties such as safety, high energy density and so on, compared to batteries that use organic liquid electrolytes. Among solid electrolytes, SiS_2-Li_2S-Li_4SiO_4 glassy electrolytes are known to exhibit high lithium ion conductivity ($\sim 10^{-3}$ S cm^{-1}) [3,4]. In ordinary batteries that use liquid electrolyte, an SEI (solid electrolyte interface) layer is formed at the interface between electrode and electrolyte and plays an important role in reversible cycles. However, the interfacial reaction of all-solid-state lithium ion batteries has not yet been elucidated. In this chapter, we will present our recent results concerning the interfacial reaction, with emphasis on the interfacial reaction products. The products at the interface were determined by using X-ray absorption spectroscopy (XAS).

Charge-discharge measurements were carried out for both all-solid-state cells and conventional coin-typed cells. An all-solid-state cell used in this study is shown in Fig. 4.2.21. $LiCoO_2$ and $LiMn_2O_4$ were used as cathode materials, with Li and In foil used as anodes. Cathode active materials were synthesized with a solid-state reaction method. Stoichiometric quantities of Li_2CO_3 and $CoC_2O_4 \cdot 2H_2O$ were used as starting materials for $LiCoO_2$, and Li_2CO_3 and Mn_2O_3 were used as those for $LiMn_2O_4$. The mixtures were heated at 600 °C for 12 hr followed by annealing at a cooling rate of 2 °C min^{-1}. The glass electrolyte, $95(0.6Li_2S \cdot 0.4SiS_2) \cdot 5Li_4SiO_4$, used in this study was synthesized by means of a twin-roller rapid quenching and mechanical milling device [2–4] which was supplied by the Minami-Tatsumisago group at Osaka Prefecture University. The structural unit of the glassy electrolytes is presented in Fig. 4.2.22 [51].

4.2 Electrode/electrolyte interfaces in all-solid-state lithium ion batteries 127

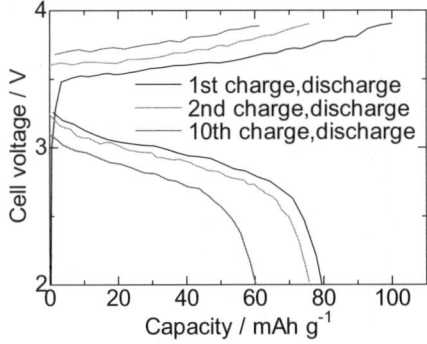

Fig. 4.2.23. Charge and discharge profiles of In /glass electrolyte/LiCoO$_2$ cell.

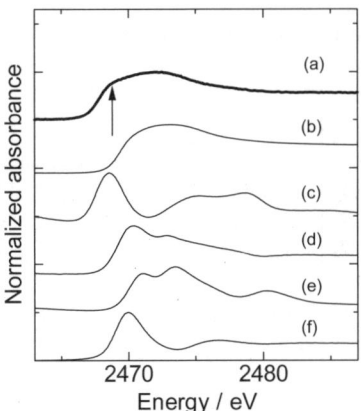

Fig. 4.2.24. S K-edge XAS spectra and (a):interface between LiCoO$_2$ and glass electrolyte(kept charged after 10th charge-discharge cycle), (b):glass electrolyte, (c):CoS, (d):SiS$_2$, (e):Li$_2$S, (f):S.

From charge discharge measurements, it was found that the cells in the solution system showed reversible charge discharge capacity when both LiCoO$_2$ and LiMn$_2$O$_4$ were used as cathode materials. On the other hand, the In/glass electrolyte/LiCoO$_2$ cell showed only the 1st charge capacity. The cell using LiCoO$_2$ showed reversibility. Figure 4.2.23 shows voltage profiles of the charge/discharge reaction in the In/glass electrolyte/LiCoO$_2$ cell. The cell with a Li anode did not show any capacity at all. In order to clarify the causes underlying these results, we examined the interface between LiCoO$_2$ or LiMn$_2$O$_4$ and the glassy electrolytes of the all solid-state cells.

The interfacial products between LiCoO$_2$ and glassy electrolytes were examined. The cell consisting of LiCoO$_2$ and glassy electrolyte was cycled 10 times before XAS measurement. Figure 4.2.24 shows S K-edge XAS spectra. The glassy electrolyte shows a broadened peak compared to its components, SiS$_2$ and Li$_2$S, which

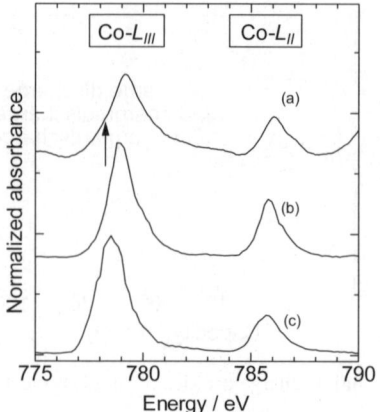

Fig. 4.2.25. Co L_{II} and Co L_{III} -edge XANES spectra and (a): interface of $LiCoO_2$ glass electrolyte(kept charged after 10th charge-discharge cycle), (b): $LiCoO_2$, (c): CoS.

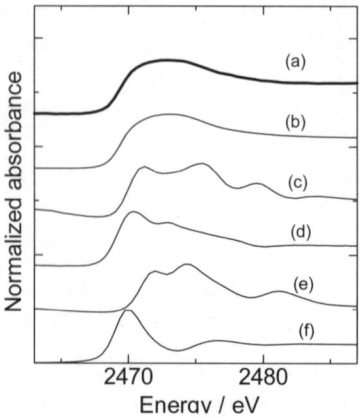

Fig. 4.2.26. S K-edge XAS spectra and (a):interface between $LiMn_2O_4$ and glass electrolyte(kept charged), (b):glass electrolyte, (c):MnS, (d):SiS_2, (e):Li_2S, (f):S.

indicates that the atoms surrounding sulfur have various electron states in the glassy electrolyte. Also, the peak position is similar to that of Li_2S. This means that the electron states of sulfur in the glassy electrolyte and Li_2S are similar to each other. Fig. 4.2.24 shows that the spectrum of the interface has a shoulder at the same position as CoS (indicated by an arrow in the figure), though the configuration and peak position are the same as for the glassy electrolyte. This indicates that CoS or some compound whose electron state is similar to that of CoS was produced at the interface.

Then we measured Co L-edge XAS spectra with the same cell (Fig. 4.2.25). The peak positions of three samples shift toward the positive energy side in the order: CoS, $LiCoO_2$ and the interfacial product. Moreover, we calculated the ratios of peak

4.2 Electrode/electrolyte interfaces in all-solid-state lithium ion batteries

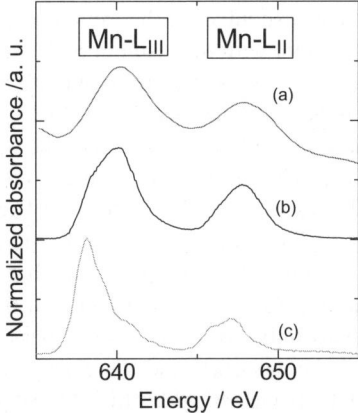

Fig. 4.2.27. Mn L_{II} and Mn L_{III}-edge XAS spectra and (a): interface between Mn L_{II} and Mn L_{III}-edge XAS spectra and (a): interface between $LiMn_2O_4$ and glass electrolyte (kept charged), (b): $LiMn_2O_4$, (c): MnS.

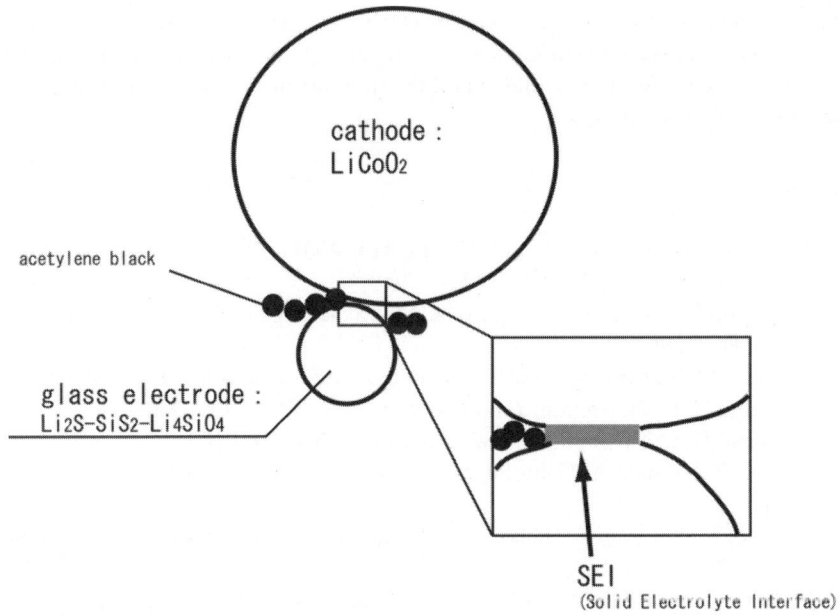

Fig. 4.2.28. Schematic diagram at $LiCoO_2$ cathode and the glass electrolyte.

intensity of L_{III} to that of L_{II}. The ratio of $LiCoO_2$ was 2.04 while that of interfacial product was 2.17, meaningfully larger than $LiCoO_2$. These phenomena indicate that the Co ion in $LiCoO_2$ in the interfacial region was oxidized through charging. Thus we can confirm that lithium ion transferred from $LiCoO_2$. In addition, the spectrum of the interfacial product has a shoulder at the same position as CoS (arrow in Fig. 4.2.25). This behavior consistent with the result of S K-edge XAS spectra.

S K-edge and Mn L-edge XAS spectra are shown in Fig. 4.2.26 and Fig. 4.2.27, respectively. We examined the interface between $LiMn_2O_4$ and the glassy electrolyte of the cell, which was charged at 4.0 V. The ratios of the peak intensity of $LiMn_2O_4$ and that of the interfacial product evaluated from the results of Mn L-edge XAS spectra were 1.77 and 1.75, respectively. From this result and Fig. 4.2.26 and 4.2.27, the presence of product at the interface was not established when a $LiMn_2O_4$ cathode and In anode were used. As a result, we suggest that what makes it possible for all-solid-state lithium ion batteries to conduct lithium ion at the interface is SEI, which stabilizes the interface between the electrode and the glassy electrolyte.

The interface between the electrode and the glassy electrolyte of an all-solid-state cell was investigated. Only the cell with a $LiCoO_2$ cathode and In anode showed good cycling ability, and if a $LiMn_2O_4$ cathode was used, reversible charge-discharge capacity was not obtained. This phenomenon was verified by XAS measurements. A schematic diagram of the $LiCoO_2$ cathode and the glassy electrolyte is shown in Fig. 4.2.28. It was revealed that some compound such as CoS was produced at the interface between $LiCoO_2$ and glassy electrolyte, while no compound was found at the interface between $LiMn_2O_4$ and glassy electrolyte. Thus, we found that what makes it possible for an all-solid-state lithium ion battery to conduct lithium ion is the interfacial product, SEI.

References

1. J. M. Tarascon and M. Armand, *Nature*, **414** (2001) 359.
2. B. Scrosati, *Nature*, **373** (1995) 557.
3. M. Tatsumisago, K. Hirai, T. Hirata, M. Takahashi and T. Minami, *Solid State Ionics*, **86** (1996) 487.
4. K. Hirai, M. Tatsumisago and T. Minami, *Solid State Ionics*, **78** (1995) 259.
5. J. Xu and G. C. Farirngton, *J. Electrochem. Soc.*, **142** (1995) 3303.
6. J. Xu and G. C. Farrington, *Solid State Ionics*, **74** (1994) 125.
7. Y. Kato, T. Ishihara, Y. Uchimoto and M. Wakihara, *J. Phys. Chem. B*, **108** (2004) 4794.
8. Y. Kato, S. Yokoyama, H. Ikuta, Y. Uchimoto and M. Wakihara, *Electrochem. Comm.*, **3** (2001) 128.
9. Y. Kato, K. Hasumi, S. Yokoyama, T. Yabe, H. Ikuta, Y. Uchimoto and M. Wakihara, *Solid State Ionics*, **150** (2002) 355.
10. Y. Kato, K. Hasumi, S. Yokoyama, T. Yabe, H. Ikuta, Y. Uchimoto and M. Wakihara, *Journal of Thermal Analysis and Calorimetry*, **69** (2002) 889.
11. Y. Kato, K. Suwa, S. Yokoyama, T. Yabe, H. Ikuta, Y. Uchimoto and M. Wakihara, *Solid State Ionics*, **152–153** (2002) 155.
12. Y. Kato, K. Suwa, H. Ikuta, Y. Uchimoto, M. Wakihara, S. Yokoyama, T. Yabe and M. Yamamoto, *J. Mater. Chem.*, **13** (2003) 280.
13. M. Saito, H. Ikuta, Y. Uchimoto, M. Wakihara, S. Yokoyama, T. Yabe and M. Yamamoto, *J. Electrochem. Soc.*, **150** (2003) A477.

14. M. Saito, H. Ikuta, Y. Uchimoto, M. Wakihara, S. Yokoyama, T. Yabe and M. Yamamoto, *J. Electrochem. Soc.*, **150** (2003) A726.
15. M. Saito, H. Ikuta, Y. Uchimoto, M. Wakihara, S. Yokoyama, T. Yabe and M. Yamamoto, *J. Phys. Chem. B*, **107** (2003) 11608.
16. Y. Kato, T. Ishihara, Y. Uchimoto, H. Ikuta and M. Wakihara, *Angew. Chem. Int. Ed.*, **43** (2004) 1966.
17. R. A. Marcus and H. Sumi, *J. Electronanal. Chem.*, **204** (1986) 59.
18. H. Sumi and R. A. Marcus, *J. Chem. Phys.*, **84** (1986) 4272.
19. H. Sumi and R. A. Marcus, *J. Chem. Phys.*, **84** (1986) 4894.
20. A. J. Bard and L. R. Faulkner, *Electrochemical Methods*, Wiley, New York, (2001).
21. W. R. Fawcett and C. A. Foss Jr., *J. Electronanal. Chem.*, **270** (1989) 103.
22. M. Maroncelli, J. MacInnis and C. R. Fleming, *Science*, **243** (1989) 1674.
23. M. J. Weaver, *Chem. Rev.*, **92** (1992) 463.
24. S. Schantz, J. Sandahl, L. Borjessn, L. M. Torell and J. R. Stevens, *Solid State Ionics*, **28–30** (1988) 1047.
25. M. Kakihana, S. Schantz and L. M. Toell, *J. Chem. Phys.*, **92** (1990) 6271.
26. W. Huang, R. Frech and R. A. Wheeler, *J. Phys. Chem.*, **98** (1994) 100.
27. F. M. Gray, *Polymer Electrolytes, RSC Monographs*, The Royal Society of Chemistry, London, (1997).
28. J. R. MacCallum and C. A. Vincent, *Polymer Electrolyte Reviews 1 and 2*, Elesevier, London, (1987–1989).
29. H. Fröhlich, *Theory of Dielectrics*, Oxford University Press, New York, (1949).
30. C. P. Smyth, *Dielectric Behavior and Structure*, McGraw-Hill, New York, (1955).
31. W. Harrer, G. Grampp and W. Jenicke, *J. Electronanal. Chem.*, **209** (1986) 223.
32. M. A. Mehta and T. Fujinami, *Chem. Lett.*, (1997) 915.
33. M. A. Mehta, T. Fujinami and T. Inoue, *J. Power Sources*, **81–82** (1999) 724.
34. M. J. Frisch, G. W. Trucks, H. B. Schlegel, G. E. Scuseria, M. A. Robb, J. R. Cheeseman, V. G. Zakrzewski, J. A. Montgomery Jr., R. E. Stratmann, J. C. Burant, S. Dapprich, J. M. Millam, A. D. Daniels, K. N. Kudin, M. C. Strain, O. Farkas, J. Tomasi, V. Barone, M. Cossi, R. Cammi, B. Mennucci, C. Pomelli, C. Adamo, S. Clifford, J. Ochterski, G. A. Petersson, P. Y. Ayala, Q. Cui, K. Morokuma, P. Salvador, J. J. Danneberg, D. K. Malick, A. D. Rabuck, K. Raghavachari, J. B. Foresman, J. Cioslowski, J. V. Ortiz, A. G. Baboul, B. B. Stefanov, G.. Liu, A. Liashenko, P. Piskorz, L. Komaromi, R. Gomperts, R. L. Martin, D. J. Fox, T. Keith, M. A. Al-Laham, C. Y. Peng, A. Nanayakkara, M. Challacombe, P. M. W. Gill, B. Johnson, W. Chen, M. W. Wong, J. L. Andres, C. Gonzalez, M. Head-Gordon, E. S. Replogle and J. A. Pople, Gaussian Inc., Pittsburgh PA, (1998).
35. A. D. Becke, *J. Chem. Phys.*, **98** (1993) 5648.
36. C. Lee, W. Yang and R. G. Parr, *Phys. Rev. B*, **37** (1988) 785.
37. A. Killis, J. F. LeNest, H. Cheradame and A. Gandini, *Makromol. Chem.*, **183** (1982) 2835.
38. M. Watanabe, M. Itoh, K. Sanui and N. Ogata, *Macromolecules*, **20** (1987) 569.
39. D. Baril, C. MIchot and M. Armand, *Solid State Ionics*, **94** (1997) 35.

40. L. M. Carvalho, P. Gu?gan, H. Cheradame and A. S. Gomes, *European Polymer J.*, **36** (2000) 401.
41. M. L. Williams, R. F. Landel and J. D. Ferry, *J. Am. Chem. Soc.*, **77** (1955) 3701.
42. K. M. Abraham, Z. Jiang and B. Carroll, *Chem. Mater.*, **9** (1997) 1978.
43. M. Armand, W. Gorecki and R. Andreani, *Proceedings of 2nd Intern. Symp. Polymer Electrolytes*, Ed. B. Scrosati, Elsevier Applied Science, London, (1989).
44. A. Valle, S. Besner and J. Prud'homme, *Electrochim. Acta*, **37** (1992) 1579.
45. M. Watanabe and A. Nishimoto, *Solid State Ionics*, **79** (1995) 306.
46. R. G. Parr, R. A. Donnelly, M. Levy and W. E. Palke, *J. Chem. Phys.*, **68** (1978) 3801.
47. R. G. Parr and R. G. Pearson, *J. Am. Chem. Soc.*, **105** (1983) 7512.
48. R. G. Pearson,. *J. Chem. Educ.*, **45** (1968) 581and 643.
49. R. G. Parr and W. Yang, *Density-Functional Theory of Atoms and Molecules*, Oxford University Press, (1989).
50. R. G. Pearson, *J. Am. Chem. Soc.*, **85** (1963) 3533.
51. R. Komiya, A. Hayashi, H. Morimoto, M. Tatsumisago and T. Minami, *Solid State Ionics*, **140** (2001) 83.

5

Construction of solid/solid interface between hydrogen storage alloy electrode and solid electrolyte for battery application

5.1 Introduction

This chapter describes the preparation, characterization and application of polymer gel electrolytes and inorganic solid electrolytes, mainly focusing on the construction of hydrogen storage alloy / solid electrolyte interfaces for battery applications.

The application of a solid or gel electrolyte, instead of a liquid electrolyte, to electrochemical devices such as batteries, fuel cells and capacitors has attracted the increasing attention especially in terms of their reliability, safety, flexibility and processibility [1–4]. For realizing such all-solid-state electrochemical devices, the electrolyte would be required to have a good electrode/electrolyte interface (compatibility with the positive and negative electrodes), high ionic conductivity, electrochemical stability, thermal stability and mechanical strength. Up to now, several attempts have been made for nickel-metal hydride (Ni/MH) and related batteries [5–14]; however, there have been only a few reports on a solid or gel electrolyte with high ionic conductivity for use in such batteries.

Mohri *et al.* [5, 6] have first reported Ni/MH-related batteries using $Sb_2O_5 \cdot xH_2O$ as a solid electrolyte, *e.g.*, $TiNi:MmH_x(Mm=1\ wt\%)/Sb_2O_5 \cdot xH_2O/MnO_2$. Kuriyama *et al.* [7–9] have investigated Ni/MH and related batteries using tetramethylammonium hydroxide pentahydrate, $(CH_3)_4NOH \cdot 5H_2O$ as a solid electrolyte with proton conductivity, *e.g.*, $LaNi_{2.5}Co_{2.4}Al_{0.1}H_x/(CH_3)_4NOH \cdot 5H_2O/NiOOH$ or MnO_2. Recently Ni/MH batteries with an alkaline solid polymer electrolyte based on poly(ethylene oxide) (PEO), KOH and water have been reported by Vassal *et al.* [10, 11], *e.g.*, $LaMmNi_{3.55}Al_{0.3}Mn_{0.4}Co_{0.75}H_x/PEO-KOH-H_2O/NiOOH$. These batteries showed a long charge-discharge cycle life but only fairly small current density could be drawn because of the low ionic conductivity of the electrolytes (*ca.* 10^{-3} S cm^{-1} at room temperature), compared with aqueous electrolyte-based Ni/MH batteries. Very recently, alkaline polymer electrolytes using Poly(vinyl alcohol) (PVA) were reported. [12–14]. PVA-KOH-H$_2$O polymer electrolyte show an ionic conductivity of *ca.* 5×10^{-2} S cm^{-1} at room temperature, and a Ni/MH cell with the polymer electrolyte could be reversibly charged and discharged at a C/5 rate [13].

Crosslinked poly(acrylate) is a well-known polymer that has a high water-absorbing capacity, a high water-holding capacity, a high gel strength and a relatively low cost. So, in our first study, an alkaline polymer gel electrolyte, called a "polymer hydrogel electrolyte" by us, was prepared from the potassium salt of crosslinked poly(acrylic acid) (PAAK) and KOH aqueous solution in order to investigate the applicability of the polymer hydrogel electrolyte to alkaline secondary batteries such as the Ni/MH battery [15–19]. An experimental Ni/MH cell for a preliminary investigation was assembled using the polymer hydrogel electrolyte, and its charge-discharge characteristics and charge retention were tested under several conditions. The main purpose of this study is to compare roughly the electrochemical characteristics between a polymer hydrogel electrolyte-based cell and a KOH aqueous solution-based cell under the same conditions. Moreover, as one of the attempts to apply the polymer hydrogel electrolyte to other electrical devices, electric double layer capacitors (EDLCs) were also assembled using the polymer hydrogel electrolyte, and their electrochemical characteristics were investigated in comparison with those of an EDLC using a KOH aqueous solution [20–22].

Our second study concerns the preparation of a proton-conducting electrolyte and its application to EDLCs. A variety of polymeric complexes swollen with organic electrolyte solutions have been proposed as solid gel electrolytes of electrochemical devices such as batteries and fuel cells. It has so far been demonstrated that polymeric gels containing alkylammonium salts are applicable to EDLCs [23, 24]. Proton-conducting systems will also function as efficient electrolytes of so-called hybrid electrochemical capacitors (HECCs) [25]. The HECC system involves Faradaic and a non-Faradaic reactions, both of which need proper proton-transport steps to compensate the charge stored in the electrodes.

Regarding the polymeric proton conductor, the first example was reported for a complex of PVA and H_3PO_4 [26], broadening the scope of the basic and applied research on polymer ionic conductors. In the solid cation conducting polymer electrolyte, e.g., PEO complexed with $LiCF_3SO_3$, the salt is dissolved in the polymer and the ion transport takes place basically in the amorphous regions of the complex. Thus, the cation transport is directly linked to the flexibility of the polymer chains: no conductivity is observed below the glass transition temperature (T_g). For proton conducting polymers, on the other hand, the membranes consist of a polymer matrix swollen with water and/or electrolyte solutions, and the ion transport takes place primarily in the electrolyte that is entrapped in the polymer matrix. The conductivity is thus not directly linked to the polymer dynamics and, as a consequence, high conductivity is possible below the T_g of the polymer [27, 28].

Our third study concerns the preparation of proton-conducting inorganic solid electrolytes and their application to all-solid-state nickel-metal hydride batteries. Quite a few inorganic solid electrolytes with conductivities over 10^{-3} S cm^{-1} at room temperature have been found so far. This is the most serious problem for realizing the all-solid-state batteries. Phosphoric acid-doped silica gels [29–33] and heteropolyacid hydrates such as $H_3PMo_{12}O_{40} \cdot 20H_2O$ and $H_3PW_{12}O_{40} \cdot 15H_2O$ [34,35] are fascinating candidates as a solid electrolyte for use in all-solid-state batteries because they show the proton conductivities of $10^{-1} - 10^{-2}$ S cm^{-1} even at room temperature. For these electrolytes, the proton conductivity depends on the water content. The water content influences the acidity, which is responsible for the corrosion of electrode materials. The control of the water content and the use of corrosion-resistant electrode materials will become important for good battery performance. Therefore, all-solid-state batteries with the phosphoric acid-doped silica gels and heteropolyacid hydrates as an electrolyte were fabricated and their battery performance was investigated [36–38].

References

1. J.-F. Fauvarque, S. Guinot, N. Bouzir, E. Salmon and I.-F. Penneau, *Electrochim. Acta*, **40** (1995) 2449.
2. S. Guinot, E. Salmon, J. F. Penneau and J.-F. Fauvarque, *Electrochim. Acta*, **43** (1998) 1163.
3. A. M. Grillone, S. Panero, B. A. Retamal and B. Scrosati, *J. Electrochem. Soc.*, **146** (1999) 27.
4. M. Kono, E. Hayashi and M. Watanabe, *J. Electrochem. Soc.*, **146** (1999) 1626.
5. M. Mohri, Y. Tajima, H. Tanaka, T. Yoneda and M. Kasahara, *Sharp Tech. J. (Tokyo, Jpn.)*, **34** (1986) 97.
6. T. Yoneda, S. Satoh, M. Mohri,T. Yoneda, S. Satoh and M. Mohri, *Sharp Tech. J. (Tokyo, Jpn.)*, **38** (1987) 55.
7. N. Kuriyama, T. Sakai, H. Miyamura, A. Kato and H. Ishikawa, *Denki Kagaku* (presently *Electrochemistry, Tokyo, Jpn.*), **58** (1990) 89.
8. N. Kuriyama, T. Sakai, H. Miyamura, A. Kato and H. Ishikawa, *J. Electrochem. Soc.*, **137** (1990) 355.
9. N. Kuriyama, T. Sakai, H. Miyamura, A. Kato and H. Ishikawa, *Solid State Ionics*, **40/41** (1990) 906.
10. N. Vassal, E. Salmon and J.-F. Fauvarque, *J. Electrochem. Soc.*, **146** (1999) 20.
11. N. Vassal, E. Salmon and J.-F. Fauvarque, *Electrochim. Acta*, **45** (2000) 1527.
12. C.-C. Yang, *J. Power Sources*, **109** (2002) 22.
13. C.-C. Yang, S.-J. Lin and S.-T. Hsu, *J. Appl. Electrochem.*, **33** (2003) 777.
14. A.A. Mohamad, N.S. Mohamed, Y. Alias and A.K. Arof, *J. Alloys Comp.*, **337** (2002) 208.
15. C. Iwakura, N. Furukawa, T. Onishi, K. Sakamoto, S. Nohara and H. Inoue, *Electrochemistry (Tokyo, Jpn.)*, **69** (2001) 659.
16. C. Iwakura, S. Nohara, N. Furukawa and H. Inoue, *Solid State Ionics*, **148** (2002) 487.
17. C. Iwakura, K. Ikoma, S. Nohara, N. Furukawa and H. Inoue, *J. Electrochem. Soc.*, **150** (2003) A1623.
18. C. Iwakura, K. Ikoma, S. Nohara, N. Furukawa and H. Inoue, *Electrochem. Solid-State Lett.*, in press.
19. H. Wada, M. Horiuchi, S. Nohara, N. Furukawa, H. Inoue and C. Iwakura, *ITE Lett.*, **5**, (2004) 348.
20. C. Iwakura, H. Wada, S. Nohara, N. Furukawa, H. Inoue and M. Morita, *Electrochem. Solid-State Lett.*, **6** (2003) A37.

21. S. Nohara, H. Wada, N. Furukawa, H. Inoue, M. Morita and C. Iwakura, *Electrochim. Acta*, **48** (2003) 749.
22. H. Wada, S. Nohara, N. Furukawa, H. Inoue, N. Sugoh, H. Iwasaki, M. Morita and C. Iwakura, *Electrochim. Acta*, **49** (2004) 4871.
23. M. Ishikawa, M. Ihara, M. Morita, Y. Matsuda, *Electrochim. Acta*, **40** (1995) 2217.
24. M. Ishikawa, L. Yamamoto, M. Morita, Y. Ando, *Electrochemistry*, **69** (2001) 437.
25. J. H. Park, O. O. Park, K. H. Shin, C. S. Jin, J. H. Kim, *Electrochem. Solid-State Lett.*, **5** (2002) H7.
26. K.-C. Gong, H. Shou-Cai, *Mater. Res. Soc. Symp.Proc.*, **135** (1985) (Solid State Ionics), 377.
27. A. J. Polak, S. Petty-Weeks, A. J. Beuhler, *Sens. Actuators*, **9** (1986) 1.
28. S. Petty-Weeks, A. J. Polak, *Sens. Actuators*, **11** (1987) 377.
29. A. Matsuda, T. Kanzaki, M. Kotani, M. Tatsumisago and T. Minami, *Solid State Ionics*, **139** (2001) 113.
30. A. Matsuda, T. Kanzaki, K. Tadanaga, M. Tatsumisago and T. Minami, *Electrochim. Acta*, **46** (2001) 939.
31. A. Matsuda, H. Honjo, M. Tatsumisago and T. Minami, *Chem. Lett.*, **1989**, 153.
32. K. Hirata, A. Matsuda, T. Hirata, M. Tatsumisago and T. Minami, *J. Sol.-Gel. Technol.*, **17** (2000) 61.
33. A. Matsuda, T. Kanzaki, K. Tadanaga, M. Tatsumisago and T. Minami, *Solid State Ionics*, **154-155** (2002) 687.
34. O. Nakamura, T. Kodama, I. Ogino, Y. Miyake, *Chem. Lett.*, **1979**, 17.
35. O.Nakamura *et al.*, *Report of the Government Industrial Research Institute, Osaka*, No.360 (1982).
36. C. Iwakura, K. Kumagae, K. Yoshiki, S. Nohara, N. Furukawa, H. Inoue, T. Minami, M. Tatsumisago and A. Matsuda, *Electrochim. Acta*, **48** (2003) 1499.
37. K. Hatakeyama, H. Sakaguchi, K. Ogawa, H. Inoue, C. Iwakura and T. Esaka, *J. Power Sources*, **124** (2003) 559.
38. K. Hatakeyama, H. Sakaguchi, T. Yamaguchi, H. Inoue, C. Iwakura and T. Esaka, *Electrochemistry*, **72** (2004) 697.

5.2 Hydrogen storage alloy electrode/polymer hydrogel electrolyte interface

5.2.1 Preparation and characterization of polymer hydrogel electrolyte

The polymer hydrogel electrolyte was prepared in the following manner [1, 2]. One gram of the potassium salt of poly(acrylic acid) having a lightly crosslinked network (Aldrich, #43532-5) was added to 0.01 dm^3 of KOH aqueous solution at a concentration of 2 ~ 20 M (M = mol dm^{-3}) with stirring in a beaker at room temperature. In a typical case using a 7.3 M KOH aqueous solution, the polymer hydrogel electrolyte was composed of 7 wt.% PAAK, 27 wt.% KOH and 66 wt.%

5.2 Hydrogen storage alloy electrode/polymer hydrogel electrolyte interface

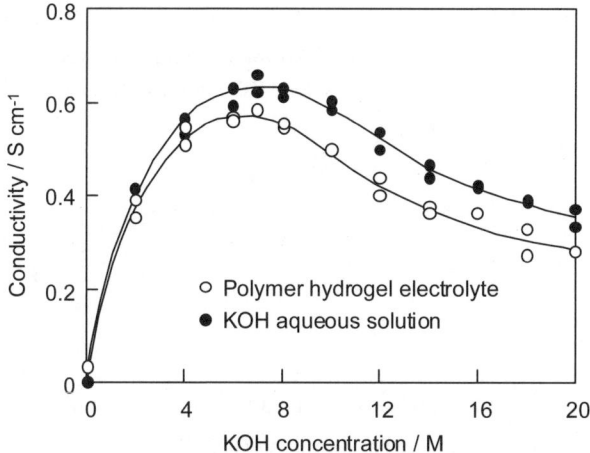

Fig. 5.2.1. Effects of the KOH concentration on the conductivity of the polymer hydrogel electrolytes and KOH aqueous solutions.

H_2O. The resulting gel was allowed to stand for over 70 h at 25°C in an airtight glass vessel to avoid water evaporation and absorption of carbon dioxide. After the gelation was completed, air bubbles in the gel were removed under vacuum.

Figure 5.2.1 shows the relationship between the conductivity and KOH concentration of polymer hydrogel electrolytes and KOH aqueous solutions at 25°C. As can be seen from this figure, the conductivity of the polymer hydrogel electrolytes is very high, compared with the PEO-KOH-H_2O (typically 60-30-10 wt.%) electrolyte reported by Vassal *et al.* [3, 4] This is because the content of the KOH aqueous solution in the polymer hydrogel electrolytes is high due to the high water-absorbing capacity of PAAK. It is thought that the conducting behavior of the polymer hydrogel electrolytes quite resembles that of KOH aqueous solutions. Maximal conductivity can be seen at intermediate KOH concentration in both electrolytes. This phenomenon is qualitatively interpreted as follows. The conductivity decreases due to the low number of charge carriers at low KOH concentrations, whereas it decreases due to restriction of the ionic mobility at high KOH concentrations. The conductivity of the polymer hydrogel electrolyte prepared using a 7.3 M KOH aqueous solution was *ca.* 0.6 S cm^{-1} at 25°C, nearly comparable to that of a 7.3 M KOH aqueous solution.

Arrhenius plots of the conductivity of the polymer hydrogel electrolyte and a 6 M KOH aqueous solution are shown in Fig. 5.2.2. An approximately linear dependency was observed over the temperature range of −40 ~ 80°C for each electrolyte. Activation energies for ionic conduction in the polymer hydrogel electrolyte and a 6 M KOH aqueous solution were calculated to be 13.7 and 13.4 kJ mol^{-1}, respectively. These data indicate that the ion-conducting mechanism of both electrolytes is quite similar over the range of KOH concentration and temperature measured in this

138 5 Construction of solid / solid interface for battery application

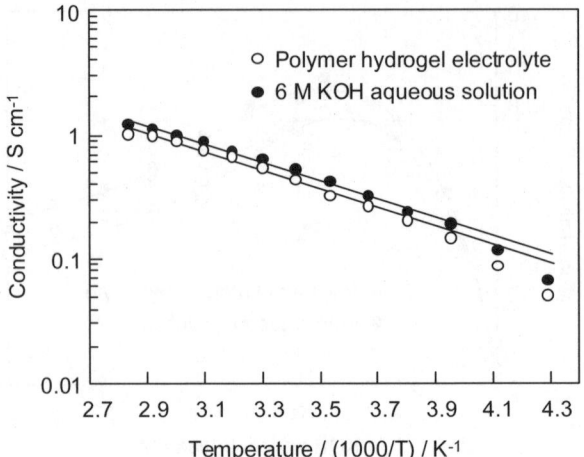

Fig. 5.2.2. Arrhenius plots of the conductivity of the polymer hydrogel electrolyte and a 6 M KOH aqueous solution.

study. No deterioration or decomposition of the polymer hydrogel was found even in concentrated alkaline media.

The ion transport number of the electrolyte is one of the most important factors to characterize the electrolyte, especially in case of all-solid-state devices. It is ideally preferable that the transport number of the OH$^-$ ion is close to 1 because the degradation of positive and negative electrodes occurs due to the appearance of the concentration gradient of OH$^-$ ions when the transport number is lower. The transport number of the OH$^-$ ion in the polymer hydrogel electrolyte was measured by the Hittorf's method [5, 6]. Electrolysis was carried out at a constant current of 20 mA. After the electrolysis, amounts of the OH$^-$ ion in positive and negative electrode compartments were measured by titration using a 0.1 M HCl aqueous solution. The transport number of the OH$^-$ ion in the polymer hydrogel electrolyte was evaluated using the following equation;

$$t_+ = Z_+ \Delta n_a / (Q/F) \qquad (5.2.1)$$

$$t_- = 1 - t_+ \qquad (5.2.2)$$

where t_+ is the cation transport number, t_- the anion transport number, Z_+ the charge number of the cation, Δn_a the change in the quantity of the KOH, Q the quantity of electricity and F the Faraday constant.

Figure 5.2.3 shows changes in the quantity of KOH in the positive and negative electrode compartments by the electrolysis. In both compartments, the changes in the quantity of KOH are almost proportional to the quantity of electricity. In addition, the absolute value of the decrease in the quantity of KOH in the positive electrode compartment is almost equal to that of the increase in the quantity of KOH in the negative electrode compartment.

5.2 Hydrogen storage alloy electrode/polymer hydrogel electrolyte interface

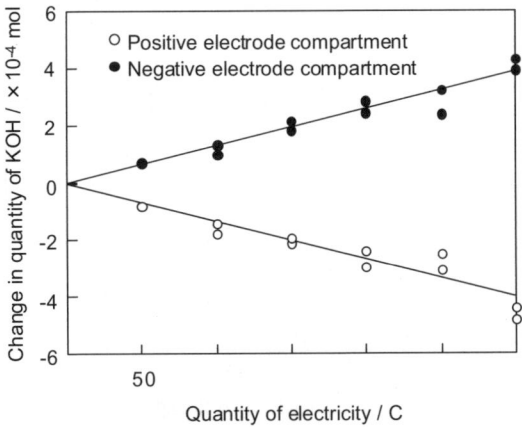

Fig. 5.2.3. Relationship between quantity of electricity and change in quantity of KOH in the positive and negative electrode compartments.

Figure 5.2.4 shows transport numbers of the OH^- ion in the polymer hydrogel electrolyte calculated using Eqs. 5.2.1 and 5.2.2. From this figure, the transport number of OH^- ion can be evaluated to be approximately *ca.* 0.85. This indicates that the charge is mainly carried by the OH^- ions in the polymer hydrogel electrolyte. Moreover, the ionic mobilities of the OH^- ion and K^+ ion at infinite dilution are 20.5×10^{-8} and 7.62×10^{-8} m s^{-1}(V m^{-1})$^{-1}$, respectively, according to the literature [7]. The transport number of the OH^- ion in a KOH aqueous solution at infinite dilution calculated from these values is 0.73. Considering that the interaction between ions is weak in a 6 M KOH aqueous solution, the transport number of the OH^- ion in a 6 M KOH aqueous solution is also expected to be *ca.* 0.73. Therefore, it can be concluded that the transport number of the OH^- ion in the polymer hydrogel electrolyte is higher than that in the KOH aqueous solution. Judging from this result, the polymer hydrogel electrolyte might be effective against the degradation of the electrode because the concentration gradient at the electrode/electrolyte interface is restrained.

In order to investigate the electrochemical behavior of the polymer hydrogel electrolyte, cyclic voltammetry was carried out using a smooth platinum electrode in the polymer hydrogel electrolyte and a 6 M KOH aqueous solution under Ar atmosphere at a scan rate of 100 mV s^{-1} at 25°C. As can be seen from Fig. 5.2.5, well-known reaction currents in the KOH aqueous solution were observed in the potential range of $-1.0 \sim +0.6$ V *vs.* Hg/HgO: *i.e.*, formation of adsorbed hydrogen, oxidation of the adsorbed hydrogen, formation of adsorbed oxygen or platinum oxide layer, and reduction of the oxide. Similar reaction currents were also observed in the cyclic voltammogram for the polymer hydrogel electrolyte; however, no significant currents for extra reactions were seen. From these results, it is found that no decomposition of the polymer hydrogel electrolyte occurs in the potential range ($-1.0 \sim +0.5$ V *vs.* Hg/HgO) for the charge and discharge of the Ni/MH cell.

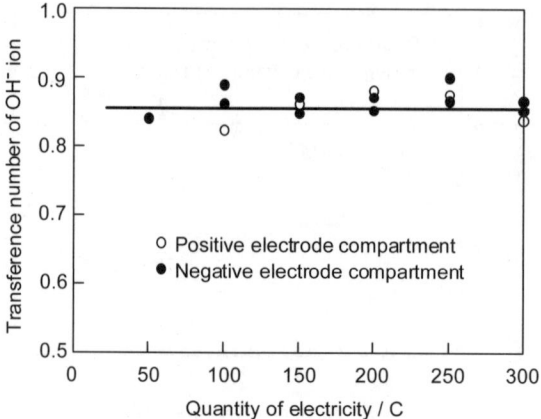

Fig. 5.2.4. Transference number of OH- ion in the polymer hydrogel electrolyte for different quantities of electricity.

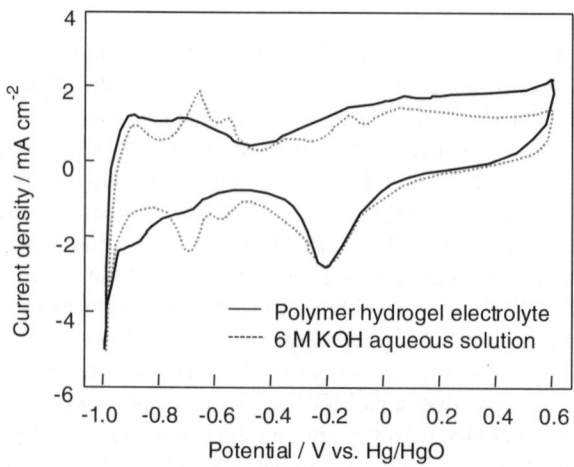

Fig. 5.2.5. Cyclic voltammograms of Pt electrode in the polymer hydrogel electrolyte and a 6 M KOH aqueous solution. Scan rate: 100 mV s^{-1}.

Because this polymer hydrogel electrolyte has high conductivity and a rather wide potential window, it could become a potential candidate for an electrolyte in alkaline secondary batteries such as a Ni/MH battery.

In the commercial sealed-type Ni/MH cell, the capacity of the cell is controlled by that of the nickel positive electrode so that oxygen evolves on the positive electrode during overcharging. The oxygen diffuses through a separator impregnated with an alkaline solution to the negative electrode where it is consumed by the following reaction with the metal hydride.

$$4MH + O_2 \rightarrow 4M + 2H_2O \quad (5.2.3)$$

5.2 Hydrogen storage alloy electrode/polymer hydrogel electrolyte interface

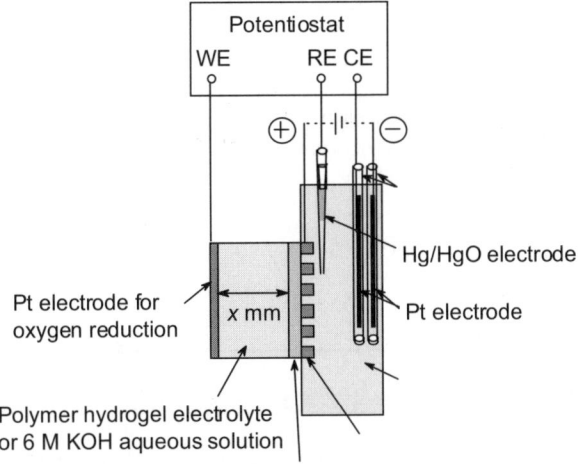

Fig. 5.2.6. Schematic representation of the cell for measuring oxygen permeation time.

This prevents an internal pressure rise in the sealed-type Ni/MH cell. Therefore, it is essential to investigate the oxygen permeability of the polymer hydrogel electrolyte. In this study, the oxygen permeability of the polymer hydrogel electrolyte was electrochemically evaluated compared with a KOH aqueous solution.

The measurement of oxygen permeability of the polymer hydrogel electrolyte was electrochemically carried out at 25°C using an experimental cell as schematically shown in Fig 5.2.6. The cell consists of two compartments separated with a platinum mesh electrode for oxygen evolution and a sulfonated polypropylene separator (120 μm nominal thickness) impregnated with the electrolyte. In one compartment filled with a 6 M KOH aqueous solution, oxygen is evolved on the platinum mesh electrode by galvanostatic electrolysis of water. In the other one, the evolved oxygen permeates through the polymer hydrogel electrolyte with a thickness of x mm, and the oxygen reduction current is potentiostatically detected on a platinum electrode for oxygen reduction at –0.8 V vs. Hg/HgO. After the galvanostatic electrolysis of water started, the times until the oxygen reduction current appeared and until the current reached a saturated value were defined as t_1 and t_2, respectively.

Figure 5.2.7 shows the oxygen permeation times t_2 as a function of electrolyte thickness for the polymer hydrogel electrolyte and the 6 M KOH aqueous solution. The t_2 of the 6 M KOH aqueous solution was very small over a wide range of thickness. On the other hand, the t_2 of the polymer hydrogel electrolyte with a thickness under 2 mm was very small, as was that of the 6 M KOH aqueous solution, and then increased exponentially with increasing electrolyte thickness over 2 mm. Because the thickness of the separator used in practical batteries is much smaller than 2 mm, say 100–150 μm, it could be expected that the polymer hydrogel electrolyte would exhibit excellent oxygen permeability in practical use. Moreover, the results of detailed investigation in the range of electrolyte thickness under 2 mm are shown

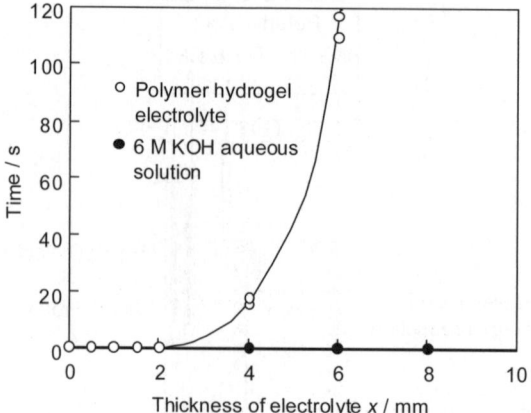

Fig. 5.2.7. Effect of thickness of the polymer hydrogel electrolyte and a 6 M KOH aqueous solution on oxygen permeation time t_2.

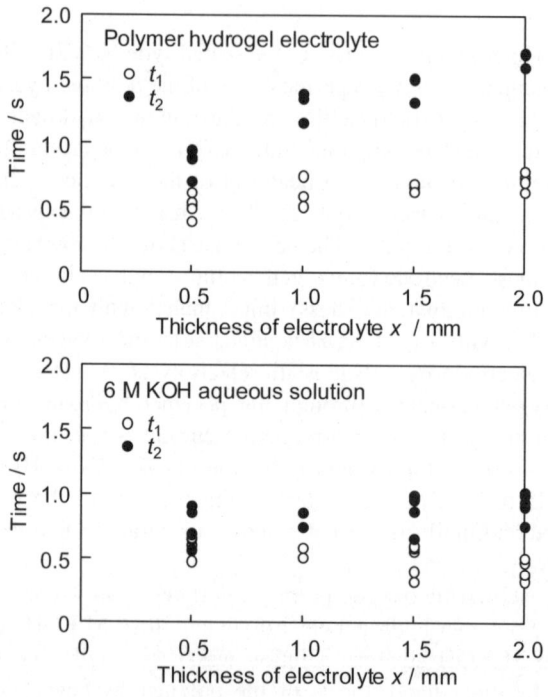

Fig. 5.2.8. Effect of thickness of the polymer hydrogel electrolyte and a 6 M KOH aqueous solution on oxygen permeation times t_1 and t_2.

in Fig. 5.2.8. Both t_1 and t_2 for the polymer hydrogel electrolyte became closer to those of the 6 M KOH aqueous solution with decreasing electrolyte thickness.

5.2 Hydrogen storage alloy electrode/polymer hydrogel electrolyte interface

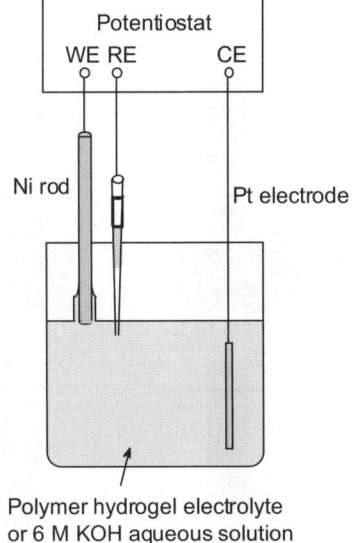

Fig. 5.2.9. Schematic representation of the cell for the electrochemical creepage tests.

From this figure, it is clear that t_1 and t_2 for the thin polymer hydrogel electrolyte with a thickness under ca. 500 μm are almost the same as those for the 6 M KOH aqueous solution. Considering the thickness of a practical separator as shown in Fig. 5.2.6, the net t_1 and t_2 for these thin electrolytes would be very close to zero. This implies that oxygen permeability of the polymer hydrogel electrolyte is excellent at a practical thickness level, almost comparable to that for the 6 M KOH aqueous solution. Consequently, it is strongly suggested that even if the polymer hydrogel electrolyte is utilized in practical sealed-type Ni/MH cells, an internal pressure rise in the cell during overcharging can be prevented without problems.

Considerable attention has been paid to sealing the Ni/MH cells to avoid electrolyte leakage. Especially, creepage of an alkaline electrolyte easily occurs along a metal sealing surface connected to the negative electrode [8–11]. The following oxygen reduction proceeds on the negatively polarized metal surface covered with a thin creepage electrolyte layer.

$$O_2 + 2H_2O + 4e \rightarrow 4OH \qquad (5.2.4)$$

This reaction causes a difference in alkali concentration between the electrolyte bulk and the thin electrolyte layer, followed by movement of water from the bulk to the thin layer. This is the reason that electrolyte creepage is facilitated on the metal surface connected to the negative electrode. Utilization of the polymer hydrogel electrolyte might suppress the electrolyte creepage because of the high water-holding capacity of PAAK. Therefore, the creeping behavior of the polymer hydrogel electrolyte along metal surfaces was investigated and compared with those of a 6 M KOH aqueous solution.

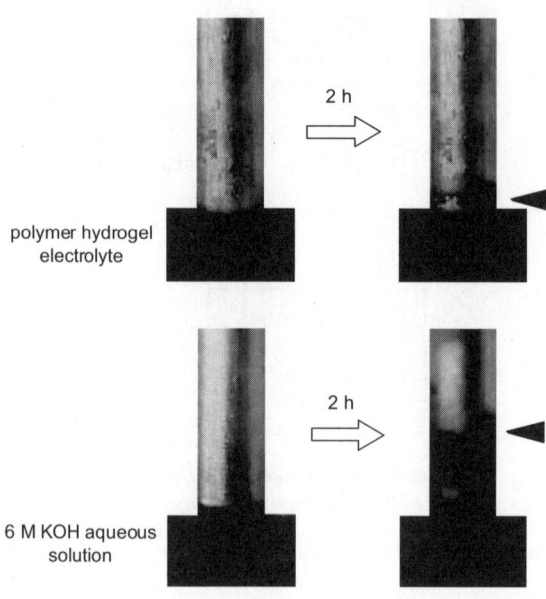

Fig. 5.2.10. Photographs of copper rods in contact with the polymer hydrogel electrolyte and a 6 M KOH aqueous solution.

Creep tests of the electrolytes were performed in the following two manners at 25°C [6]. First, a nonpolarized copper rod (diameter: 6 mm) was placed in vertical contact with the surface of the electrolyte colored with Eriochrome black T ($C_{20}H_{12}N_3Na_7O_7S$), and then the creeping behavior of the electrolyte along the rod was measured in the air. Second, as schematically shown in Fig. 5.2.9, a nickel rod (diameter: 5 mm) in vertical contact with the electrolyte surface was polarized to –0.9 V *vs.* Hg/HgO, and the oxygen reduction current was measured with time in air and under an Ar rich atmosphere.

Figure 5.2.10 shows the creeping behavior of the two electrolytes along a nonpolarized copper rod. In case of the 6 M KOH aqueous solution, the solution crept slightly just after contact with the copper rod, and creepage was clearly observed at a high level after 2 h as can be seen in Fig. 5.2.10. The KOH aqueous solution creeps easily along the copper rod surface because of the strong surface tension of water. On the other hand, in case of the polymer hydrogel electrolyte, creepage was not observed just after contact and was hardly observed even after 2 h. Utilization of the polymer hydrogel electrolyte instead of a KOH aqueous solution fairly suppressed the electrolyte creepage. It is strongly suggested that the high water-holding capacity of PAAK led to the suppression of electrolyte creepage.

5.2 Hydrogen storage alloy electrode/polymer hydrogel electrolyte interface

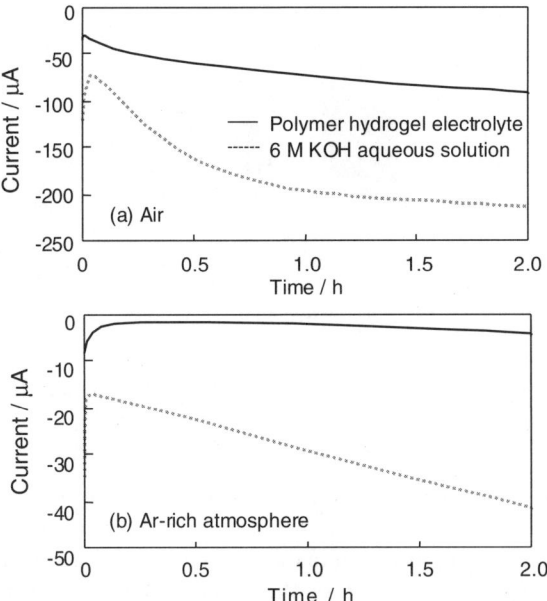

Fig. 5.2.11. Oxygen reduction currents at -0.9 V vs. Hg/HgO for the polymer hydrogel electrolyte and a 6 M KOH aqueous solution (a) in air and (b) under an Ar-rich atmosphere.

Figure 5.2.11 shows the time courses of oxygen reduction currents for the two electrolytes in contact with a negatively polarized nickel rod. In air (Fig. 5.2.11 (a)), a remarkable oxygen reduction current was observed for the 6 M KOH aqueous solution. The oxygen reduction current should be mainly proportional to the area of the thin electrolyte layer due to the electrolyte creepage. According to Eq. 5.2.4, the electrode creepage easily proceeds with the oxygen reduction in air. However, the oxygen current is fairly small for the polymer hydrogel electrolyte, indicating that the electrochemical creepage is largely suppressed by using the polymer hydrogel electrolyte. Moreover, even under an Ar-rich atmosphere (Fig. 5.2.11 (b)), the oxygen reduction current was clearly observed for the 6 M KOH aqueous solution, and it considerably increased with time. It is suggested that residual oxygen in the gas phase was dissolved into the thin electrolyte layer covering the electrode and was then reduced on the electrode surface. On the other hand, the oxygen reduction current for the polymer hydrogel electrolyte was quite small, and the rate of an increase in the current was also markedly suppressed by using the polymer hydrogel electrolyte. The current after 2 h for the polymer hydrogel electrolyte was *ca.* one-tenth that for the 6 M KOH aqueous solution. This implies that expansion of the thin electrolyte layer, *i.e.*, electrolyte creepage, along the negatively polarized metal surface was largely suppressed. The PAAK with high a water-holding capacity could serve effectively in suppressing electrolyte creepage even if the creepage was facilitated by negative polarization of the electrode.

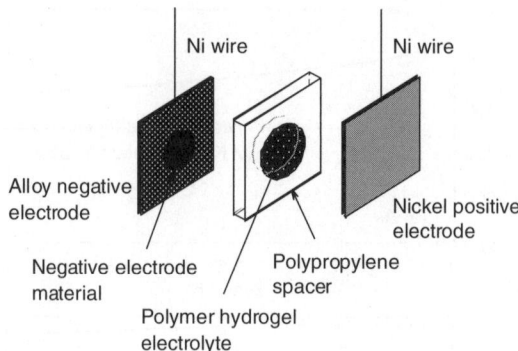

Fig. 5.2.12. Experimental cell assembly for the polymer hydrogel electrolyte-based Ni/MH cell.

Based on the results of the creep tests in the present study, it can be expected that the suppression of creepage by the polymer hydrogel electrolyte leads to reduction of sealing materials or improvement in the reliability of the prevention of electrolyte leakage. The suppressed creepage of the polymer hydrogel electrolyte can be a very significant advantage in applying the electrolyte to practical sealed-type Ni/MH batteries.

5.2.2 Application of polymer hydrogel electrolyte to nickel-metal hydride batteries

A simple-type experimental cell for the polymer hydrogel electrolyte-based Ni/MH cell was assembled as shown in Fig. 5.2.12 to obtain basic electrochemical data [1, 2]. For comparison, a similar type of Ni/MH cell was also assembled using a 6 M KOH aqueous solution as an electrolyte. The negative and positive electrodes were activated in a 6 M KOH aqueous solution in a glass beaker at 30°C before placing the electrodes in the test cell. The paste-type negative electrode ($MmNi_{3.6}Mn_{0.4}Al_{0.3}Co_{0.7}H_x$), a disc of polymer hydrogel electrolyte (3 mm thick, 1.54 cm^2), a positive electrode (NiOOH) used in commercial batteries and a polypropylene sheet with a circular hole (3 mm thick, 1.54 cm^2) as a spacer were stacked alternatively as shown in the figure. No separator was used in this cell for simplification. The polymer hydrogel electrolyte consisting of 7 wt.% PAAK, 23 wt.% KOH and 70 wt.% H_2O was used as the electrolyte for the test cell. The cell was deaerated under vacuum and then stored in an Ar atmosphere taking care of avoiding water evaporation. The total thickness of the cell in this study was *ca.* 5 mm.

In charge-discharge cycle tests, the negative electrode was charged at 50 mA g^{-1} for 4 h (200 mAh g^{-1}) and discharged at various currents to a cutoff voltage of 0.8 V at 25°C. After each charging, the circuit was kept open for 10 min. Figure 5.2.13 shows typical charge-discharge curves (1st, 10th and 50th cycles) of the Ni/MH cells with the polymer hydrogel and a 6 M KOH aqueous solution. Well-defined

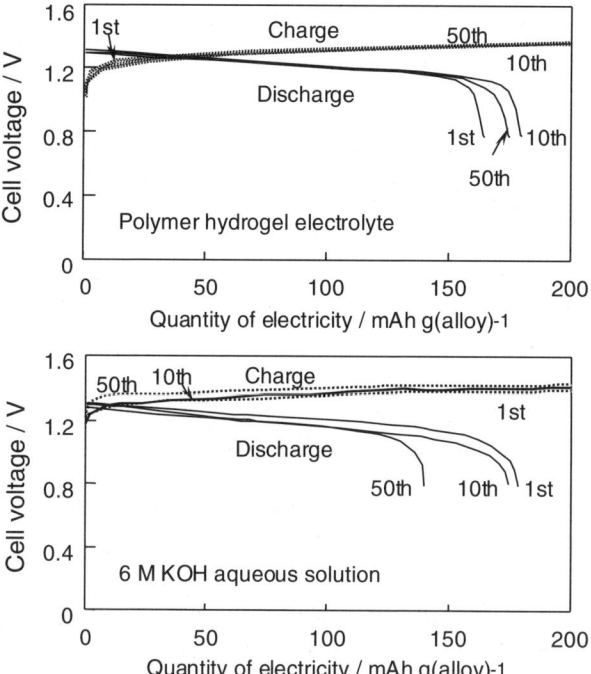

Fig. 5.2.13. Typical charge-discharge curves (1st, 10th and 50th cycles) of the Ni/MH cells with the polymer hydrogel electrolyte and a 6 M KOH aqueous solution.

charge and discharge curves were obtained in either case. A plateau of the cell voltage is seen in the discharge curves ranging from 1.3 to 1.1 V. At the first cycle, the discharge capacity was *ca.* 160 mAh g^{-1}. After the second cycle, a stationary discharge capacity of *ca.* 170–180 mAh g^{-1} was obtained which corresponds to *ca.* 85–90% of the charged quantity of electricity (200 mAh g^{-1}) of the negative electrode.

The voltage at the last stage of discharge for the Ni/MH cell with the polymer hydrogel electrolyte is rather high, compared to the Ni/MH cell with a 6 M KOH aqueous solution. In the Ni/MH cell of this study, the polymer hydrogel electrolyte and the 6 M KOH aqueous solution were 3-mm thick and no separator was used. Because the difference in the conductivity between both electrolytes was so small (*ca.* 0.6 S cm^{-1}), the difference in the voltage drops across the hydrogel and aqueous electrolytes of 3-mm thickness can be neglected. Although the active materials were impregnated in a foamed nickel substrate under high pressure, the active materials might be partially released from the electrode during the charge-discharge process. However, as a whole, the charge-discharge characteristics of both Ni/MH cells are quite similar.

The relationships between the charged quantity of electricity and the discharge capacity are shown in Fig. 5.2.14 for the Ni/MH cells with the polymer hydrogel

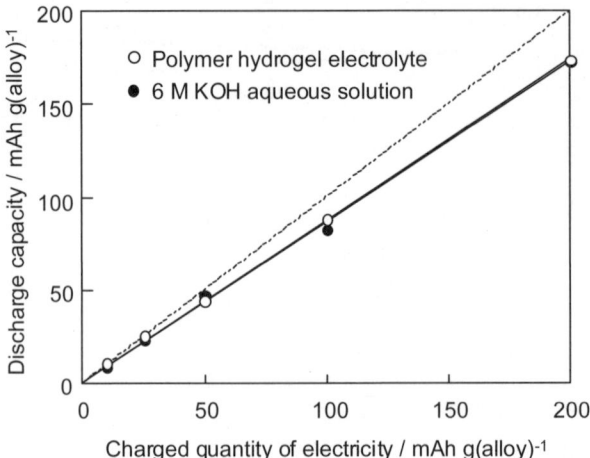

Fig. 5.2.14. Relationship the between charged quantity of electricity and the discharge capacity for the Ni/MH cells with the polymer hydrogel electrolyte and a 6 M KOH aqueous solution.

electrolyte and a 6 M KOH aqueous solution. The discharge capacity was 85 ~ 90% of the charged quantity of electricity. An electrochemical reaction of the PAAK did not occur during the charging process. The decrease in discharge capacity from the nominal capacity might be caused by hydrogen release from the negative electrode because the experimental cell was an open system.

Figure 5.2.15 shows the discharge capacities as a function of cycle number for the Ni/MH cells with the polymer hydrogel electrolyte and a 6 M KOH aqueous solution. As can be seen from this figure, the cycle life of the cell with the polymer hydrogel electrolyte was prolonged, maintaining a discharge capacity of over 85% even at the 50th cycle. The discharge capacity was *ca.* 70% of the charged quantity of electricity at the 300th cycle. Under the present conditions, there may be no problems at the electrode/electrolyte interface, and the charge and discharge reactions might then proceed smoothly. These features resulted from the high water-holding capacity of the PAAK. The polymer hydrogel electrolyte might hold enough water for smooth progress of the charge-discharge reactions at the positive and negative electrodes. The polymer hydrogel electrolyte used in this study was found to be very stable during charge-discharge cycling in a concentrated KOH aqueous solution. It showed a strong adhesive force onto the electrode materials, and thus contact discontinuity between electrode and polymer hydrogel electrolyte did not occur during the cycle test.

Figure 5.2.16 shows the capacity retention observed from the storage test for twelve days of the Ni/MH cells with the polymer hydrogel electrolyte and a 6 M KOH aqueous solution at 25°C. The capacity retention of these experimental cells is lower than that of the commercial Ni/MH batteries. This is probably because these experimental cells were the open systems assembled without using a separator for simplification. However, it is clear from this figure that the capacity retention of

5.2 Hydrogen storage alloy electrode/polymer hydrogel electrolyte interface 149

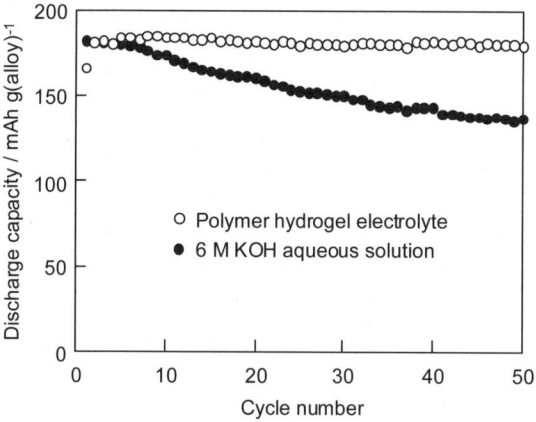

Fig. 5.2.15. Discharge capacities as a functon of cycle number for the Ni/MH cells with the polymer hydrogel electrolyte and a 6 M KOH aqueous solution.

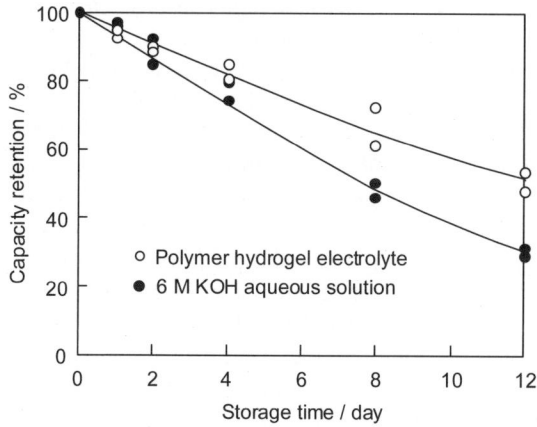

Fig. 5.2.16. Capacity retention as a function of storage time for the Ni/MH cells with the polymer hydrogel electrolyte and a 6 M KOH aqueous solution.

the Ni/MH cell with the polymer hydrogel electrolyte was much higher than that of the Ni/MH cell with the 6 M KOH aqueous solution. Table 5.2.1 shows the surface composition of alloys in the negative electrodes stored for eight days (Fig. 5.2.16) in the polymer hydrogel electrolyte and the 6 M KOH aqueous solution. The composition of the alloys was determined by EPMA. No significant change was observed for the contents of Mm, Ni and Co of the alloys before and after the storage test. However, the contents of Mn and Al decreased for the alloys after the storage test in the polymer hydrogel electrolyte and the 6 M KOH aqueous solution. This result is attributed to the dissolution of those metals in the electrolytes during storage of the cells, affecting the discharge capacity of the cells. It should be noted that the

150   5 Construction of solid / solid interface for battery application

Table 5.2.1. Surface composition of alloys in negative electrodes stored for eight days in the polymer hydrogel electrolyte and a 6 M KOH aqueous solution.

Negative electrode	atom%				
	Mm	Ni	Mn	Al	Co
As prepared	18.6	60.8	5.3	4.0	11.2
Stored in the polymer hydrogel electrolyte	18.2	61.6	5.1	2.8	11.2
Stored in a 6M KOH aqueous solution	18.8	62.0	4.6	2.3	11.5

dissolution rates of those elements are larger in the 6 M KOH aqueous solution than in the polymer hydrogel electrolyte. This is probably due to the low solubility of alloy constituents in the polymer hydrogel electrolyte. It seems that inhibition of the dissolution of those metals resulted in the high capacity retention of the Ni/MH cell with the polymer hydrogel electrolyte. Study along this line is in progress.

In our previous study, negative electrode capacity-controlled Ni/MH cells were assembled and tested as mentioned above [1, 2]. As a result, the charge-discharge reactions were found to occur smoothly at the interface between the negative electrode and the polymer hydrogel electrolyte as well as in the KOH aqueous solution. Nevertheless, in commercial Ni/MH batteries, the capacity is generally controlled by that of the nickel positive electrode. In addition, a disc of the polymer hydrogel electrolyte loaded in a perforated polypropylene sheet (3-mm thick) as a spacer was used for the experimental cell assembly in the previous study. In order to investigate the applicability of the polymer hydrogel electrolyte to the commercial Ni/MH batteries, it is essential that the experimental cell be improved further and characterized under the conditions limited by the positive electrode. In this study, positive electrode capacity-controlled Ni/MH cells were assembled using sulfonated polypropylene separators (120 μm nominal thickness) impregnated with the polymer hydrogel electrolyte and their electrochemical characteristics were investigated in detail for practical use [12, 13].

An experimental Ni/MH cell was assembled using a sulfonated polypropylene unwoven cloth separator (120 μm nominal thick) impregnated with the polymer hydrogel electrolyte, a sintered commercial $Ni(OH)_2$ positive electrode and two paste-type $MmNi_{3.6}Mn_{0.4}Al_{0.3}Co_{0.7}$ alloy negative electrodes as shown in Fig. 5.2.17. The sulfonated polypropylene separator served not only as a separator but also as a support material for the polymer hydrogel electrolyte. The negative electrodes were prepared in the same manner as previously reported [1, 2]. The nickel positive electrode was covered with the sulfonated polypropylene separator and stacked between the two negative electrodes. The total capacity of the two negative electrodes was *ca.* 4 times that of the positive electrode. Charge and discharge reserves were provided in the negative electrodes by partially charging

5.2 Hydrogen storage alloy electrode/polymer hydrogel electrolyte interface

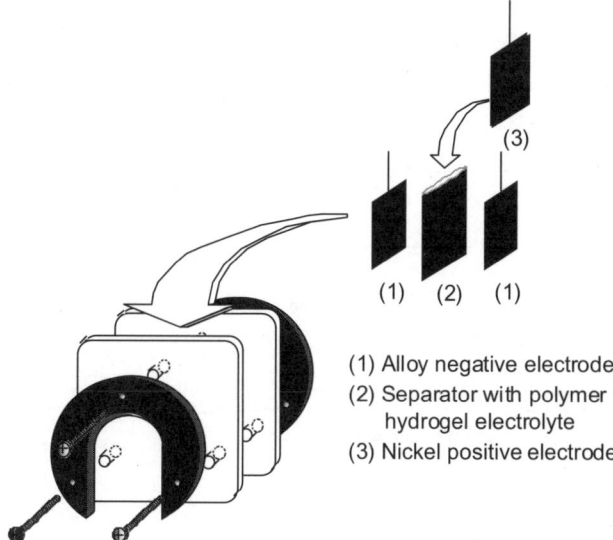

(1) Alloy negative electrode
(2) Separator with polymer hydrogel electrolyte
(3) Nickel positive electrode

Fig. 5.2.17. Experimental cell assembly for the polymer hydrogel electrolyte-based Ni/MH battery.

before assembling the cell as in case of the commercial Ni/MH battery. In this manner, the capacity of the cell was controlled by that of the positive electrode. A similar type cell of the Ni/MH battery was also assembled using a 7.3 M KOH aqueous solution as an electrolyte for comparison.

In the charge-discharge cycle test, the experimental cell was charged at 145 mA $g[Ni(OH)_2]^{-1}$ for 2.4 h and discharged at 145 mA $g[Ni(OH)_2]^{-1}$ to a cell voltage of 0.9 V at 25°C. After each charging, the circuit was opened for 10 min. The discharge capacity of the cell is described as the capacity (mAh) per mass of $Ni(OH)_2$ (g) because of the positive electrode capacity-controlled cell. High-rate chargeability (HRC) and high-rate dischargeability (HRD) were obtained at various current densities as described later. The capacity retention characteristics of the cell were also evaluated in the manner as described later.

Figure 5.2.18 shows typical charge-discharge curves (10th cycle) of the Ni/MH cell and the negative and positive electrodes with the polymer hydrogel electrolyte and the 7.3 M KOH aqueous solution. As can be seen from Fig. 5.2.18 (a), both the Ni/MH cells showed well-defined charge-discharge curves and the discharge curves had plateaus ranging from 1.3 to 1.1 V. The charge-discharge characteristics of the cell with polymer hydrogel electrolyte are quite similar to those with the 7.3 M KOH aqueous solution. Based on this experimental condition limited by the positive electrode, it is clear that the charge and discharge reactions smoothly proceeded at the interface between the polymer hydrogel electrolyte and the nickel positive electrode as well as the alloy negative electrode as mentioned above. Moreover, the discharge capacity of the cell with the polymer hydrogel electrolyte was almost the same as

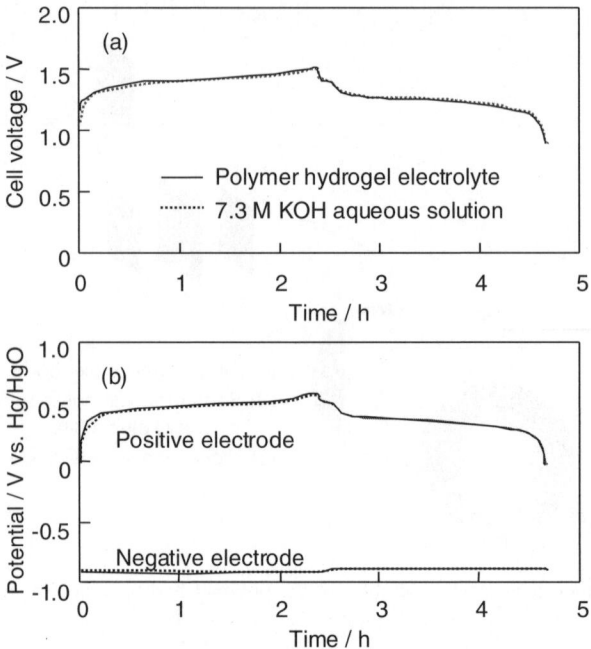

Fig. 5.2.18. Typical charge-discharge curves (10th cycle) of (a) the Ni/MH cells and (b) positive and negative electrodes with the polymer hydrogel electrolyte and a 7.3 M KOH aqueous solution.

the theoretical capacity of the nickel positive electrode calculated based on $Ni(OH)_2$, 289.1 mAh g[$Ni(OH)_2$]$^{-1}$. This fact implies that almost all of the active material in the positive electrode were available even when using the polymer hydrogel electrolyte. Furthermore, the potential changes in the negative and positive electrodes during the charge-discharge cycle test were measured using a Hg/HgO reference electrode as shown in Fig. 5.2.18 (b). In both cases of the two electrolytes, the potential of the negative electrode hardly changed during charging and discharging of the cells, while that of the positive electrode changed as did the charge-discharge curves of the cells. This is evidence that the experimental cells used in this study were controlled by the positive electrodes.

The relationships between the quantity of charged electricity (Q_{charge}) and the discharge capacity (C) for the Ni/MH cells with the polymer hydrogel electrolyte and the KOH aqueous solution are shown in Fig. 5.2.19. The Cs for the two cells are very similar in the range up to Q_{charge} of 400 mAh g[$Ni(OH)_2$]$^{-1}$. Under this experimental condition, the C/Q_{charge} ratios for both cells apparently deviated from 100% at a Q_{charge} of more than ca. 250 mAh g[$Ni(OH)_2$]$^{-1}$. This deviation can be ascribed to the evolution of oxygen on the positive electrode. Furthermore, C in each case approached the theoretical capacity of $Ni(OH)_2$ with increasing Q_{charge}, but did

5.2 Hydrogen storage alloy electrode/polymer hydrogel electrolyte interface

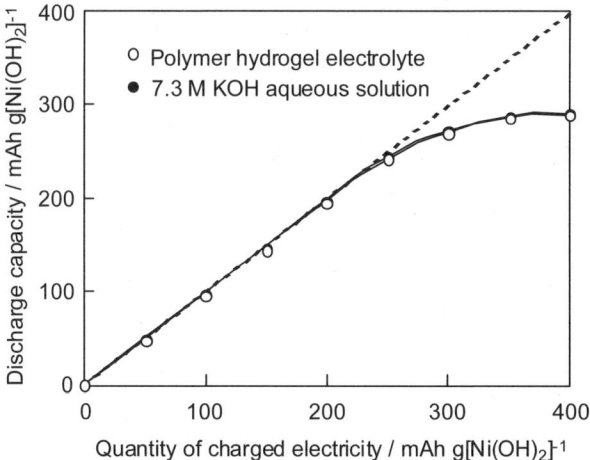

Fig. 5.2.19. Relationship between quantity of charged electricity and discharge capacity for the Ni/MH cells with the polymer hydrogel electrolyte and a 7.3 M KOH aqueous solution.

not exceed that value. This naturally results from the fact that the cells controlled by the positive electrode were used in this study.

Figure 5.2.20 shows the discharge capacities as a function of cycle number for the Ni/MH cells with the polymer hydrogel electrolyte and the 7.3 M KOH aqueous solution. Even in the case of the polymer hydrogel electrolyte, the discharge capacity hardly changed with increasing cycle number, and almost the same discharge capacity as the theoretical value was retained at the 50th cycle. This indicates that the charge and discharge reactions reversibly proceed at the electrode/electrolyte interface under the present conditions. The polymer hydrogel electrolyte used in this study was also found to be stable during charge-discharge cycling in a concentrated KOH aqueous solution. In addition, it may have a strong adhesive force onto the electrode materials, maintaining uniform contact between the electrodes and polymer hydrogel electrolyte during charge-discharge cycling. These advantages of the polymer hydrogel electrolyte may contribute to the stable charge-discharge cycle performance.

High-rate chargeability (HRC) and high-rate dischargeability (HRD) were investigated for the Ni/MH cell with the polymer hydrogel electrolyte at 25°C, compared to that with the 7.3 M KOH aqueous solution. HRC was evaluated by discharging at 145 mA g[Ni(OH)$_2$]$^{-1}$ after charging at various current densities to 120% of the theoretical capacity of Ni(OH)$_2$. HRC was calculated by the following equation.

$$\mathrm{HRC}(\%) = (C_i/C_{145}) \times 100 \quad (5.2.5)$$

where C_i and C_{145} are the discharge capacities after charging at i and 145 mA g[Ni(OH)$_2$]$^{-1}$, respectively. HRD was also evaluated in the following manner. The cells were charged at 145 mA g[Ni(OH)$_2$]$^{-1}$ for 2.4 h and then discharged at different

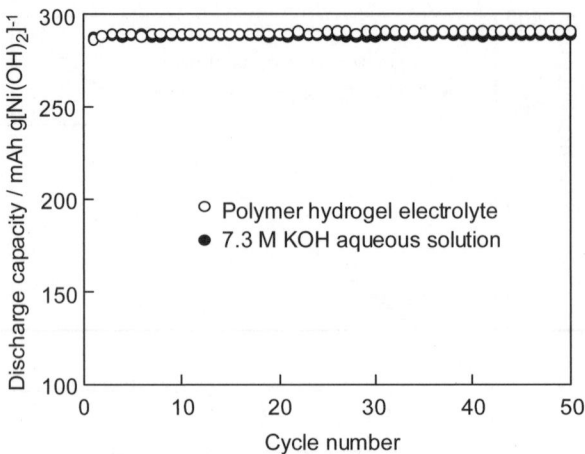

Fig. 5.2.20. Discharge capacities as a function of cycle number for the Ni/MH cells with the polymer hydrogel electrolyte and a 7.3 M KOH aqueous solution.

current densities to a cell voltage of 0.8 V, followed by continuous discharging at 25 mA g[Ni(OH)$_2$]$^{-1}$ without charging. HRD was calculated by the following equation.

$$\text{HRD}(\%) = C_i/(C_i + C_{25}) \times 100 \tag{5.2.6}$$

where C_i is the discharge capacity at a current density of i mA g[Ni(OH)$_2$]$^{-1}$ and C_{25} the discharge capacity when the cell is discharged at 25 mA g[Ni(OH)$_2$]$^{-1}$ just after discharging at i mA g[Ni(OH)$_2$]$^{-1}$.

Figure 5.2.21 show the HRC and HRD for the cells with the polymer hydrogel electrolyte and the 7.3 M KOH aqueous solution. As can be seen from Fig. 5.2.21 (a), both types of cells maintained an HRC of more than 90% even at a high charge current density such as *ca.* 1500 mA g[Ni(OH)$_2$]$^{-1}$. Moreover, an HRD of over 65% was retained at a high discharge current density of *ca.* 1500 mA g[Ni(OH)$_2$]$^{-1}$ even in case of the polymer hydrogel electrolyte as well as the 7.3 M KOH aqueous solution as shown in Fig. 5.2.21 (b). These results strongly indicate that the HRC and HRD of the polymer hydrogel electrolyte-based cell are excellent and almost the same as that of a KOH aqueous electrolyte-based cell. The excellent high-rate capability can be ascribed to the high conductivity of the polymer hydrogel electrolyte and good contact between the electrolyte and electrodes.

The capacity of the Ni/MH battery as well as other batteries generally decays with storage time on open circuit due to self-discharge. Especially, the capacity decay is serious at relatively high temperature. Therefore, from a practical viewpoint, it is essential that the capacity retention characteristics should be investigated in detail for the Ni/MH cell with the polymer hydrogel electrolyte. In this study, the capacity retention characteristics of the experimental Ni/MH cell with the polymer hydrogel electrolyte were examined, compared to those with the 7.3 M KOH aqueous solution

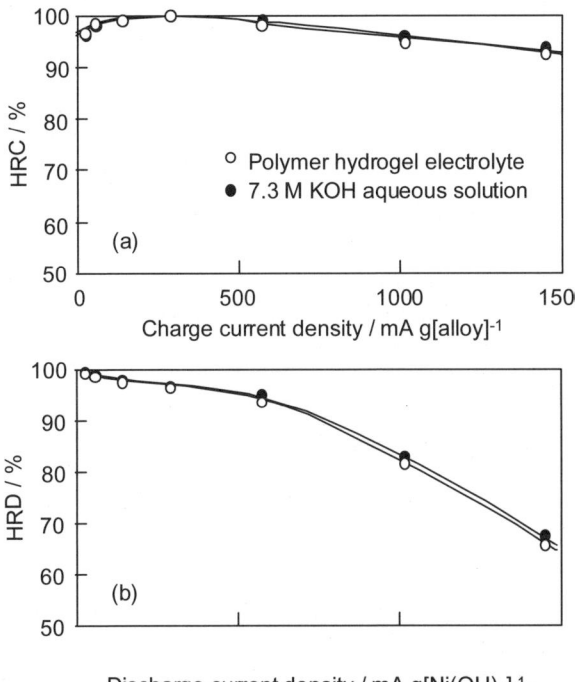

Fig. 5.2.21. (a) High-rate chargeability (HRC) and (b) high-rate dischargeability (HRD) for the Ni/MH cells with the polymer hydrogel electrolyte and a 7.3 M KOH aqueous solution.

at various temperatures, and the reversible and irreversible capacity losses were also examined.

The storage tests of the experimental cell were carried out to evaluate the capacity retention characteristics as follows. The experimental cell was charged at 145 mA g[Ni(OH)$_2$]$^{-1}$ for 2.4 h, allowed to rest for 10 min and then discharged at 145 mA g[Ni(OH)$_2$]$^{-1}$ to a cell voltage of 0.9 V. After 3 charge-discharge cycles and charging, the cell was stored on open circuit for various times and then discharged. The ratio of discharge capacity after the storage to that before the storage was evaluated as capacity retention. Moreover, the cell was charged and discharged again after the storage tests to estimate the reversible and irreversible capacity losses. For the capacity loss due to storage, the capacities recovered and unrecovered by recharging were defined as the reversible and irreversible capacity losses, respectively. The storage tests of the experimental cell were carried out at several temperatures from 0 to 55°C.

Figure 5.2.22 shows capacity retention characteristics as a function of storage time at 25°C for the Ni/MH cells with the polymer hydrogel electrolyte and the 7.3 M KOH aqueous solution. For both electrolytes, the discharge capacities before storage were very close to the theoretical capacity of the nickel positive electrode calculated

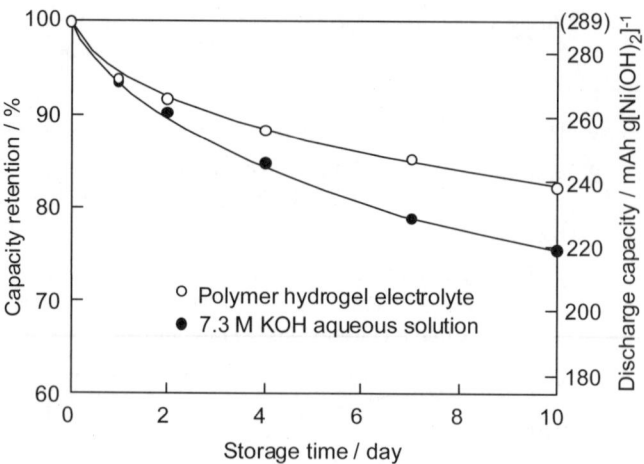

Fig. 5.2.22. Capacity retention as a function of storage time for the Ni/MH cells with the polymer hydrogel electrolyte and a 7.3 M KOH aqueous solution at 25°C.

based on $Ni(OH)_2$, 289.1 mAh $g[Ni(OH)_2]^{-1}$. As can be seen from this figure, in case of the 7.3 M KOH aqueous solution, the discharge capacity was decayed with storage time, and the discharge capacity was decreased to ca. 76% of that before the storage after storage for 10 days. On the other hand, in case of the polymer hydrogel electrolyte, the capacity decay with storage time was clearly suppressed, and the capacity retention after the storage for 10 days was ca. 83%. This implies that the polymer hydrogel electrolyte could improve the capacity retention characteristics of Ni/MH cells.

Discharge curves of the Ni/MH cells before and after the storage test for 10 days at 25°C are shown in Fig. 5.2.23. In each case, a well-defined discharge curve with a plateau ranging from 1.3 to 1.1 V was observed, indicating that the discharge reaction smoothly proceeded. From curves a and b in Fig. 5.2.23 as well as in Fig. 5.2.22, it is clear that the capacity decay due to storage is much smaller in case of the polymer hydrogel electrolyte. Moreover, the cell after the storage test was charged again and discharged (curve c) to estimate the reversible and irreversible capacity losses. In the case of the 7.3 M KOH aqueous solution, the discharge capacity was recovered to some degree on recharging, and the reversible capacity loss can be estimated at ca. 17% of the discharge capacity before the storage test. However, ca. 7% was not recovered under the present test conditions. On the other hand, in the case of the polymer hydrogel electrolyte, the reversible and irreversible capacity losses were 14 and 3%, respectively, which are quite small, compared to those for the KOH aqueous solution.

The reversible capacity loss can be ascribed to self discharge due to reduction of NiOOH to $Ni(OH)_2$ by water and hydrogen released at the negative electrode[14, 15]. Because the reversible capacity loss was depressed by using the polymer hydrogel electrolyte as can be seen from Fig. 5.2.23, it is inferred that the spontaneous

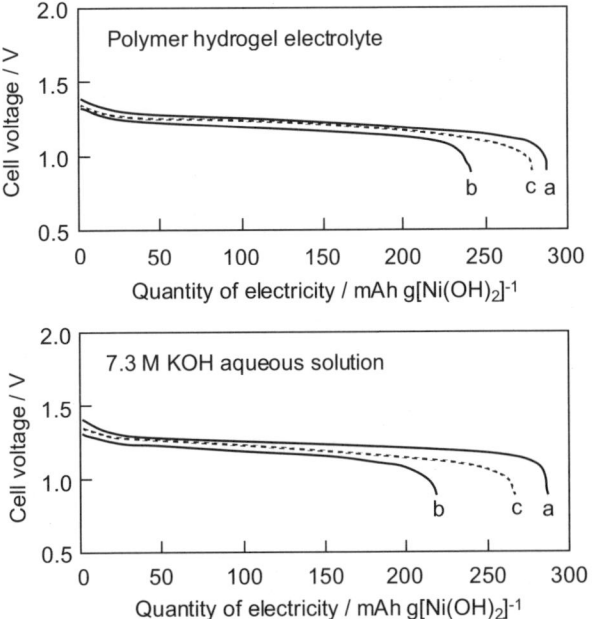

Fig. 5.2.23. Discharge curves of the Ni/MH cells with the polymer hydrogel electrolyte and a 7.3 M KOH aqueous solution (a) before and (b) after the storage test for 10 days and (c) after the storage test for 10 days, discharging and charging at 25°C.

reduction of NiOOH was suppressed. In addition, the irreversible capacity loss may be due to degradation of both the positive and negative electrodes and/or some other factors resulting from the open cell condition. As mentioned above, dissolution and diffusion of negative electrode constituents such as Al and Mn were suppressed by using the polymer hydrogel electrolyte. The inhibitory action against the dissolution and diffusion of the electrode constituents would work well in the polymer hydrogel electrolyte. The negative electrode constituents dissolved into the electrolyte might diffuse to the positive electrode across the electrolyte and exert a bad influence on the positive electrode performance. The polymer hydrogel electrolyte would inhibit this influence. Furthermore, the polymer hydrogel electrolyte also might suppress the dissolution of the positive electrode constituents. Thus, it is suggested that these effects may result in suppression of the degradation of the positive electrode, leading to a decrease in the irreversible capacity loss.

Figure 5.2.24 shows the capacity retention characteristics as a function of temperature (storage time: 10 days) for the Ni/MH cells. Over a wide temperature range of 0–55°C, the polymer hydrogel electrolyte exhibited better capacity retention characteristics than the 7.3 M KOH aqueous solution. Especially, at relatively high temperature, the discharge capacity largely decayed due to storage for 10 days in the case of the 7.3 M KOH aqueous solution. It is assumed that increasing temperature promotes both the spontaneous reduction of NiOOH and dissolution of the electrode

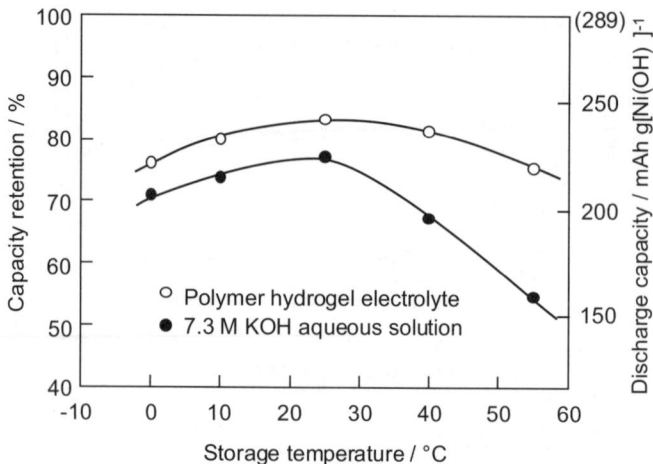

Fig. 5.2.24. Capacity retention as a function of storage temperature for the Ni/MH cells with the polymer hydrogel electrolyte and a 7.3 M KOH aqueous solution after the storage test for 10 days.

constituents as mentioned above. However, using the polymer hydrogel electrolyte, the capacity loss due to storage was markedly suppressed. Under the present test conditions, capacity retentions at 55°C for the 7.3 M KOH aqueous solution and polymer hydrogel electrolyte were 54 and 76%, respectively. The effects of the polymer hydrogel electrolyte on charge retention characteristics were much more remarkable at 55°C than at 25°C.

Discharge curves before and after the storage test at 55°C are shown in Fig. 5.2.25. Although a well-defined discharge curve with a plateau was observed in each case, the capacity loss due to storage was quite large for the 7.3 M KOH aqueous solution. In the case of the cell after the storage test and recharging (curve c) with the KOH aqueous solution, the reversible and irreversible capacity losses can be estimated to be 28 and 18%, respectively, of the discharge capacity before the storage. In particular, the irreversible capacity at 55°C was *ca.* 2.6 times that at 25°C, suggesting that degradation of the positive electrode was largely promoted. On the other hand, in case of the polymer hydrogel electrolyte, the reversible and irreversible capacity losses were 17 and 7%, respectively. These results clearly indicate that both reversible and irreversible capacity losses were remarkably suppressed. In addition, it seems that the polymer hydrogel electrolyte is especially effective in suppressing the irreversible capacity loss. This suggests that dissolution of the electrode constituents were fairly inhibited by using the polymer hydrogel electrolyte. From all the data on capacity retention characteristics as mentioned above, it was clarified that the polymer hydrogel electrolyte had the significant advantage of suppressing self-discharge of the Ni/MH cell. Along this line, a further detailed investigation is now in progress.

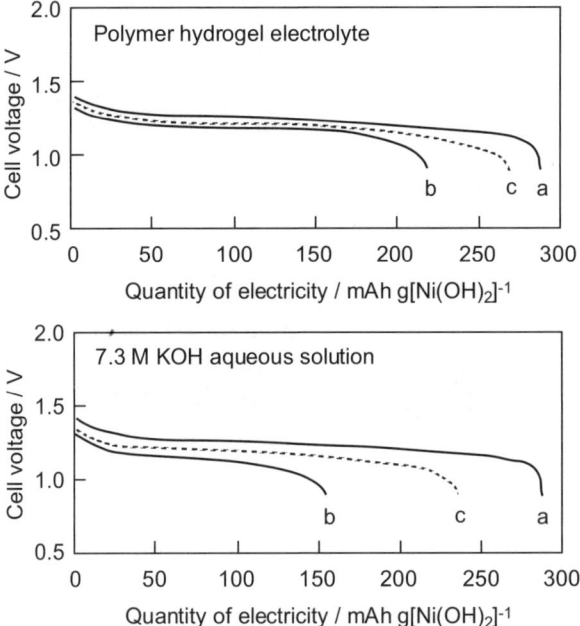

Fig. 5.2.25. Discharge curves of the Ni/MH cells with the polymer hydrogel electrolyte and a 7.3 M KOH aqueous solution (a) before storage, (a) before and (b) after the storage test for 10 days and (c) after the storage test for 10 days, discharging and charging at 55°C.

The capacity decay of the experimental Ni/MH cell with storage time on open circuit was clearly suppressed by using the polymer hydrogel electrolyte at 25°C. From the results of recharging of the cell after the storage test, it was found that the capacity loss due to storage consisted of reversible and irreversible capacity losses and that both of them were depressed by using the polymer hydrogel electrolyte. Moreover, the polymer hydrogel electrolyte exhibited excellent capacity retention characteristics over a wide temperature range of 0–55°C, and the effects of the polymer hydrogel electrolyte were more remarkable at relatively high temperature, *e.g.*, 55°C, than at 25°C. In particular, the irreversible capacity loss was fairly depressed by using the polymer hydrogel electrolyte, leading to improvement in the capacity retention characteristics.

5.2.3 Application of polymer hydrogel electrolyte to electric double-layer capacitors

The polymer hydrogel electrolyte was prepared from crosslinked potassium poly(acrylate) (PAAK) and a 10 M KOH aqueous solution in the same manner as described in the previous section (see 5.2.1). The polymer hydrogel electrolyte was composed of 7 wt.% PAAK, 36 wt.% KOH and 57 wt.% H_2O. As shown in Fig. 5.2.26, an experimental EDLC cell was assembled using a sulfonated polypropylene

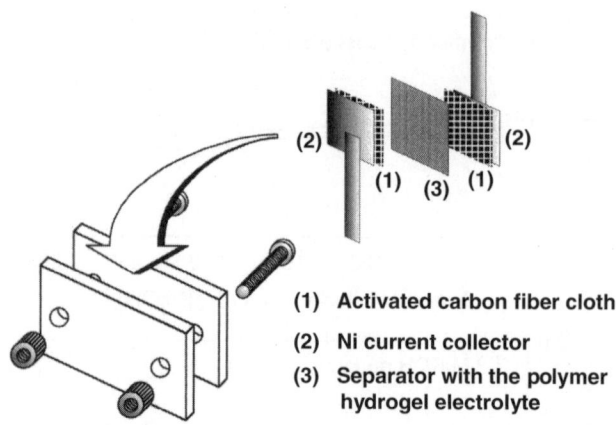

Fig. 5.2.26. Schematic representation of the experimental cell assembly.

nonwoven cloth separator impregnated with the polymer hydrogel electrolyte and two activated carbon cloths (specific surface area: *ca.* 2000 m^2 g^{-1}) as electrode materials [16].

Figure 5.2.27 shows cyclic voltammograms at 25°C for the activated carbon electrodes in a 10 M KOH aqueous solution and the polymer hydrogel electrolyte. A voltammogram without visible peaks due to redox reactions, close to the ideal rectangular shape, was observed for each electrolyte. This indicates that the two electrolytes were electrochemically stable under this condition and that charge and discharge reversibly occur at the electrode/electrolyte interface. From the voltammogram, the capacitances of the electrodes in the polymer hydrogel electrolyte and the KOH aqueous solution were evaluated to be 108 F g^{-1} and 104 F g^{-1}, respectively. The value in the polymer hydrogel electrolyte was slightly higher than that in the KOH aqueous solution. These results were confirmed by several further experiments under the same conditions.

Charge-discharge curves for the experimental EDLC cells with a 10 M KOH aqueous solution and the polymer hydrogel electrolyte are shown in Fig. 5.2.28. Almost linear curves are observed for each electrolyte, indicating that good electrode/electrolyte interfaces were formed and the experimental cells successfully functioned as EDLCs. Moreover, an IR drop was hardly observed for both charge and discharge curves under this condition. The polymer hydrogel electrolyte with high ionic conductivity would have good contact with the electrode as well as the KOH aqueous solution. From the discharge curves, the discharge capacitances of the electrodes were determined to be 110 F g^{-1} and 106 F g^{-1} for the polymer hydrogel electrolyte and the KOH aqueous solution, respectively, very close to those based on the cyclic voltammograms. Moreover, the discharge capacitance for the polymer hydrogel electrolyte was slightly higher than that for the KOH aqueous solution. In more than 50 experiments performed by changing the concentration of the KOH aqueous solution and using various kinds of carbon cloths, the cell with the

5.2 Hydrogen storage alloy electrode/polymer hydrogel electrolyte interface 161

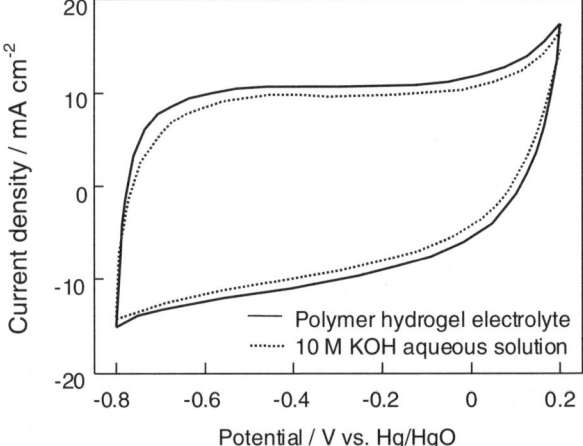

Fig. 5.2.27. Cyclic voltammograms for the activated carbon electrodes of the EDLC cells with the polymer hydrogel electrolyte and a 10 M KOH aqueous solution. Scan rate: 10 mV s^{-1}.

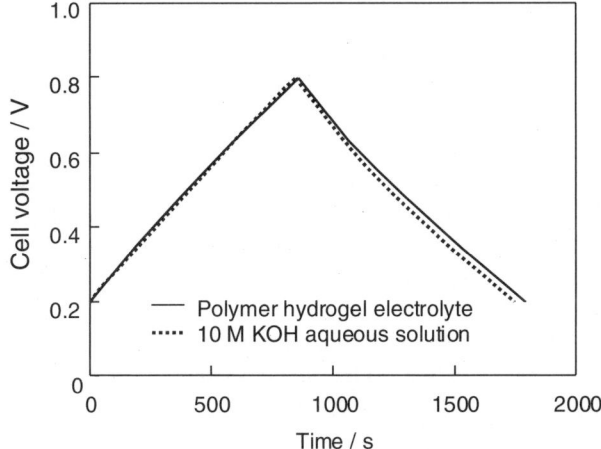

Fig. 5.2.28. Charge-discharge curves (10th cycle) for the EDLC cells with the polymer hydrogel electolyte and a 10 M KOH aqueous solution. Charge-discharge current density: 1 mA cm^{-2}.

polymer hydrogel electrolyte showed a slightly higher capacitance than that with the KOH aqueous solution. Therefore, it can be concluded that the difference in capacitance is small, and it is valid. AC impedance measurements suggested that the difference in capacitance was ascribed to the pseudocapacitance caused by a certain part of the PAAK [17, 18]. Charge-discharge cycle performance was also investigated. Discharge capacitance of the cell with the polymer hydrogel electrolyte was 82% of the initial value at the 20,000th cycle, while it was 72% in the case of the KOH aqueous solution. This result clearly indicates that the cell assembled using the

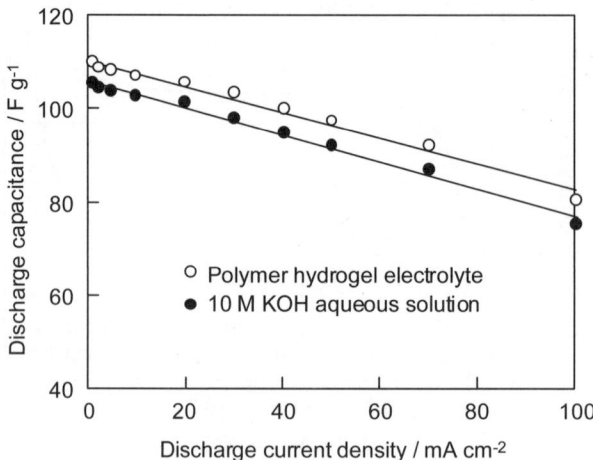

Fig. 5.2.29. Discharge capacitance as a function of the discharge current density for the EDLC cells with the polymer hydrogel electrolyte and a 10 M KOH aqueous solution.

polymer hydrogel electrolyte exhibited better cycle performance, compared with the case of the KOH aqueous solution. Utilization of the polymer hydrogel electrolyte prepared from PAAK with a high water-holding capacity may be more effective in maintaining a good electrode/electrolyte interface.

High-rate dischargeability of the EDLC cells using the polymer hydrogel electrolyte was also evaluated at 25°C, compared to the case of a KOH aqueous solution. As shown in Fig. 5.2.29, it was found that the cell with the polymer hydrogel electrolyte had excellent high-rate dischargeability as did that with the KOH aqueous solution. Even at high discharge current densities such as 100 mA cm^{-2}, the cell with the polymer hydrogel electrolyte exhibited higher discharge capacitance than that with the KOH aqueous solution. The excellent high-rate dischargeability can be mainly ascribed to the high ionic conductivity of the polymer hydrogel electrolyte almost comparable to the KOH aqueous solution. The results of electrochemical characterization as mentioned above strongly indicate that the polymer hydrogel electrolyte can be potentially applied to the EDLCs as an electrolyte with good performance.

5.2.4 Preparation and characterization of proton-conducting polymeric gel electrolytes

Proton (H^+)-conducting solid electrolytes have attracted much attention because of their potential application to such electrochemical devices as fuel cells, batteries, sensors, electrochromic displays, photo-electrochemical solar cells and capacitors. Among them, polymeric complexes that consist of organic polymers containing dissolved electrolytic salts can provide not only high ionic conductivity but also mechanical flexibility under ambient temperature conditions. In this work, a new

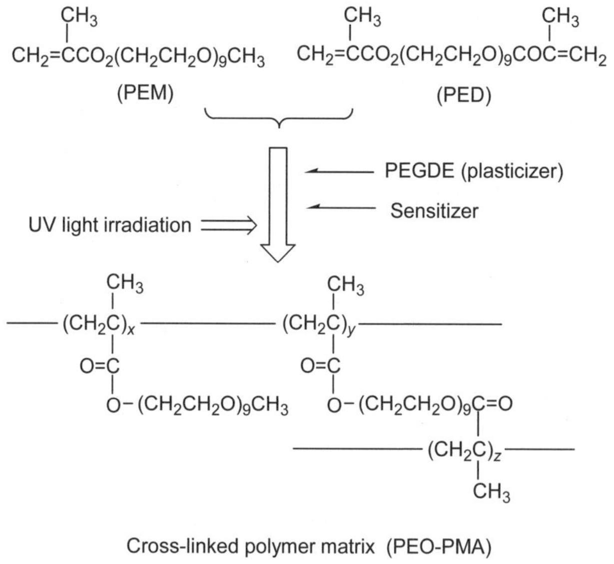

$$CH_2=\overset{\overset{\displaystyle CH_3}{|}}{C}CO_2(CH_2CH_2O)_9CH_3 \quad CH_2=\overset{\overset{\displaystyle CH_3}{|}}{C}CO_2(CH_2CH_2O)_9\overset{\overset{\displaystyle CH_3}{|}}{C}OC=CH_2$$

(PEM) ... (PED)

⟵ PEGDE (plasticizer)
⟵ Sensitizer

UV light irradiation ⟹

Cross-linked polymer matrix (PEO-PMA)

Fig. 5.2.30. A preparation scheme for the polymeric complex based on the cross-linked PEO-PMA.

type of polymeric proton conductor that can be used as a solid electrolyte for such electrochemical energy devices has been developed. The system is based on a novel concept of polymeric complexes doped with organic or inorganic acids. Cross-linked polymethacrylate which had originally been reported as a polymer matrix for a lithium ion (Li^+) conductor [19] was adopted. Different preparation procedures have been examined for the polymeric electrolytes, and the electrochemical properties of the resulting polymeric complexes have been investigated experimentally.

Two kinds of macro-monomers, poly(ethylene oxide) monomethacrylate (PEM) and poly(ethylene oxide) dimethacrylate (PED), were used for the matrix formation [19]. Poly(ethylene glycol) dimethylether (PEGDE) as a plasticizer and 2,2-dimethoxy-2-phenylacetophenone as a photo-sensitive initiator were added into the mixed macro-monomers. The resulting mixture was exposed to UV light to yield a cross-linked polymer matrix, PEO-PMA (Fig. 5.2.30). The prepared polymeric membrane was then swollen in aqueous solutions of inorganic or organic acids (H_3PO_4, HCl, CH_3COOH, etc.) with different concentrations and immersion times. In this stage, the electrolytes were doped in the polymeric membrane. The composition of the resulting gel is formulated to be (PEO-PMA)/PEGDE/HX (x mol dm^{-3}), where x denotes the acid concentration in the doping solution [20, 21].

The ionic conductivity of the gel membranes containing H_3PO_4, (PEO-PMA)/PEGDE/H_3PO_4, depended on the film preparation conditions. Figure 5.2.31 shows an example of the ionic conductivity data as a function of the acid concentration in the doping solution. Because the (PEO-PMA)/PEGDE matrix has no conductance (ionic nor electronic) at an ambient temperature range, the

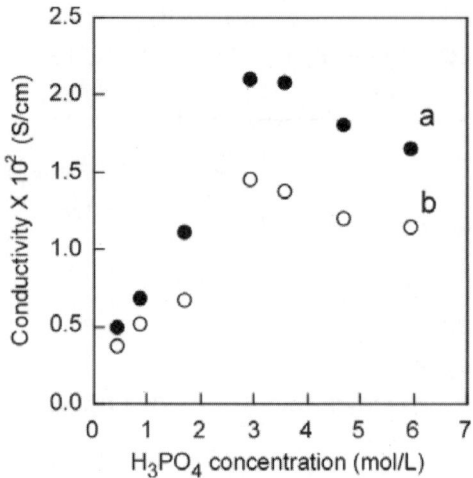

Fig. 5.2.31. Conductivity changes with the H_3PO_4 concentration, at immersion time of 2 h. a: (PEO-PMA)/PEGDE = 50/50 (in mass), b: (PEO-PMA)/PEGDE = 38/62 (in mass).

conductance observed for the polymeric membrane swollen with H_3PO_4 solution was certainly based on the ion transport in the gel system. The conductivity once increased with the acid concentration to the maximum at around 3 mol dm^{-3} while decreased in the higher acid concentration region. The maximum conductivity at room temperature was *ca.* 2×10^{-2} S cm^{-1}, which is comparable to that for an aqueous liquid H_3PO_4 solution. The ionic conductivity also depended on the PEGDE content in the polymer matrix. It increased with an increase in the PEGDE content, which suggests that the local viscosity of the gel plays an important role in the ionic conduction of the gel.

The amounts of the dopant acid (H_3PO_4) and the water uptake in the resulting gel were determined by mass changes before and after the sequence of acid doping and drying [21]. The acid content in the gel increased almost linearly with the concentration in the solution up to 5 mol dm^{-3}, while the water content in the gel decreased with the acid concentration in the solution. This result explains the conductance behavior shown in Fig. 5.2.31. That is, the decrease in the ionic conductivity for the gel doped with high acid concentration is due to the depression of the water in the gel, where both the ionic dissociation and the ion transport become unfavorable.

The temperature dependence of the ionic conductivity was investigated in a temperature range of 20–70°C, where the acid concentration in the doping solution was used as the parameter [21, 22]. Most samples showed Arrhenius-type variations, except for highly doped gel electrolytes. The maximum conductivity was *ca.* 2.8×10^{-2} cm^{-1} at 70°C. The conductivity increased monotonously with the acid concentration but decreased at higher acid concentration and higher temperature.

5.2 Hydrogen storage alloy electrode/polymer hydrogel electrolyte interface

This was due to the insufficient amount of water in the gel composition with high acid contents.

The apparent activation energy for ionic conduction was determined from the slope of the curves in the linearly increasing region of the Arrhenius plots. Around 10 kJ mol^{-1} or lower values were obtained as the activation energy for samples doped with moderate acid concentrations. Thus, both the Grotthus- and Vehicle-type mechanisms [23, 24] exist in the conducting process for the present (PEO-PMA)/PEGDE/H$_3$PO$_4$ system. The ionic motion in the present gel system seems to be highly decoupled from the segmental motion of the polymer matrix, the conductivity then being rather close to that of the liquid electrolyte system [25]. The PEO-PMA matrix, in fact, plays an essential role in enclosing the otherwise free phosphoric acid within a solid framework that provides a large number of charge carriers such as H$_2$PO$_4^-$ and even H$_4$PO$_4^+$.

The effects of the dopant acid have also been examined on the conductance behavior of the gel. When such strong acids as sulfuric and hydrochloric acids were used as the dopant, a higher conductivity than 10^{-2} S cm^{-1} was obtained. The doping of weak acids, acetic (HAc) and succinic acids (HSc), gave lower conductivity even when a high concentration was adopted for the doping solution. Interactions between the dopant acids and the polymer matrix were investigated by FT-IR spectra and XRD patterns for the gels doped with different acids. When strong acids such as sulfuric and phosphoric acids were doped, a rather strong interaction by hydrogen-bonding was observed, as shown by the broadening of the carbonyl group band. On the other hand, rather weak intermolecular forces were suggested for the gels doped with weak organic acids such as HAc and HSc. These were consistent with the conductance behavior. That is, the degree of ionic dissociation of the acid seems to be directly reflected in the molecular interactions and hence the ionic conductivity of the resulting gels.

The electrochemical activity of the gel electrolyte system was also examined. Figure 5.2.32 compares cyclic voltammograms of the ferric/ferrous redox system in an aqueous solution and in the present gel electrolyte. The upper voltammogram was measured in aqueous solution and the bottom was for the redox system in the gel. The gel electrolyte has a wide electrochemical window and sufficient activity of the electrochemical redox couple, suggesting that the present gel system will be applicable to the proton transferring processes, as in electrochemical capacitors.

The presence of water in the electrolyte generally causes some problems in the devices containing moisture-sensitive materials. Also, degradation of the polymer will occur under strong acidic conditions with considerable amounts of water. It has been demonstrated that dissolution of H$_3$PO$_4$ in poly(methylmethacrylate) plasticized by dimethylformamide (DMF) or poly(glycidylmethacrylate) with DMF or propylene carbonate (PC) produces so-called polymeric gel electrolytes with high proton conductivities around 10^{-4} S cm^{-1} at ambient temperatures [26–28]. Because these polymeric gels are synthesized using anhydrous solvents under a dry atmosphere, the above-mentioned drawbacks can be avoided. Here is reported a novel nonaqueous polymeric gel electrolyte that consists of poly(ethylene oxide)-modified poly(methacrylate) (PEO-PMA) containing an organic plasticizer,

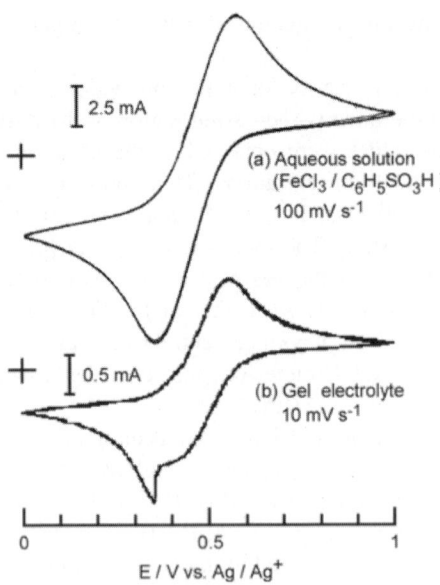

Fig. 5.2.32. Cyclic voltammograms for Pt in (a) 1 M $FeCl_3/C_6H_5SO_3H$ and (b) (PEO-PMA)/PEGDE doped with 1 M $FeCl_3/C_6H_5SO_3H$.

poly(ethylene glycol) dimethyl ether) (PEGDE) or DMF, in which anhydrous H_3PO_4 was chosen as a proton donor [29–31].

Anhydrous H_3PO_4 was dissolved in PEGDE, or a mixture of PEGDE and DMF. PEM/PED (3:1 by molar ratio) and a radical initiator were then added to the H_3PO_4/PEGDE (or H_3PO_4/PEGDE+DMF) solution. The resulting mixture was then developed on an Al plate and exposed to UV light for polymerization to yield H_3PO_4-doped poly(ethylene oxide)-modified poly(methacrylate) polymeric gel : (PEO-PMA)/PEGDE(or PEGDE+DMF)/H_3PO_4 (see Fig. 5.2.30). All steps of the preparation procedure were carried out under a dry Ar atmosphere.

Optically transparent and uniform polymeric gel membranes were obtained over a wide range of component concentrations, an H_3PO_4 content of 13–52 mass% and a PEGDE content of 0–77 wt%. The membranes have sufficient mechanical strength to withstand measurement of their electrochemical properties.

Figure 5.2.33 shows the temperature dependence of the ionic conductivity obtained for the gel membranes with different compositions. A linear relation is observed on every $\log \sigma$ vs. $1/T$ plot. The slope of the linear plot gives apparent activation energy (E_a) for the ionic conduction. The E_a value increased with an increase in the H_3PO_4 content in the gels plasticized either with PEGDE or mixed PEGDE+DMF. The addition of DMF as the plasticizer tended to decrease the activation energy, but the difference became small when the polymeric gel complexes contained high acid contents. This suggests that the activation process of the conduction becomes similar in the gels with high H_3PO_4 contents. The E_a values

5.2 Hydrogen storage alloy electrode/polymer hydrogel electrolyte interface

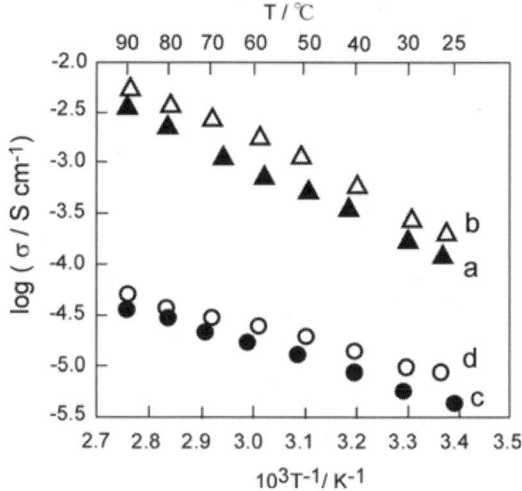

Fig. 5.2.33. Temperature dependence of the ionic conductivity for PEO-PMA-based polymeric gel complexes with different acid contents. a, c: PEGDE as the plasticizer; (PEO-PMA)/PEGDE = 38/62 (in mass). b, d: PEGDE+DMF as the plasticizer; (PEO-PMA)/PEGDE/DMF = 38/47/15 (in mass). H_3PO_4 content (mass %): 13 (c, d); 52 (a, b).

in the present nonaqueous gels were generally higher than those observed for the hydrogels containing H_3PO_4 or acetic acid, where around 10 kJ mol^{-1} was obtained.

In Fig. 5.2.34, possible mechanisms for the ionic conduction in the present nonaqueous gel system are schematically shown. The proton conduction proceeds mainly through the molecular coordination of protons with ethylene oxide (EO) units in the polymer matrix and/or the amide group of the plasticizing DMF molecule. The proton conduction mechanism appears to change with the increase in the H_3PO_4 content. That is, for the samples with high H_3PO_4 contents, a vehicle-type mechanism dominates the ionic conduction, whereas for the gels with low acid contents and/or high solvent contents, a Grotthus-type mechanism appears to be the primary route of the conduction. This is further supported by the experimental results of FT-IR and thermal analyses [30].

The applicability of the PEO-PMA-based nonaqueous polymeric gel was examined as a solid electrolytes for all-solid electrochemical devices. Here, some preliminary results are shown as the electrolyte for an EDLC. An activated carbon fiber (ACF) cloth of 1300 m^2 g^{-1} BET surface area was used as the test electrode of the EDLC with the gel electrolyte. In Fig. 5.2.35, a test cell for evaluating the capacitor performance is schematically shown. A piece of ACF cloth (26.5 ± 0.5 mg) was adhered to a Pt sheet current collector with conductive carbon paste. The cell has a stacked structure with two ACF electrodes between which the polymeric gel film is sandwiched as the solid electrolyte. For comparison, aqueous H_3PO_4 solution (2.0 mol dm^{-3}) impregnated in a polypropylene separator was used as the liquid electrolyte of the EDLC. The EDLC performances were evaluated by charge and

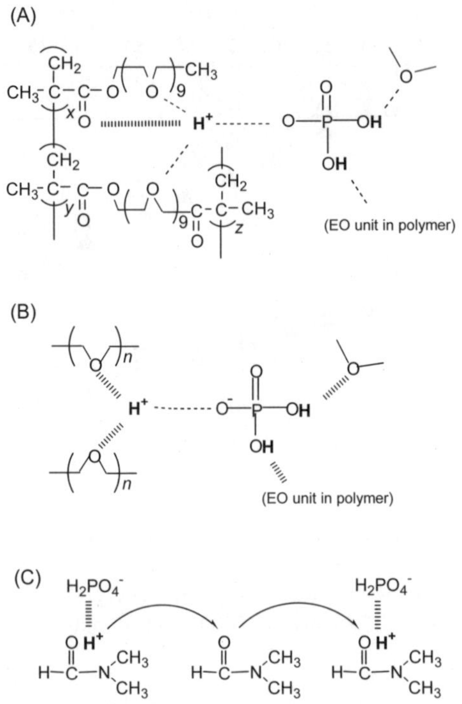

Fig. 5.2.34. Schematic diagrams for proton conduction in the nonaqueous polymeric gel. (A) Conduction in PEO-PMA/H$_3$PO$_4$ system, (B) in PEO-PMA/PEGDE/H$_3$PO$_4$ system, H$_3$PO$_4$ system, H$_3$PO$_4$ system.

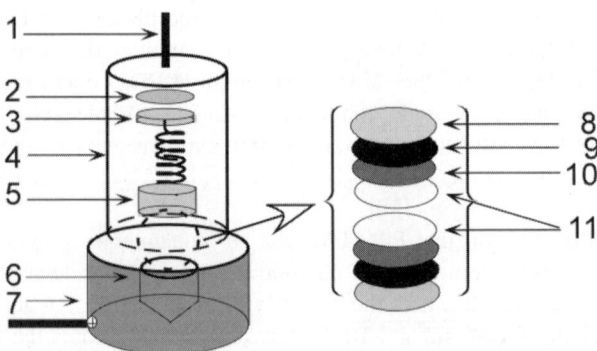

Fig. 5.2.35. Schematic diagram of the test cell for the capacitor performance. 1: Ni lead wire, 2,3: Ni disks, 4: Teflon sleeve, 5, 6: stainless steel electrodes, 7: stainless steel terminal, 8: Pt sheet current collector, 9: carbon paste, 10: ACF cloth, 11: polymeric gel.

5.2 Hydrogen storage alloy electrode/polymer hydrogel electrolyte interface 169

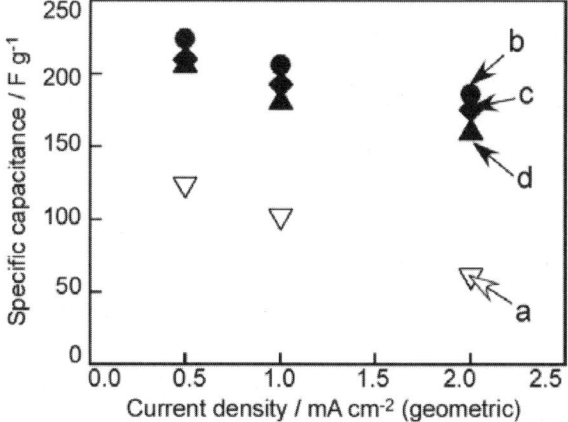

Fig. 5.2.36. Variations in the specific capacitance of ACF with current density for the test cells using polymeric gel electrolytes at 90°C. a: without plasticizer, b: PEGDE, c: PEGDE+DMF, d: DMF.

discharge cycling under constant current conditions in the temperature range between 30 and 90°C, where the capacitor was first charged to 1.0 V and then discharged to 0 V at a constant current from 0.13 to 2.6 mA.

In Fig. 5.2.36, the specific capacitance obtained at 90°C is plotted against the cycling current density. Here, the *iR*-free values of the capacitance are shown to discuss the rate performances. When is cycled the cell at a relatively low rate (0.5 mA cm^{-2}), *ca.* 120 F g^{-1} of the discharge capacity was obtained for the cell with the plasticizer-free electrolyte. This value was about half the capacitance obtained for a cell using aqueous liquid electrolyte. For the cells with plasticized nonaqueous electrolyte, almost the same level of capacity was achieved as that observed in the liquid electrolyte. In general, the specific capacitance decreased with an increase in the cycling current, but the rate dependence was not significant up to 2.0 mA cm^{-2}, which corresponds to several "*C* rate" for conventional battery systems. The rate capability shown in Fig. 5.2.36 corresponds to a capacitor performance (power density) of *ca.* 200 W kg^{-1} at an energy density of *ca.* 10 Wh kg^{-1}, when the calculation is based on the mass of the ACF electrodes.

In summary, the present nonaqueous polymeric gel was found to be applicable to practical capacitor systems, even for the limited gel composition and in a limited temperature range. Improvements in the low temperature properties (conductivity and elasticity) of the gel would enable realizing capacitors with high rate capability in a low temperature region.

References

1. C. Iwakura, N. Furukawa, T. Onishi, K. Sakamoto, S. Nohara and H. Inoue, *Electrochemistry (Tokyo, Jpn.)*, **69** (2001) 659.

2. C. Iwakura, S. Nohara, N. Furukawa and H. Inoue, *Solid State Ionics*, **148** (2002) 487.
3. N. Vassal, E. Salmon and J.-F. Fauvarque, *J. Electrochem. Soc.*, **146** (1999) 20.
4. N. Vassal, E. Salmon and J.-F. Fauvarque, *Electrochim. Acta*, **45** (2000) 1527.
5. J. O'M. Bockris and A. K. N. Reddy, *Modern Electrochemistry 1 Ionics*, Plenum Publishing Corporation, New York, pp. 489–493 (1998).
6. H. Wada, M. Horiuchi, S. Nohara, N. Furukawa, H. Inoue and C. Iwakura, *ITE Lett.*, in press.
7. D. A. MacInnes, *The Principles of Electrochemistry*, Reinhold Publishing Corporation, New York (1939),
8. M.N. Hull and H.I. James, *J. Electrochem. Soc.*, **124** (1977) 332.
9. S.M. Davis and M.N. Hull, *J. Electrochem. Soc.*, **125** (1978) 1918.
10. L.M. Baugh, J.A. Cook and J.A. Lee, *J. Appl. Electrochem.*, **8** (1978). 253
11. L.M. Baugh, J.A. Cook and F.L. Tye, *J. Power Sources*, **7** (1978) 519.
12. C. Iwakura, K. Ikoma, S. Nohara, N. Furukawa and H. Inoue, *J. Electrochem. Soc.*, **150** (2003) A1623.
13. C. Iwakura, K. Ikoma, S. Nohara, N. Furukawa and H. Inoue, *Electrochem. Solid-State Lett.*, in press.
14. C. Iwakura, Y. Kajiya, H. Yoneyama, T. Sakai, K. Oguro and H. Ishikawa, *J. Electrochem. Soc.*, **150** (1989) 1351.
15. T. Sakai, M. Matsuoka and C. Iwakura, *Handbook on the Physics and Chemistry of Rare Earths*, K. A. Gschneidner, Jr. and L. Eyring (Eds.), Elsevior Science, Amsterdam, Vol. 21, pp. 133–178 (1995).
16. H. Wada, S. Nohara, N. Furukawa, H. Inoue, N. Sugoh, H. Iwasaki, M. Morita and C. Iwakura, *Electrochim. Acta*, in press.
17. C. Iwakura, H. Wada, S. Nohara, N. Furukawa, H. Inoue and M. Morita, *Electrochem. Solid-State Lett.*, **6** (2003) A37.
18. S. Nohara, H. Wada, N. Furukawa, H. Inoue, M. Morita and C. Iwakura, *Electrochim. Acta*, **48** (2003) 749.
19. M. Morita, T. Fukumasa, M. Motoda, H. Tsutsumi, Y. Matsuda, T. Takahashi and H. Ashitaka, *J. Electrochem. Soc.*, **137** (1990) 3401.
20. J.-L. Qiao, N. Yoshimoto and M. Morita, *J. Power Sources*, **105** (2002) 45.
21. J.-L. Qiao, N. Yoshimoto, M. Ishikawa and M. Morita, *Electrochim. Acta*, **47** (2002) 3447.
22. J.-L. Qiao, N. Yoshimoto, M. Ishikawa and M. Morita, *Solid State Ionics*, **156** (2003) 415.
23. J. B. Goodenough, *Solid State Electrochemistry*, P. G. Bruce (Ed.), Cambridge University Press, Cambridge, Chapter 3 (1995).
24. H. Ericson, C. Svanberg, A. Brodin, A. M. Grillone, S. Panero, B. Scrosati and P. Jacobsson, *Electrochim. Acta*, **45** (2000) 1409.
25. A. M. Grillone, S. Panero, B. A. Retamal and B. Scrosati, *J. Electrochem. Soc.*, **146** (1999) 27.
26. J. R. Stevens, W. Wieczorek, D. Raducha and K. R. Jeffrey, *Solid State Ionics*, **97** (1997) 347.

27. D. Raducha, W. Wieczorek, Z. Florjanczyk and J. R. Stevens, *J. Phys. Chem.*, **100**, (1996) 20126.
28. A. M. Grillone, S. Panero, B. A. Retamal and B. Scrosati, *J. Electrochem. Soc.*, **146** (1999) 27.
29. M. Morita and J.-L. Qiao, *J. Korean Electrochem. Soc.*, **6** (2003) 141.
30. J.-L. Qiao, N. Yoshimoto, M. Ishikawa and M. Morita, *Chem. Mater.*, **15** (2003) 2005.
31. J.-L. Qiao, N. Yoshimoto, M. Ishikawa and M. Morita, *Electrochim. Acta*, **50** (2004) 837.

5.3 Construction of hydrogen storage alloy electrode/inorganic solid electrolyte interface

5.3.1 Preparation, characterization and application of phosphoric acid-doped silica gel electrolytes

Inorganic solid electrolytes for nickel/metal hydride (Ni/MH) batteries have scarcely been investigated in recent years because of their low proton conductivity. Phosphoric acid-doped (P-doped) silica gels with high proton conductivity of *ca.* 10^{-2} S cm^{-1} at room temperature were lately prepared by a sol-gel method [1–5]. The P-doped silica gels, therefore, are interesting materials in terms of a new solid electrolyte for use in Ni/MH batteries.

A typical preparation scheme for a P-doped silica gel (molar ratio of H_3PO_4 to SiO_2: 1.5) is shown in Fig. 5.3.1. The P-doping was basically carried out before gelation. The proton conductivity of P-doped silica gels increases with the increase in concentration of the doped phosphoric acid and humidity. The acidity of the P-doped silica gels also increases with the humidity, which is likely to cause corrosion of negative and/or positive electrode materials. The reduced water content in the P-doped silica gels will suppress corrosion and reduce proton conductivity. To determine the maximal water content must be a key point in realizing all-solid-state Ni/MH batteries with the P-doped silica gel electrolytes.

The drying in vacuum at different temperatures for the P-doped silica gel (P/Si = 1.5) led to a change in its crystal structure [6]. In the X-ray diffraction patterns shown in Fig. 5.3.2, at a drying temperature of 60°C or less, only one broad diffraction peak was observed, suggesting an amorphous structure. Over 60°C, the initial broad peak steadily disappeared and new sharp peaks assigned to crystalline $Si_5O(PO_4)_6$ appeared. The conductivity of the P-doped silica gel increased with the decreasing drying temperature (Fig. 5.3.3), suggesting that the water content in the P-doped silica gels dried at different temperatures influences the conductivity. The water content was determined by thermal analyses. In the thermogravimetry (TG) curves for the P-doped silica gel dried in vacuum at different temperatures (Fig. 5.3.4), the endothermic weight loss was ascribed to the evaporation of water from the gel. In each case, the evaporation process proceeds in two steps, *i.e.*, a first evaporation process at lower than 140°C and a second one higher than 140°C. The

172   5 Construction of solid / solid interface for battery application

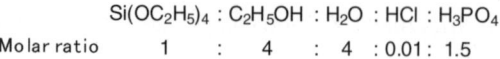

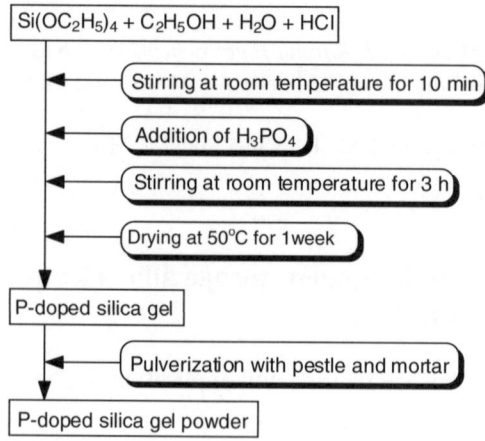

Fig. 5.3.1. A typical preparation scheme for P-doped silica gel (moler ratio of H_3PO_4 to SiO_2: 1.5).

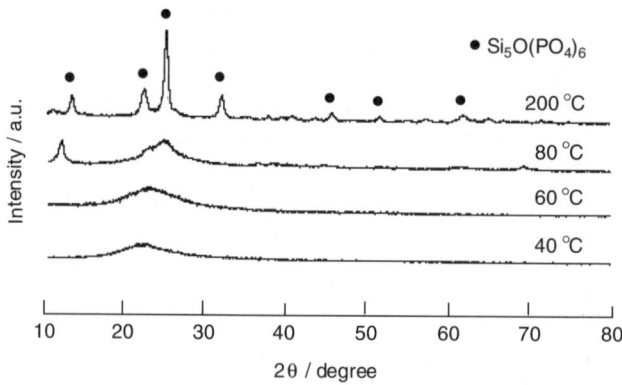

Fig. 5.3.2. X-ray diffraction patterns for P-doped silica gels dried in vacuum at different temperatures.

evaporation at the lower temperature is assigned to that of physically adsorbed water, while the evaporation at the higher temperature is assigned to that of water strongly adsorbed on phosphoric acid [5]. The weight loss in each TG curve corresponds to the percentage of water evaporated from the gel, *i.e.*, the water content of the gel. The weight loss in each step and the total weight loss were determined from the TG curves (Fig. 5.3.5). The weight loss at the lower temperature slightly increased with the decreasing drying temperature, whereas that at the higher temperature

5.3 Construction of hydrogen storage alloy electrode / inorganic solid electrolyte interface   173

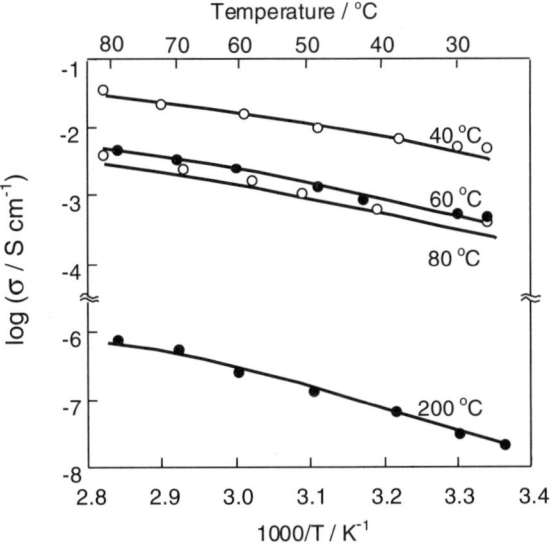

Fig. 5.3.3. Temperature dependence of logarithmic conductivity for P-doped silica gel pellets dried in vacuum at different temperatures. Weight of gel pellet: 0.3 g.

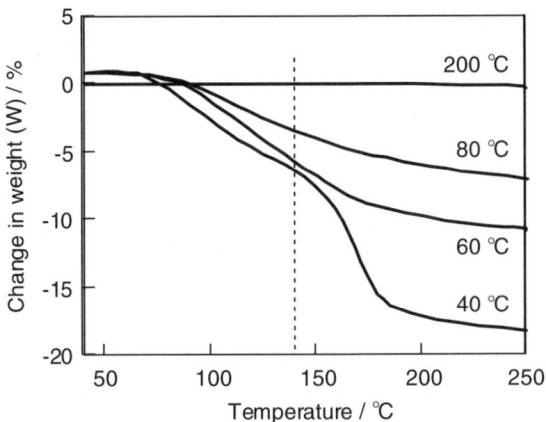

Fig. 5.3.4. TG and DTA curves of P-doped silica gels dried in vacuum at different temperatures. Heating rate: 10 K min^{-1}.

significantly increased. The logarithmic conductivity at 30°C showed dependence of the drying temperature similar to the weight loss at the higher temperature. This strongly suggests that water adsorbed in three-dimensionally continuous micropores of the gel significantly influences the proton conductivity.

The preparation of negative and positive electrodes and the fabrication of all-solid-state Ni/MH batteries were carried out according to Fig. 5.3.6. The charge-discharge cycle performance for the all-solid-state Ni/MH batteries is shown in Fig.

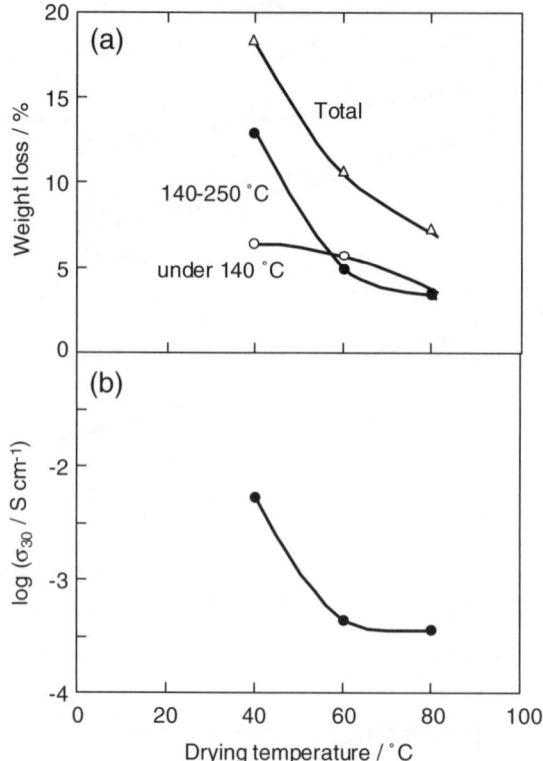

Fig. 5.3.5. Weight loss (a) and logarithmic conductivity at 30°C (b) for P-doped silica gels as a function of drying temperature.

5.3.7. The batteries using the gels dried at 40°C, 60°C and 80°C worked well. The best cycle performance was observed in a battery using the gel dried at 60°C.

Phosphoric acid doped in the gel can influence not only proton conductivity but also the corrosion rate, which should influence the cycle performance of the all-solid-state Ni/MH batteries. Therefore, P-doped silica gels with different concentrations of phosphoric acid (P/Si = 1.5, 1.0, 0.5) were prepared. The XRD spectra for gels with different P/Si ratios showed that the crystallization began at lower drying temperatures for the gels with a higher P/Si ratio. Regardless of the P/Si ratio, two kinds of weight losses were observed in any TG curve. The weight loss at the lower temperature showed a minimum for the gel with P/Si = 1.0, while that at the higher temperature showed a maximum. In contrast, conductivity was increased with the P/Si ratio (Fig. 5.3.8) due to the increase in the concentration of doped phosphoric acid or proton concentration.

The charge-discharge cycle performance was markedly improved by using an equimolar mixture of an alloy and its hydride as a negative electrode material and nickel hydroxide and nickel oxyhydroxide as a positive electrode material (Fig.

5.3 Construction of hydrogen storage alloy electrode / inorganic solid electrolyte interface 175

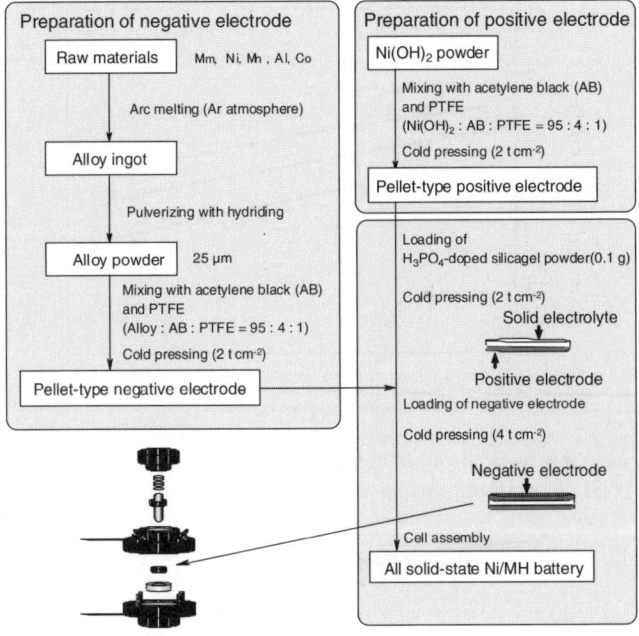

Experimental cell assembly

Fig. 5.3.6. A scheme for the preparation of negative and positive electrodes and the fabrication of all-solid state Ni/MH batteries using the P-doped silica gels dried in vacuum at different temperatures.

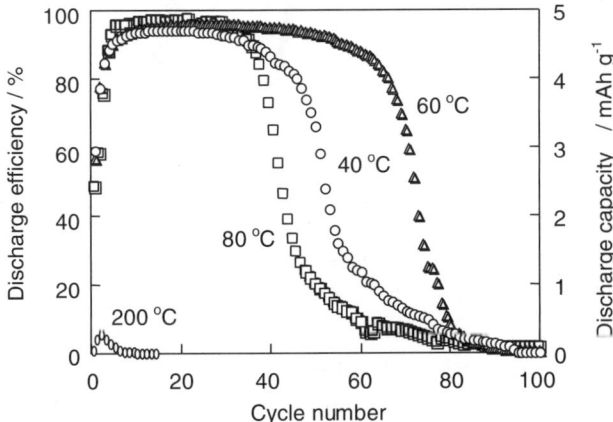

Fig. 5.3.7. Discharge efficiency and discharge capacities of all solid-state Ni/MH batteries using P-doped silica gels dried in vacuum at different temperatures as a function of cycle number. Charge condition: 5 mA g(alloy)$^{-1}$, 1 h. Discharge condition: 5 mA g(alloy)$^{-1}$, 0.5 V. Rest time: 10 min.

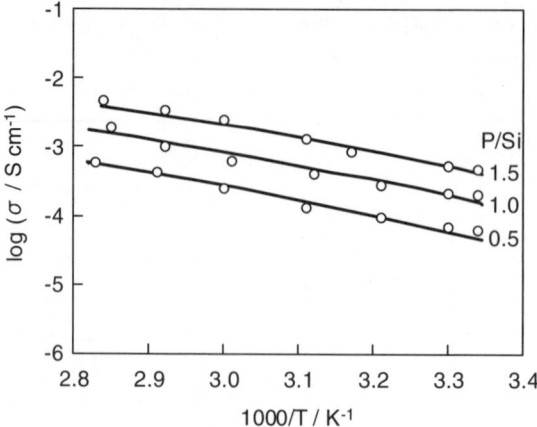

Fig. 5.3.8. Temperature dependence of logarithmic conductivity for P-doped silica gel pellets with different P/Si ratios. Drying condition: 60°C in vacuum, 1h.

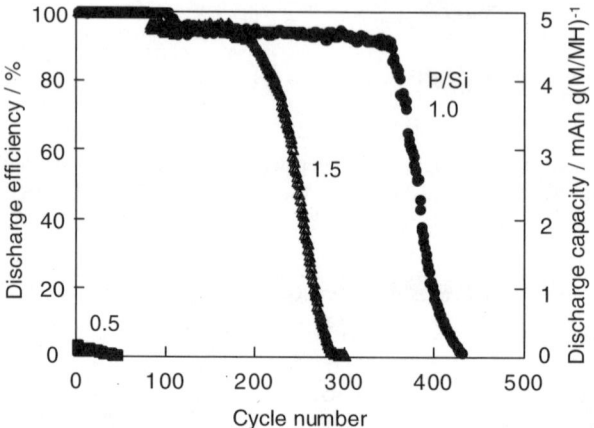

Fig. 5.3.9. Discharge efficiency and discharge capacities of all solid state Ni/MH batteries using P-doped silica gels with different P/Si ratios as a function of cycle number. Drying condition of gels: 60°C in vacuum, 1 h. Charge condition: 5 mA g(M/MH)$^{-1}$, 1 h. Discharge condition: 5 mA g(M/MH)$^{-1}$, 0.5 V. Rest time: 10 min.

5.3.9). The battery using the gel with P/Si = 0.5 showed a poor cycle performance probably due to low conductivity of the gel. The decay of discharge efficiency and discharge capacity for the gel with P/Si = 1.5 or 1.0 proceeded in two steps, a small decay from 100% to around 95% and a large one from around 95% to 0%. During the former decay, the plateau voltage in the discharge curve hardly lowered, and the plateau region shortened due to the very slow diffusion of H$^+$ in NiOOH [7]. On the other hand, during the latter decay, the plateau voltage steadily lowered and then the plateau region inclined, indicating an increase in the charge transfer resistance. The

5.3 Construction of hydrogen storage alloy electrode / inorganic solid electrolyte interface 177

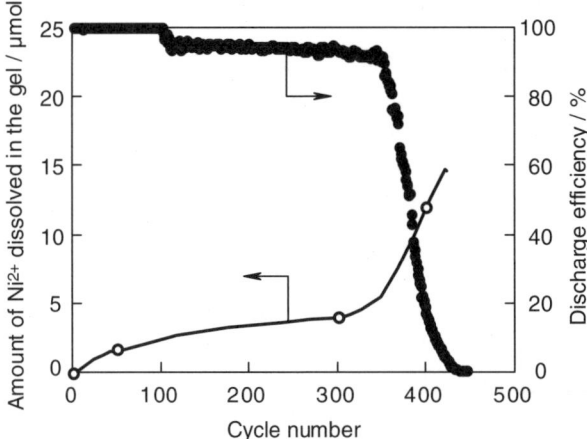

Fig. 5.3.10. Amount of Ni^{2+} dissolving from both electrode materials into the gel with P/Si=1.0 and discharge efficiency as a function of cycle number.

cycle life was prolonged in the order of the P/Si ratio of 0.5 < 1.5 < 1.0, suggesting that the cycle life was influenced not only by the conductivity of the gels but also by the change in the interfaces of electrode material/gel during the continuous charging and discharging. The interfacial change caused by corrosion and the vaporization of water from the gels is likely to cause serious problems in charge and discharge cycles. Ni^{2+} dissolved out in the gel with P/Si = 1.0 during the charge-discharge cycles (Fig. 5.3.10). This can be ascribed to the increase in the charge transfer resistance between the negative or positive electrode and the gel originating from the corrosion of the active materials.

5.3.2 Preparation, characterization and application of inorganic oxide solid electrolytes

There have been only a few reports for all-solid-state batteries using inorganic solid electrolytes, as pointed out in section 5.3.1. The performance of solid-state batteries consisting of inorganic solid electrolytes was inferior to that of polymer systems because of their lower ionic (OH^- or H_3O^+) conductivities. However, it is well-known that heteropolyacid hydrates, such as $H_3PW_{12}O_{40} \cdot 29H_2O$ and $H_3PMo_{12}O_{40} \cdot 29H_2O$, show higher proton conductivities at room temperature than other proton conductors [8]. Therefore, the fabrication of Ni/MH and the related batteries has been attempted using these electrolytes [9, 10]. Because heteropolyacid hydrates exhibit strong acidity, the suitability of the electrode materials to the hydrate electrolyte was also tested.

Heteropolyacid hydrates of $H_3PW_{12}O_{40} \cdot nH_2O$ and $H_3PMo_{12}O_{40} \cdot nH_2O$ (Wako Pure Chemical Industries, Ltd.) were used as electrolytes of the metal hydride batteries without any pretreatment. Before assembling the battery, the hydration

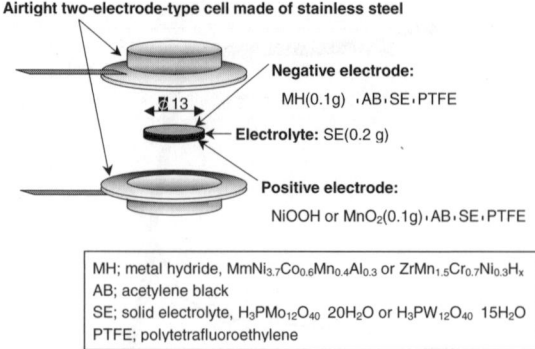

Fig. 5.3.11. A schematic diagram for the construction of negative and positive electrodes and the setup of solid-state MH batteries using the heteropolyacid hydrates.

number of n was determined to be 21 and 29 for $H_3PW_{12}O_{40} \cdot nH_2O$ and $H_3PMo_{12}O_{40} \cdot nH_2O$, respectively, using a thermobalance. These values decreased to 15 and 20, respectively, when the samples were left in the electrochemical test cell overnight at 30°C. According to the relationship between temperature and humidity investigated by Nakamura et al. [11], the humidity appears to be around 65 % in the test cell.

AC conductivity was measured at 30°C by an impedance meter. In this case, the heteropolyacid hydrate powders were isotropically pressed to obtain dense pellets, which were placed in a gas-tight stainless steel cell to prevent the loss of water content from the hydrates. The conductivity of $H_3PMo_{12}O_{40} \cdot nH_2O$ was estimated to be *ca.* 3.3×10^{-2} S cm^{-1}, which was nearly equal to the value of $H_3PMo_{12}O_{40} \cdot 18H_2O$ in the literature [11]. Thus, the hydration number estimated from thermogravimetry almost coincided with that from conductivity. As for the conductivity of $H_3PW_{12}O_{40} \cdot 15H_2O$, it is presumed to be *ca.* 1×10^{-2} S cm^{-1} according to the hydration number obtained. The difference in the conductivities between the two heteropolyacids would be essentially due to the hydration numbers.

The construction of negative and positive electrodes and the setup of solid-state MH batteries are shown in Fig. 5.3.11. Figure 5.3.12 shows the charge-discharge curves of the test cell consisting of $MmNi_{3.6}Al_{0.4}Mn_{0.3}Co_{0.7}H_x$ / $H_3PW_{12}O_{40} \cdot 15H_2O$/ NiOOH (Mm: misch metal). In a few initial cycles, the cell could be discharged starting at around 1.3 V which is the theoretical voltage of the Ni/MH battery, indicating that the heteropolyacid hydrate can serve as an electrolyte for the Ni/MH battery. Considering that the cell discharging stopped after several cycles, the charge-discharge cyclability was very poor. This poor cyclability would be due to the corrosion of the negative electrode material by the electrolyte with a strong acidic character. Consequently, the misch metal-based alloy was replaced with a zirconium-based alloy which is comparatively resistant to acids. The cyclability was somewhat improved as compared with the previous cell. However, the performance was still insufficient.

5.3 Construction of hydrogen storage alloy electrode / inorganic solid electrolyte interface 179

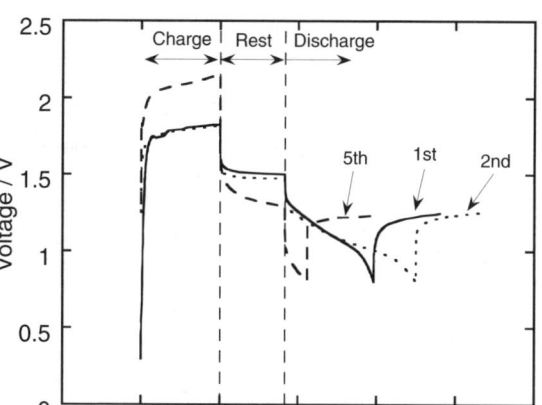

Fig. 5.3.12. Charge-discharge curves for the battery, $MmNi_{3.6}Al_{0.4}Mn_{0.3}Co_{0.7}H_x / H_3PW_{12}O_{40} \cdot 15H_2O / NiOOH$. Charge and discharge current densities are 50 mA g(alloy)$^{-1}$ and 20 mA g(alloy)$^{-1}$, respectively.

Based on the demonstration by Mohri et al., the cycle life performance was improved by using manganese dioxide for the positive electrode in the solid-state Ni/MH battery employing acidic $Sb_2O_5 \cdot nH_2O$ as an electrolyte [12, 13]. Manganese dioxide, MnO_2, was then employed in the cell instead of nickel oxyhydroxide, NiOOH. As a result, a long life of over 200 cycles was achieved with the resulting cell (charge-discharge current density was 5 mA g^{-1}). However, the discharge efficiency was low and charging was impossible at higher current densities. This indicates that the high rate charge and discharge performance depends on the rate of reaction in the positive active materials.

Considering the higher ionic conductivity of 12-molybdophosphoric acid hydrate under the electrochemical test conditions, the battery characteristics were investigated using the hydrate as an electrolyte. Based on the results for the battery using 12-tungstophosphoric acid hydrate, $ZrMn_{1.5}Cr_{0.7}Ni_{0.3}$ and MnO_2 were employed as the negative and positive electrode materials. Figure 5.3.13 illustrates the charge-discharge curves obtained with the test cell consisting of $ZrMn_{1.5}Cr_{0.7}Ni_{0.3}H_x / H_3PMo_{12}O_{40} \cdot 20H_2O / MnO_2$ at a current density of 5 mA g^{-1}. During the initial several cycles, the cell exhibited relatively low voltage profiles due to insufficient activation of the electrodes. Afterwards, up to the 100th charge-discharge, the voltage profiles appeared to be steady. The open circuit voltages of the cell after charging were lower than those of the conventional Ni/MH battery because of the lower electrode potential of MnO_2 than NiOOH. Gentle voltage drops were observed just after the beginning of discharge, suggesting that $H_3PMo_{12}O_{40} \cdot 20H_2O$ would be a desirable electrolyte for solid-state metal hydride batteries. However, the overpotential increased somewhat after the 100th cycle.

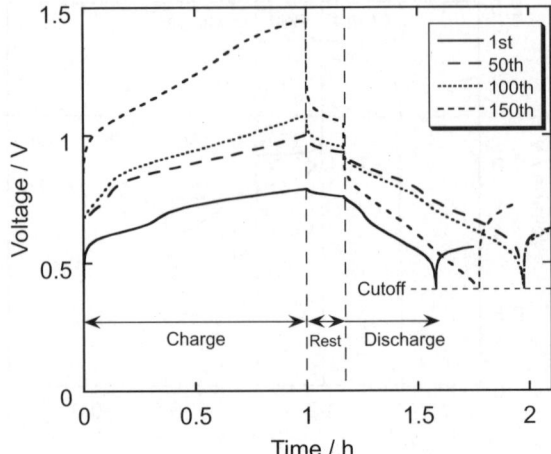

Fig. 5.3.13. Charge-discharge curves for the cell, $ZrMn_{1.5}Cr_{0.7}Ni_{0.3}H_x$ / $H_3PMo_{12}O_{40} \cdot 20H_2O$ / MnO_2. Charge and discharge current densities are 5 mA g(alloy)$^{-1}$.

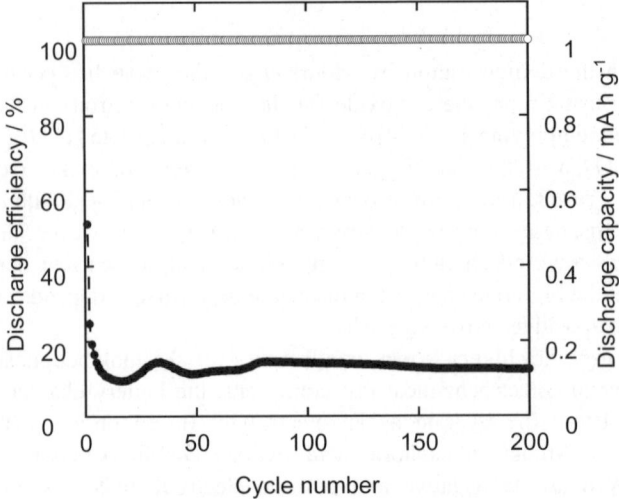

Fig. 5.3.14. Discharge efficiency and capacity variations with the number of charge-discharge cycle for the cells using 12-molybdophosphoric acid hydrate and 12-tungstophosphoric acid hydrate. Charge and discharge current densities are 1 mA g(alloy)$^{-1}$. ○ ; $ZrMn_{1.5}Cr_{0.7}Ni_{0.3}H_x$ / $H_3PMo_{12}O_{40} \cdot 20H_2O$ / MnO_2, ● ; $ZrMn_{1.5}Cr_{0.7}Ni_{0.3}H_x$ / $H_3PW_{12}O_{40} \cdot 15H_2O$ / MnO_2.

Figure 5.3.14 displays the discharge efficiency and capacity variations with the number of charge-discharge cycle for the cells using 12-molybdophosphoric acid and 12-tungstophosphoric acid at a current density of 1 mA g^{-1}, where the discharge efficiency is defined as the ratio of measured discharge capacity (right axis in the figure) to the quantity of charged electricity. The test cell using 12-

5.3 Construction of hydrogen storage alloy electrode / inorganic solid electrolyte interface 181

molybdophosphoric acid could be discharged up to 200 cycles, and the discharge efficiency held at *ca.* 100%. On the other hands, the cell using 12-tungstophosphoric acid did not show such a superior performance. As expected above, the higher conductivity of the electrolyte would induce a good performance.

Some improvements in electrochemical properties, such as coulomb efficiency and high rate and cycle life performances, have been achieved by increasing the electrode-electrolyte interface and/or employing electrolytic manganese dioxide as the positive electrode material instead of the reagent grade manganese dioxide used previously [10]. Figure 5.3.15 comparatively shows the 50th charge-discharge curves for the MnO_2 (reagent grade, Wako Pure Chemical Industries, Ltd.) / MH test cells obtained by two different pressing methods. The one-step press means that the compacted powders of electrode and electrolyte materials were pressed at 1×10^3 kgf cm^{-2} at a time to obtain a pellet. This process was completely different from the previous way of repeatedly loading and pressing each component three times in turn (three-steps press). The cell using the one-step press in the present study showed a lower overvoltage and longer discharge time compared with the previous battery prepared by the three-step press process. This would be due to the fact that the cell constructed in this study had a rough electrode/electrolyte interface to smooth the electrode reaction. The cycle life performance was, however, still not sufficient, though the discharge efficiency increased by *ca.* 5 %.

The poor cyclability of the cell might be due to the cathode material. Considering that the following reaction occurs on the MnO_2 electrode in acid electrolyte,

$$MnO_2 + H_3O^+ + e^- \rightarrow MnOOH + H_2O \qquad (5.3.1)$$

the low reactivity with protons and the subsequent low proton diffusivity would cause severe problems. Thus, the cathode material was changed from reagent grade manganese dioxide to electrolytic manganese dioxide used for commercial batteries (Mitsui Mining & Smelting Co., Ltd.). Figure 5.3.16 shows the variations in the discharge efficiency with charge-discharge cycling. It is noteworthy that the discharge efficiency increases to *ca.* 90% and the cell cycle life is extended to *ca.* 450 cycles. Figure 5.3.17 shows the 50th and 450th charge-discharge curves of the present cell. In the previous cell using reagent grade manganese dioxide, the terminal voltage exceeded 1.4 V at the end of the 150th charge. In the present cell using the electrolytic manganese dioxide, the polarization was not very high even at 450 cycles. These facts indicate a smooth electrode reaction on the cathode and suggest that the diffusion of protons into the electrode bulk is very important to improve the cycle life performance. Considering that the average valence states of manganese were 3.75 in the reagent grade manganese dioxide and 3.92 in the electrolytic manganese dioxide, the proton diffusion might be disturbed by manganese sesquioxide partially present in the reagent grade manganese dioxide as is generally known.

The influence of the atmospheric humidity on cell performance was additionally examined. After the usual charge-discharge test (200 cycles at a current density of 5 mA g(alloy)$^{-1}$), the test cell was opened and left in the atmosphere for 30 days. Thereafter, the battery performance was tested again. As a result, the discharge

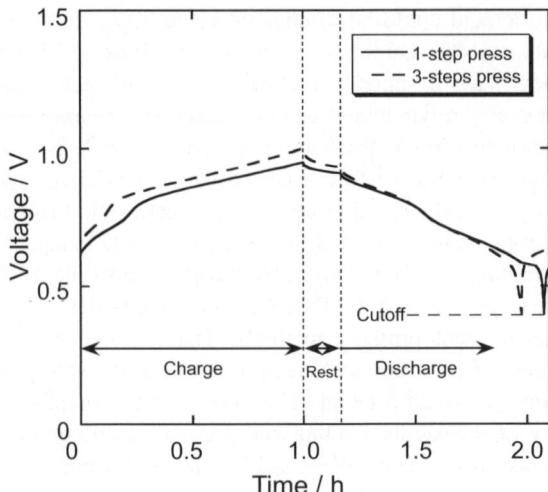

Fig. 5.3.15. The 50th charge-discharge curves for the cell, $ZrMn_{1.5}Cr_{0.7}Ni_{0.3}H_x$ / $H_3PMo_{12}O_{40} \cdot 20H_2O$ / MnO_2(regent grade), obtained in two different pressing methods. Charge and discharge current densities are 5 mA g(alloy)$^{-1}$.

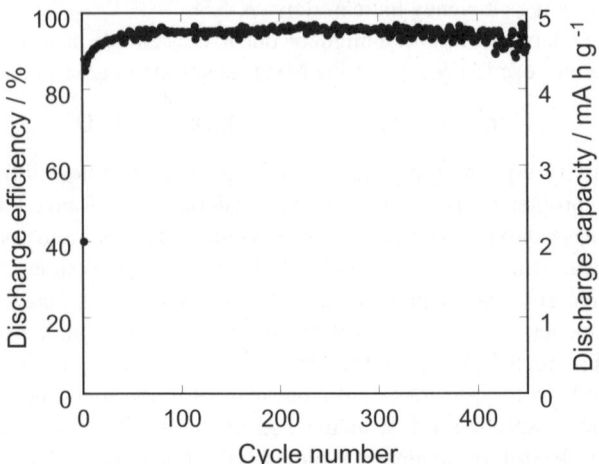

Fig. 5.3.16. Discharge efficiency variation with the cycle numbers for the battery using electrolytic manganese dioxide as a positive electrode material. Charge and discharge current densities are 5 mA g(alloy)$^{-1}$.

efficiency was almost the same as before, which showed that the battery was not influenced very much by normal humidity and temperature change. This result may be favorable for easy treatment and storage of the batteries.

Next, the cyclability of the cell was tested at a higher current density of 10 mA g(alloy)$^{-1}$. Unfortunately, the discharge efficiency decreased to 80% accompanied

5.3 Construction of hydrogen storage alloy electrode / inorganic solid electrolyte interface   183

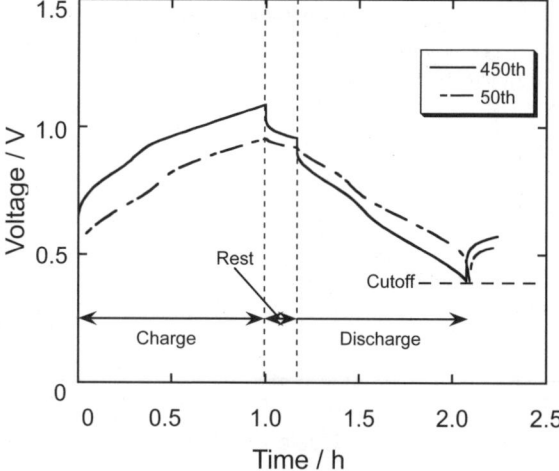

Fig. 5.3.17. Charge-discharge curves for the battery using electrolytic manganese dioxide as the positive electrode material. Charge and discharge current densities are 5 mA g(alloy)$^{-1}$.

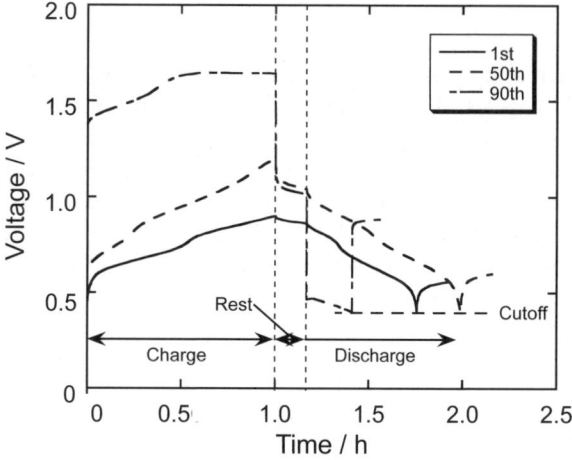

Fig. 5.3.18. Charge-discharge curves for the battery using electrolytic manganese dioxide at a higher current density of 10 mA g(alloy)$^{-1}$.

by cell life termination within 100 cycles. Figure 5.3.18 shows the 1st, 50th and 90th charge-discharge curves, where a very high overvoltage appears at the 90th charge and discharge curves. This suggests that the amount of electrode/electrolyte interface, *i.e.*, the reaction area, was not sufficient to sustain the current density.

To improve the high rate charge-discharge characteristics, the interfacial area was increased by changing the content of solid electrolyte in the positive electrode from 5 wt.% to 10, 20 and 40 wt.%. The discharge efficiencies for the cells tested at a current density of 10 mA g(alloy)$^{-1}$ are shown in Fig. 5.3.19. The cycle life is substantially

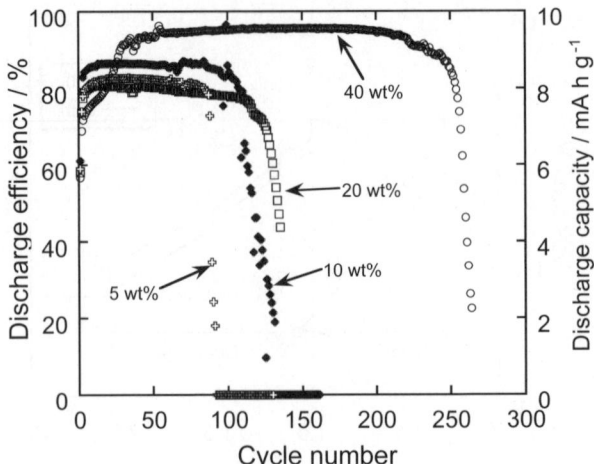

Fig. 5.3.19. Discharge efficiency variations with the cycle numbers for the battery with different electrolyte contents in the positive electrode. Charge and discharge current densities are 10 mA g(alloy)$^{-1}$.

improved by increasing the electrolyte contents up to 40 wt.%; the cell was able to operate over 200 cycles, maintaining the discharge efficiency at *ca.* 95%, and the resting voltage was almost the same as that at the end of charging. This emphasizes the importance of optimizing the electrode/electrolyte interface to improve the cell performance. Considering the above results, the present battery performance was superior to that of the MnO$_2$/MH solid-state battery using Sb$_2$O$_5$ · nH$_2$O [12, 13] or (CH$_3$)$_4$NOH · 5H$_2$O [14–16] and was comparable to that of the Ni/MH solid-state battery using H$_3$PO$_4$-doped silica gel [6].

References

1. A. Matsuda, H. Honjo, M. Tatsumisago and T. Minami, *Chem. Lett.*, **1989**, 153.
2. K. Hirata, A. Matsuda, T. Hirata, M. Tatsumisago and T. Minami, *J. Sol-Gel Sci. Technol.*, **17** (2000) 61.
3. A. Matsuda, T. Kanzaki, Y. Kotani, M. Tatsumisago and T. Minami, *Solid State Ionics*, **139** (2001) 113.
4. A. Matsuda, T. Kanzaki, K. Tadanaga, M. Tatsumisago and T. Minami, *Electrochim. Acta*, **46** (2001) 939.
5. A. Matsuda, T. Kanzaki, K. Tadanaga, M. Tatsumisago and T. Minami, *Solid State Ionics*, **154–155** (2002) 687.
6. C. Iwakura, K. Kumagae, K. Yoshiki, S. Nohara, N. Furukawa, H. Inoue, T. Minami, M. Tatsumisago and A. Matsuda, *Electrochim. Acta*, **48** (2003) 1499.
7. K. P. Ta and J. Newman, *J. Electrochem. Soc.*, **145** (1998) 3860.
8. O. Nakamura, T. Kodama, I. Ogino and Y. Miyake, *Chem. Lett.*, **1979**, 17.

9. K. Hatakeyama, H. Sakaguchi, K. Ogawa, H. Inoue, C. Iwakura and T. Esaka, *J. Power Sources*, **124** (2003) 559.
10. K. Hatakeyama, H. Sakaguchi, T. Yamaguchi, H. Inoue, C. Iwakura and T. Esaka, *Electrochemistry*, **72** (2004) 697.
11. O.Nakamura *et al.*, *Report of the Government Industrial Research Institute, Osaka*, No.360 (1982).
12. M. Mohri, Y. Tajima, H. Tanaka and T. Yoneda, *Sharp Tech. J.*, **34** (1986) 97.
13. T. Yoneda, S. Satoh and M. Mohri, *Sharp Tech. J.*, **38** (1987) 55.
14. N. Kuriyama, T. Sakai, H. Miyamura, A Kato and H. Isikawa, *Denki Kagaku*, **58** (1990) 89.
15. N. Kuriyama, T. Sakai, H. Miyamura, A Kato and H. Isikawa, *J. Electrochem. Soc.*, **137** (1990) 355.
16. N. Kuriyama, T. Sakai, H. Miyamura, A Kato and H. Isikawa, *Solid State Ionics*, **40/41** (1990) 906.

5.4 Conclusions

Information obtained from the studies described in this chapter is summarized as follows.

The polymer hydrogel electrolyte prepared from potassium salt of crosslinked poly(acrylic acid) and a KOH aqueous solution had the high ionic conductivity (*ca.* 0.6 S cm^{-1} at 25°C) and a wide potential window, nearly comparable to the KOH aqueous solution. The oxygen permeability of the polymer hydrogel electrolyte was excellent at practical thickness level. The electrolyte creepage along nonpolarized and negatively polarized metal surfaces was markedly suppressed by using the polymer hydrogel electrolyte. Experimental negative and positive electrode capacity-controlled Ni/MH cells assembled using the polymer hydrogel electrolyte exhibited equivalent or better charge-discharge characteristics, compared with those assembled using the KOH aqueous solution. Capacity retention characteristics of the cells were fairly improved by using the polymer hydrogel electrolyte, and the difference in the capacity retention between the cells with the polymer hydrogel electrolyte and the KOH aqueous solution was more remarkable at relatively high temperatures. The polymer hydrogel electrolyte used in this study therefore seems to have the applicability to alkaline secondary batteries such as a Ni/MH battery. Moreover, an experimental cell of electric double layer capacitor (EDLC) assembled using the polymer hydrogel electrolyte exhibited slightly higher capacitance and better cycle performance than that with a KOH aqueous solution. The high-rate dischargeability of the EDLC cell with the polymer hydrogel electrolyte was also excellent, comparable to that in case of the KOH aqueous solution.

For a novel polymeric gel electrolyte system, poly(ethylene oxide)-modified polymethacrylate (PEO-PMA) matrix with linear polyether (PEGDE) doped with H_3PO_4 from aqueous solutions, *ca.* 2×10^{-2} S cm^{-1} of the conductivity was obtained at room temperature. The contents of acid and water in the gel gave critical influences on the conductance behavior of the resulting hydorogel. Nonaqueous polymeric gel complexes composed of poly(ethylene oxide)-modified polymethacrylate (PEO-PMA) dissolving anhydrous H_3PO_4 gave also high ionic conductivity of $\sim 10^{-3}$ S cm^{-1} at 70°C. An EDLC test cell with the gel electrolyte showed as high capacity as that with an aqueous liquid electrolyte: high rate capability was obtained for the cell at 90°C.

The conductivity of the P-doped silica gel increases with decreasing drying temperature, and the temperature dependence of the conductivity was similar to that of total water content. The all-solid-state Ni/MH battery using the P-doped silica gel dried in vacuum at 60°C as an electrolyte worked well during *ca.* 80 charge-discharge cycles. The distribution of water molecules in the P-doped silica gels and conductivity depended on the P/Si ratio, that is H^+ concentration. The all-solid-state Ni/MH battery using the P-doped silica gel with P/Si = 1.0 as an electrolyte showed the longest cycle life (*ca.* 420 cycles) in the present study, and the decay of discharge capacity closely related to the corrosion of both electrode materials.

Heteropolyacid hydrates were found to work as electrolytes for solid-state NiOOH/MH and MnO_2/MH batteries. The electrode materials compatible with the electrolytes were manganese dioxide and zirconium-based alloy for positive and negative electrodes, respectively. The performance of MnO_2/MH battery using $H_3PMo_{12}O_{40} \cdot 20H_2O$ was enhanced by improvement of the pressing method and employment of electrolytic manganese dioxide. As a result, the discharge efficiency at 5 mA g(alloy)$^{-1}$ increased to 90% and the cell cycle life was extended to 450 cycles. Furthermore, the high rate characteristic of the battery was improved significantly by optimizing the electrolyte content in the positive electrode.

6

Polymer solid electrolytes for lithium-ion conduction

6.1 Introduction: Role of rubbery state for ionic conduction

Developments of polymeric functionality materials have been focused, taking advantages of polymeric materials in general, *e.g.* light weight and excellent processability compared with inorganic and metallic materials. This trend is recently much intensified by the coming era of nanotechnology [1–3]. Among polymeric materials (fiber, plastics and rubber), rubbery materials, which are often called elastomers, are unique due to their mechanical property *i.e.* rubber elasticity, or from thermodynamic viewpoint, entropic elasticity [4]. This unique property itself is an excellent functionality because of an extraordinarily high elongation and rapid recovery to the original size upon the release of stress. However, it is sometimes difficult to develop a device showing both rubber elasticity and a needed high functionality as shown in Table 6.1.1. Among these functionalities, electronic or ionic conduction [5, 6] and blood-compatibility [7] have been investigated on elastomeric materials. In order to preserve rubber elasticity, a certain compromise may be necessary in the design of elastomeric functionality materials.

The other but highly relevant nature of rubbers is their being amorphous both at lower and higher temperatures than glass-transition temperature (T_g). Actually this is prerequisite to the high elasticity of elastomers above T_g, which is thermodynamically due to entropy term. The fundamental key to the design of elastomeric functionality materials is to take advantage of being a soft and flexible (above T_g) amorphous matrix for a specific functionality [5–11].

In the present chapter, use of amorphous matrices from poly(oxyethylene)s and from ionene polymers consisting of oxytetramethylene units for specific functions, especially lithium ion-conducting, is described based on the results from our group during the past five or so years. The elastomers are classified into two according to how the three-dimensional network structure is organized: Crosslinked rubbers and thermoplastic elastomers (TPE). The formers are due to covalent bonding by a chemical reaction (crosslinking reaction), and when raw rubbers are crosslinked by sulfur, the resulted networks are often called vulcanizates. As the crosslinkers, use of organic peroxides is also popular other than sulfur. The network structure in the latter,

Table 6.1.1. Functionalities relevant to elastomers.

Area	Functionality
Chemical	environmental degradability, ablation, polymer catalyst, ion exchange, gas or liquid permeation, gas or ion sensor, chemomechanical, chemoluminescence
Thermal	non-flammability, low thermal expansion, high thermal conduction, pyroelectricity, thermochromism
Mechanical	vibration isolation, control of vibration, mechanochemical, piezoelectricity
Photonic	transparency, control of refractive index, non-linear optics, photochromic, photochemical, photomechanical, photoconductive
Radiation	anti-rad, photo-resist, X-ray or electron beam resist electronic ferroelectric, semi-conductive, electron- or ion-conductive, electrochromism, electroluminescence
Magnetic	ferromagnetic, organic magnet, magnetic adhesive
Biomedical	biocompatibility, antithrombogenic, polymer drug, controlled release or targeting of drug, enzyme immobilization, biosensor, biodegradability, bio-absorbability

TPE, is not due to covalent bonds but due to physical interactions, such as hydrogen bonding, the Coulombic interactions, or any other intermolecular interactions not based on chemical origins. The network in this category is more or less transient or temporary unlike the vulcanizates which are the base of recyclability of TPE as a material. In our studies, we have included network structures as possible materials for an ionic conductor: Poly(oxyethylene)s are subject to crosslinking reaction using an organic peroxide. The ionene polymers based on oxytetramethylene units are one class of TPE whose origin of network structure is aggregation of ionic groups in the ionenes. T_gs of amorphous poly(oxyethylene) and poly(oxytetramethylene) are well below room temperature. Therefore, we can take advantage of soft and flexible matrix by using them, which may support our view that elastomers can be an excellent matrix for soft functionality materials under ambient conditions [5, 6, 8].

There may be two factors for fast ionic conduction in general. Traditionally, polymeric materials have been used as electronically insulating materials. This trend is changing as suggested by the Nobel Prize Winner in Chemistry, Dr. H. Shirakawa, in 2000 for electron-conducting polymers *e.g.* poly(acetylene)s. In the studies on poly(ethylene) as an excellent insulating material, the following equation was proposed for the ionic conduction in the polymer matrix [12, 13], which was the reason of dielectric breakdown.

$$\sigma = \sigma_0 \exp[-\{\gamma v_1^*/v_f + (E + W/2\varepsilon)/kT\}] \quad (6.1.1)$$

where σ is the conductivity, σ_0 is a constant, γ is the numerical factor to correct the overlap of free volume, v_f is the free volume, v_1^* is the critical volume needed for migration of ions, E is the potential of barrier for ionic migration, W is the

6.1 Introduction: Role of rubbery state for ionic conduction

ionic dissociation energy, ε is the dielectric constant and k is Boltzmann's constant. The first term of Equation 6.1.1 is related to the free volume of polymer, which is dominating at a higher temperature region than glass-transition temperature (T_g) for ionic conduction. On the other hand, the second term means that the ionic conduction occurs between the lattice vacancies in the ionic crystal. This term is related with the ionic conductivity at the lower temperature region than T_g. Therefore, the first term and the second term provide us with the suggestions about the molecular design for an excellent polymer solid electrolyte. The polymeric material with lower T_g, *i.e.*, elastomer is preferred for the design based on the first term, and the construction of ion-conducting columns (often by crystalline structures) assisted by hopping mechanism is on the second.

We have been studying the elastomeric polymers for the preparation of ion-conducting materials, which have been conducted per the molecular design based on the first term in Equation 6.1.1. In this case, an elastomer is the best choice as a matrix for ion conduction. This is because the rubbery state (amorphous and at lower temperature than T_g) is actually a liquid state, even though it is regarded as a solid from its appearance [14]. For practical purposes, giving a network structure to the rubbery state may be useful and in many cases necessary for mechanical stability and strength [4, 8, 9]. Poly(ethylene oxide) (PEO) is one of the well known water-soluble polymers. Its chemical structure is very simple *i.e.* repeating unit is oxyethylene (CH_2CH_2O). However, depending on polymerization conditions including the catalyst used, we obtain the products of various average molar masses. Additionally, PEO is a highly crystalline polymer. Due to these factors, appearance of PEO ranges from an easily water-soluble liquid to crystalline hard solids. Oligomeric PEO whose molar mass is lower than *ca.* 1,000 is not crystalline, but PEOs of the higher molar mass are highly crystallizable. Crystalline structure of PEO was already reported as Form I monoclinic system and a 7/2 helical structure [15, 16].

For many functionality materials, rubbery matrix is of much use. Take an example of ion-conducting polymer or polymer solid electrolyte: The polymer electrolytes are to be used in all-solid-state (secondary) batteries instead of an electrolytic solution. Different from electron conduction, ion is a mass and diffusion of the mass is very slow in solid state, *i.e.*, mobility of ions in the solid is very low. The conductivity (σ) is expressed as

$$\sigma = \sum_j q_j n_j \mu_j \qquad (6.1.2)$$

where several kinds of ions are present and j means the *j*-th ion. The σ is a summation of the contribution from all kinds of ions. q_j is a charge number of ion and q_j is an e (elementary electronic charge) in the case of the univalent ion, $2e$ for divalent ones, *etc*. The n_j is a number of carriers, and μ_j is mobility of ions. For electron conductors number of the carrier (n) is important for conductivity, because electron is very small to afford very high mobility. In case of ion conductors, the mobility μ is much concerned, which is extremely low in solids. This is the reason why electrolyte solutions are still in use for batteries, which is against a strong trend in electronics industries toward all solid state.

Rubbery state is both amorphous and flexible due to the micro-Brownian segmental motion of polymer chains above T_g. Thus, it is a polymeric liquid state at a molecular level in spite of its solid-like appearance [5, 8, 14]. From this consideration, rubber matrix can be a good medium for ion diffusion, and the ion-conducting material therefrom can be of use to all solid state devices involving electrolytes. Additionally, good contact with solid electrodes may be easily realized by using soft and flexible solid electrolytes. For this end, the matrix has to be able to dissolve ions. PEO has been known for its ability to form complex with alkali cations. Amorphous PEO is, therefore, a very good matrix for lithium-ion conduction together with the fact that T_g of PEO is very low. This was reported to be the case many years ago by Wright *et al.* [17, 18] followed by the exciting suggestion of applicability of polymer electrolytes based on PEO to lithium-ion batteries[19].

Recent development of lithium-ion batteries, that are still using an electrolytic organic solution, suggests that some ion-conducting polymers can be a crucial material for light-weight, flexible and quite often hopefully small-size and thin all-solid-state devices such as a secondary battery and an electrochromic or electroluminescence display. The objective of our studies is to develop excellent polymer electrolytes for higher performance lithium-ion secondary batteries, and our results so far are described in the following sections.

References

1. K. Eric Drexler, *Engines of Creation: The Coming Era of Nanotechnology*, Doubleday, New York (1986).
2. A C&EN Special Report on Nanotechnology, *Chem. Eng. News*, Oct. 16 (2000).
3. D. Mulhall, *Our Molecular Future*, Prometheus Books, New York (2002).
4. L. R. G. Treloar, *The Physics of Rubber Elasticity*, 3rd ed., Clarendon Press, Oxford (1975).
5. S. Kohjiya and Y. Ikeda, *Recent Res. Dev. Electrochem.*, **4** (2001) 99.
6. Y. Ikeda, *J. Appl. Polym. Sci.*, **78** (2000) 1530.
7. S. Kohjiya, Y. Ikeda and S. Yamashita, *Polyurethanes in Biomedical Engineering II*, eds. H. Plank, I. Syre, M. Dauner and G. Egbers, Elsevier, Amsterdam (1987).
8. S. Kohjiya, *Proc. the 38th Japan Congress on Mater. Res.*, p.15 (1995).
9. S. Kohjiya, T. Kawabata, K. Maeda, S. Yamashita and Y. Shibata, *Second Inter. Symp. on Polym. Electrolytes*, ed. B. Scrosati, Elsevier Appl. Sci., London, p.187 (1990).
10. S. Kohjiya, K. Maeda and S. Yamashita, *J. Mater. Sci.*, **25** (1990) 3368.
11. S. Kohjiya, A. Ono, T. Kishimoto, S. Yamashita, H. Yanase and T. Asada, *Mol. Cryst. Liq. Cryst.*, **185** (1990) 183.
12. T. Miyamoto and K. Shibayama, *Kobunshi Kagaku*, **28** (1971) 797.
13. T. Miyamoto and K. Shibayama, *J. Appl. Phys.*, **44** (1973) 5372.
14. A. S. Lodge, *Elastic Liquid*, Academic Press, London (1960).
15. Y. Takahashi and H. Tadokoro, *Macromolecules*, **6** (1973) 672.
16. Y. Ikeda, Y. Wada, Y. Matoba, S. Murakami and S. Kohjiya, *Electrochim. Acta*, **45** (2000) 1167.

17. D. E. Fenton, J. M. Parker and P. V. Wright, *Polymer*, **14** (1973) 589.
18. P. V. Wright, *Br. Polym. J.*, **7** (1975) 319.
19. M. B. Armand, J. M. Chabagno and M. J. Duclot, *Fast Ion Transport in Solids*, ed. P. Vashishta, North-Holland, Amsterdam (1979).

6.2 Branched poly(oxyethylene)s as polymer solid electrolytes

6.2.1 Copolymers of ethylene oxide

In the previous chapter rubbery polymer matrix is assumed to be a very promising candidate of ion-conducting materials, which are of much use for all-solid-state, especially, lithium-ion batteries. Among many rubbery polymers, poly(ethylene oxide) (PEO, the simplest among poly(oxyethylene)s) has been the material of choice since the pioneering reports [1–3], because it is polar enough for dissolving alkali cations including lithium and its glass-transition temperature (T_g) is relatively low. However, there is a problem to be solved: PEO is highly crystalline, *i.e.*, usually it is not amorphous. How to realize amorphous state using PEO is to be considered.

One design to obtain amorphous nature is to use oligomeric PEO molecules followed by chemical attaching to a mother molecule. Because oligomers are liquid-like and have no shape-forming ability, they are to be attached to a suitable main frame. The mother is preferably elastomeric in accordance with our purpose. Poly(siloxane) grafted with oligomeric PEO as side chains (a graft copolymer) should be one of the best molecular designs, because poly(siloxane) has the lowest T_g (-120 °C) among polymers [4]. The graft copolymer formed solid solutions with lithium salt, and the observed conductivities were excellently high even at room temperature. However, its mechanical properties were not good. To improve them, crosslinking leading to a polymer network may be one solution. This subject will be discussed in the subsection 6.2.3. In case of poly(siloxane) grafted with oligomeric PEO, the crosslinking resulted in low conductivities. Also, poly(siloxane)s are not much stable when in contact with metallic lithium, which is assumed to be used as a negative electrode in a solid-state lithium-ion battery.

One of the other ways to have an amorphous PEO matrix is to use the copolymerization technique. Ethylene oxide (EO) has been copolymerized with epichlorohydrin (EH) to afford hydrin rubbers for industries. Copolymers of EO and EH of various EO contents were synthesized, and the conductivity of their complex with lithium perchlorate was evaluated [5–7]. However, in terms of ionic conductivity even the best result was not so high at room temperature (lower than 10^{-5} S cm^{-1}). In this system, amorphous phase was observed or even predominant if EO content is low, but the presence of EH units in the copolymer did not contribute to the solubilization of lithium cation. Additionally, increase of EH content resulted in the increase of T_g to give rise to a lower mobility of lithium cations.

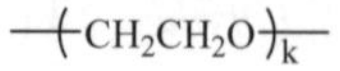

$$-(\text{CH}_2\text{CH}_2\text{O})_k-$$

PEO

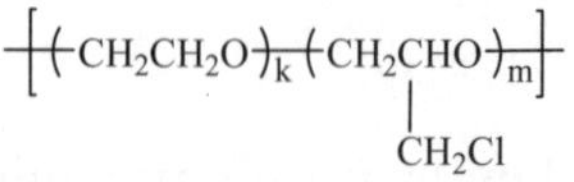

$$-[(\text{CH}_2\text{CH}_2\text{O})_k(\text{CH}_2\text{CHO})_m]-$$
$$|$$
$$\text{CH}_2\text{Cl}$$

Copolymer of EO and EH

$$-[(\text{CH}_2\text{CH}_2\text{O})_k(\text{CH}_2\text{CHO})_m]-$$
$$|$$
$$\text{CH}_2\text{O}(\text{CH}_2\text{CH}_2\text{O})_n\text{CH}_3$$

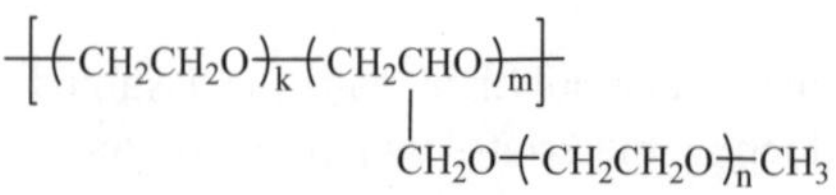

Branched PEOs: n=1, MEC; n=2, BEC; n=3, TEC

Fig. 6.2.1. Chemical structures of PEO, EO-EH copolymers and branched PEOs.

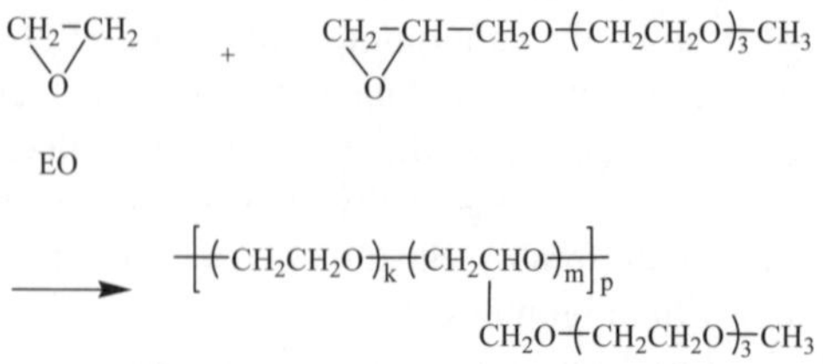

$$\underset{\text{EO}}{\overset{\text{CH}_2-\text{CH}_2}{\diagdown\diagup}} + \underset{}{\overset{\text{CH}_2-\text{CH}-\text{CH}_2\text{O}(\text{CH}_2\text{CH}_2\text{O})_3\text{CH}_3}{\diagdown\diagup}}$$

$$\longrightarrow \;-[(\text{CH}_2\text{CH}_2\text{O})_k(\text{CH}_2\text{CHO})_m]_p-$$
$$|$$
$$\text{CH}_2\text{O}(\text{CH}_2\text{CH}_2\text{O})_3\text{CH}_3$$

TEC

Fig. 6.2.2. Synthetic route of TEC [10].

6.2.2 High molar mass poly(oxyethylene)s

Taking the previous results into account, the ultimate molecular design of lithium ion conducting polymer was estimated as shown in Fig. 6.2.1 where MEC stands for PEO carrying mono(oxyethylene) side chains, BEC for di(oxyethylene), and TEC for tri(oxyethylene). Here, structures of PEO and EO-EH copolymer are also included. The proposed structure is PEO having oligo(oxyethylene) side chains. The side chain may hinder the crystallization of PEO main chain, and it may contribute to the more solubilization of lithium cation due to the same oxyethylene structure both at main chain and at side chains. By substitution reaction of chlorine atom in EH monomer by oligo(oxyethylene) group, monomers carrying mono, di, and tri(oxyethylene) units were synthesized, and subjected to copolymerization with EO. As an example, the copolymerization reaction for TEC is shown in Fig. 6.2.2. We were able to change the composition in the copolymers i.e. amount of side chains by changing the monomer feed ratio at the time of copolymerization, and the products were high molar mass branched poly(oxyethylene)s. Among the branched poly(oxyethylene)s i.e. MEC, BEC, and TEC, results on BEC [8, 9] and TEC [10–12] were already published.

6.2 Branched poly(oxyethylene)s as polymer solid electrolytes   193

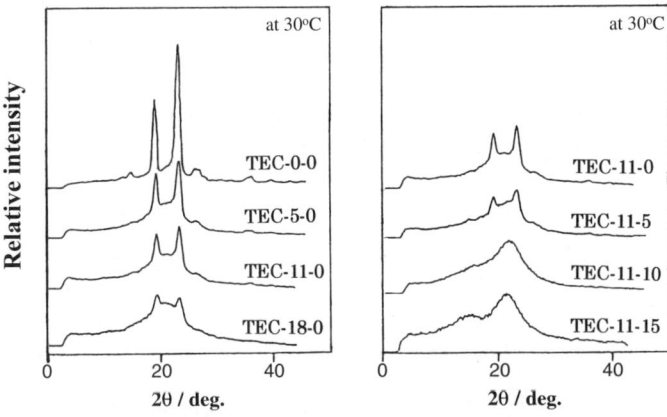

Fig. 6.2.3. WAXD intensity profiles of TECs and those of TEC-11 doped with LiClO$_4$ [11].

Figure 6.2.3 shows wide-angle X-ray diffraction (WAXD) profiles on TEC of various side chain contents without doping and with doping of lithium perchlorate relative to ether oxygen in the polymer. Hereafter, sample codes like TEC-*a-b* are used; the figure at the center (*a*) is a side chain content in molar percentage of monomeric units in the polymer, and that at the end (*b*) is amount of Li$^+$ ion doped in terms of molar ratio to ether oxygen ([Li$^+$]/[-O-]) in the polymer. By this coding, BEC-0-0 and TEC-0-0 are equal to PEO-0 or simply PEO. In case of PEO which is a homopolymer, the figure (*c*) in PEO-*c* indicates amount of Li$^+$ doped. The WAXD of PEO is fully in agreement with the reported one [13]. The peak heights, especially the larger two due to (120) and (032) reflections, decreased with increases of the side-chain content and of the doped salt. Peak intensities due to (110) and (024) were observed to decrease also, though the intensity was relatively small. WAXD on BECs gave very similar results. Therefore, the crystallization of PEO was effectively hindered by the copolymerization technique *i.e.* by the introduction of side chains onto the main chain. From these WAXD profiles the degree of crystallization was calculated. The results on PEO, MEC, BEC and TEC are shown in Fig. 6.2.4. No results on MEC of lower contents than 18 mol % of the side chain, in this case mono(oxyethylene), are available. However, comparing the effects of mono, di and tri(oxyethylene) side chains, it is concluded that very bulky side chains are not necessary to inhibit the crystallization of PEO, since BEC is more amorphous than TEC as shown in the figure. Also, with *ca.* 20 % side chain content MEC showed no crystallinity, in other words, completely amorphous. Table 6.2.1 shows DSC (differential scanning calorimeter) results on T_g and melting behaviors of PEOs doped with lithium perchlorate. TEC-18-5 was found to contain a small amount of crystalline part by DSC to give melting temperature T_m of 31.5 °C. Both MEC and BEC afforded a completely amorphous matrix as revealed by both WAXD and DSC. One or two oxyethylene units at the side chain (MEC and BEC) were enough for the purpose of disturbing the crystallization of PEO, but the amorphous

Table 6.2.1. Grass-transition and melting behaviors of poly(oxyethylene)s doped with lithium perchorate.

Sample code	T_g / °C	T_m / °C	ΔH_m / cal g^{-1}
MEC-18-5	-47.0	-	-
BEC-18-5	-48.1	-	-
TEC-18-5	-50.9	31.5	3.3

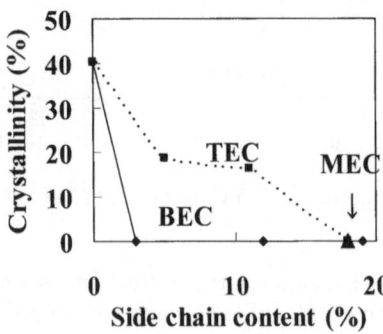

Fig. 6.2.4. Change of crystallinity at 30°C with side chain content of branched PEOs doped with 5% LiClO$_4$.

region containing tri(oxyethylene) side chains (TEC) gave the lowest T_g, though the difference was not very large. The tri(oxyethylene) segments are estimated to be more effective for enhancing the micro-Brownian motion of the main chain of PEO than the shorter side chains at amorphous states.

Ionic conductivity (σ) was measured on TEC-18 doped with lithium perchlorate by an alternating current impedance method in a temperature-controlled box. From the polymer electrolyte film, a disk was cut and was placed between two stainless steel blocking electrodes with a Teflon® ring spacer. The assembly was sealed in a Teflon® container. These procedures were conducted in a glove box under dry argon. The ionic conductivity was measured in a thermostat oven. The bulk resistance (R_b) was determined from the complex impedance (the Cole-Cole) plot, and conductivity (σ) was calculated using the equation

$$\sigma = d/(A \cdot R_b) \qquad (6.2.1)$$

where d is the thickness and A is the cross sectional area of the film. The results are shown in Fig. 6.2.5 as a function of temperature. The convex curves in the Arrhenius plot imply that the results are in accord with the Williams-Landel-Ferry (WLF) equation [14].

$$\log \sigma(T)/\sigma(T_g) = C_1(T - T_g)/(C_2 + (T - T_g)) \qquad (6.2.2)$$

The fitting of the equation with the results in Fig. 6.2.5 has produced C_1 and C_2 values as shown in Table 6.2.2, which also displays conductivity values. The

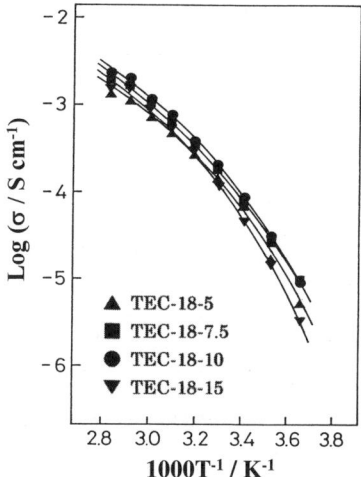

Fig. 6.2.5. Temperature dependence of ionic conductivity of TEC-18 doped with LiClO$_4$ [11].

Table 6.2.2. Conductivity and WLF parameters of TEC-18 doped with lithium perchorate [11].

Sample code	σ at 30 °C / S cm^{-1}	σ at T_g / S cm^{-1}	C_1	C_2 / K
TEC-18-5	1.5×10^{-4}	6.6×10^{-16}	14.7	24.6
TEC-18-7.5	1.8×10^{-4}	1.8×10^{-12}	11.7	36.2
TEC-18-10	2.1×10^{-4}	4.7×10^{-13}	12.2	31.7
TEC-18-15	1.2×10^{-4}	3.0×10^{-11}	10.1	32.3

observed conductivity at 30 °C is one of the highest ones for polymer electrolytes. The very low conductivity values calculated using the WLF equation at T_g and good fitting to the WLF equation strongly indicate that ion conduction depends on free volume available in the PEO matrix *i.e.* on an amorphous region. In spite of the detection of crystallite by DSC as shown in Table 6.2.1, it does not seem to interfere with the migration of the cation.

The WAXD and conductivity results on TEC are plotted against tri(oxyethylene) side chain content in Fig. 6.2.6. It is very clear that the conductivity increased and crystallinity decreased with increasing side chain contents. Again, the result supports that ion conduction is via amorphous region in PEO matrix. Similar results have also been obtained on BEC and MEC. Once the majority in PEO matrix is amorphous, the Li$^+$ conductivity is not much dependent on the sort of side chains. In other words, MEC, BEC and TEC all showed an excellent Li$^+$ conductivity at their amorphous state, though T_g differences shown in Table 6.2.1 may suggest a little difference in the micro-Brownian segmental movement at the rubbery state.

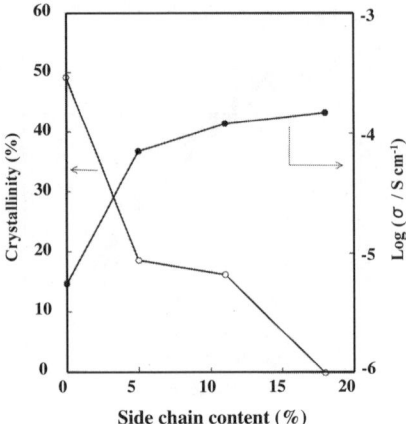

Fig. 6.2.6. Change of crystallinity and ionic conductivity of TEC doped with 5% LiClO$_4$ at 30°C with side chain content [10].

6.2.3 Poly(oxyethylene) networks

The poly(siloxane) grafted with PEO and the highly branched yet linear poly(oxyethylene)s of high molar mass showed a fair mechanical property (afforded a good stable films) as well as good lithium-ion conductivities. Yet, the mechanical properties are to be improved for the manufacturing processes and the practical applications. Equivalent to the former, we prepared copolymeric gels composed of poly(dimethyl siloxane) and PEO network chains [15]. They were both hydrophobic and hydrophilic, *i.e.*, amphiphilic, and showed characteristic solubility behaviors. Upon doping with lithium perchlorate, conductivity between 10^{-5} and 10^{-6} S cm^{-1} was observed. Again, the presence of low T_g poly(siloxane) chains was not so effective to promote ionic diffusion. Probably low polarity of siloxane polymer was not favorable to ionic carrier formation.

In order to synthesize polymer networks of branched poly(oxyethylene)s, terpolymers (copolymers from three kinds of monomers) were prepared using allyl glycidyl ether as a third comonomer [16]. Figure 6.2.7 shows the synthetic route of the terpolymer. The introduced allyl groups (-CH$_2$-CH=CH$_2$) is for a crosslinking reaction to afford a polymer network. Characterizations of the produced terpolymer are listed in Table 6.2.3. All terpolymers contained 2 mol% of crosslinking group (AGE, allyl glycidyl ether unit), and this amount of AGE limited the degree of crosslinking within the range of elastomers in terms of mechanical behaviors. Introduction of 8 mol % of di(oxyethylene) chains (see Terpolymer-8) was not enough to make the polymer completely amorphous in accordance with the previous results. Crosslinking reaction was conducted on Terpolymer-20 (side chain content was 20 mol%) at 100 °C using 0.3 wt % of benzoyl-*m*-toluoyl peroxide as a free radical initiator and various amount of N,N'-*m*-phenylene bismaleimide as a

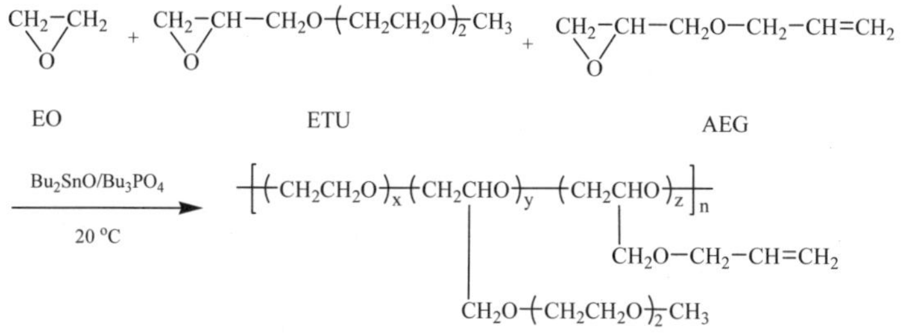

6.2 Branched poly(oxyethylene)s as polymer solid electrolytes   197

$$\underset{\text{EO}}{\overset{\text{CH}_2-\text{CH}_2}{\underset{\text{O}}{\vee}}} + \underset{\text{ETU}}{\overset{\text{CH}_2-\text{CH}-\text{CH}_2\text{O}+\text{CH}_2\text{CH}_2\text{O}\overline{)_2}\text{CH}_3}{\underset{\text{O}}{\vee}}} + \underset{\text{AEG}}{\overset{\text{CH}_2-\text{CH}-\text{CH}_2\text{O}-\text{CH}_2-\text{CH}=\text{CH}_2}{\underset{\text{O}}{\vee}}}$$

$$\xrightarrow[20\,°C]{\text{Bu}_2\text{SnO}/\text{Bu}_3\text{PO}_4} \left[\left(\text{CH}_2\text{CH}_2\text{O} \right)_x \left(\text{CH}_2\text{CHO} \right)_y \left(\text{CH}_2\text{CHO} \right)_z \right]_n$$

with branches $\text{CH}_2\text{O}-\text{CH}_2-\text{CH}=\text{CH}_2$ and $\text{CH}_2\text{O}+\text{CH}_2\text{CH}_2\text{O})_2\text{CH}_3$

Fig. 6.2.7. Synthesis of terpolymer [16].

Table 6.2.3. Characteristics of the terpolymers [16].

Sample code	Composition (mol%)			M_w	M_n	M_w/M_n	T_g/°C	T_m/°C	ΔH_m/Jg^{-1}
	[EO]	[ETU]	[AGE]						
Terpolymer-8	90	8	2	2.3×10^6	4.6×10^5	5.0	-65	43	12.3
Terpolymer-15	83	15	2	1.6×10^6	3.1×10^5	5.2	-69	39	0.7
Terpolymer-20	78	20	2	1.5×10^6	3.1×10^5	4.9	-72	38	0.3
Terpolymer-30	68	30	2	1.1×10^6	1.9×10^5	5.9	-72	38	0.1

coagent. The amount of the coagent is assumed to determine the network chain density ν (in mol cm^{-3}).

Conductivities of the terpolymers doped with three kinds of lithium salts are listed in Table 6.2.4. Among the salts, LiN(SO$_2$CF$_3$)$_2$ (LiTFSI) afforded the highest conductivity, which was 1.1×10^{-4} S cm^{-1} at 30 °C for the terpolymer containing 30 mol % of di(oxyethylene) chain, which is in accord with the previous results. Conductivities of the crosslinked polymers, *i.e.*, networks doped with three kinds of lithium salts are shown in Table 6.2.5. Here, the figure following Network is amount of the coagent in wt %. The network chain density varied from 3.1×10^{-5} to 1.02×10^{-4} mol cm^{-3} according to the amount of the coagent. The conductivity decreased a little by crosslinking: In case of LiTFSI as a dopant, Network-1 with the lowest degree of crosslinking showed almost the same conductivities as the Terpolymer-20, but Network-5 (with the highest network chain density) showed the lowest conductivity among the crosslinked ones and the precursor (Terpolymer-20). Yet, the decrease was found to be very small compared with the previous findings [4]. This result strongly suggests that the excellent amorphous state of the branched poly(oxyethylene)s was maintained in their crosslinked networks, of course, within a few mol percent of crosslinking site. In the present system, 2 mol % (amount of AGE unit, see Table 6.2.3) was the maximum of the degree of crosslinking.

The improvement of mechanical property by crosslinking is shown in Fig. 6.2.8, which displays a dynamic mechanical behavior in the form of temperature dispersion of dynamic modulus (E') of Terpolymer-20 and its crosslinked networks. E' may be understood to be equivalent to the Young's modulus in static mechanics. The present crosslinking on the terpolymer afforded a stable rubbery plateau region around a

Table 6.2.4. Ionic conductivity of the terpolymers doped with Li salts [16].

Sample code	LiClO$_4$[a]		LiBF$_4$[a]		LiTFSI[a]	
	σ at 30 °C / S cm^{-1}	σ at 60 °C / S cm^{-1}	σ at 30 °C / S cm^{-1}	σ at 60 °C / S cm^{-1}	σ at 30 °C / S cm^{-1}	σ at 60 °C / S cm^{-1}
Terpolymer-8	2.8 × 10^{-5}	2.5 × 10^{-4}	2.7 × 10^{-5}	2.5 × 10^{-4}	5.8 × 10^{-5}	7.5 × 10^{-4}
Terpolymer-15	4.7 × 10^{-5}	4.3 × 10^{-4}	4.7 × 10^{-5}	4.3 × 10^{-4}	8.7 × 10^{-5}	9.1 × 10^{-4}
Terpolymer-20	5.2 × 10^{-5}	5.4 × 10^{-4}	5.3 × 10^{-5}	5.5 × 10^{-4}	1.2 × 10^{-4}	1.2 × 10^{-3}
Terpolymer-30	5.2 × 10^{-5}	5.4 × 10^{-4}	5.4 × 10^{-5}	5.6 × 10^{-4}	1.1 × 10^{-4}	1.1 × 10^{-3}

[a] [Li]/[-O-] = 0.06.

Table 6.2.5. Characteristics of the network polymer electrolytes doped with Li salts [16].

Sample code	BM[a] (wt%)	LiClO$_4$[b]			LiBF$_4$[b]			LiTFSI[b]			ν /×10^{-5} molcm^{-3}
		T_g /°C	σ at 30°C / S cm^{-1}	σ at 60°C / S cm^{-1}	T_g /°C	σ at 30°C / S cm^{-1}	σ at 60°C / S cm^{-1}	T_g /°C	σ at 30°C / S cm^{-1}	σ at 60°C / S cm^{-1}	
Network-1	1	-47	5.1×10^{-5}	5.4×10^{-4}	-50	5.3×10^{-5}	5.4×10^{-4}	-60	1.2×10^{-4}	1.2×10^{-3}	3.1
Network-2	2	-47	5.1×10^{-5}	5.4×10^{-4}	-50	5.3×10^{-5}	5.4×10^{-4}	-60	1.0×10^{-4}	1.1×10^{-3}	4.8
Network-3.5	3.5	-46	5.0×10^{-5}	5.3×10^{-4}	-49	5.2×10^{-5}	5.3×10^{-4}	-57	1.0×10^{-4}	1.1×10^{-3}	6.5
Network-5	5	-43	3.5×10^{-5}	5.3×10^{-4}	-45	3.6×10^{-5}	3.4×10^{-4}	-53	8.5×10^{-5}	8.4×10^{-4}	10.2

[a] N,N'-m-phenylenebismaleimide
[b] [Li]/[-O-]=0.06.

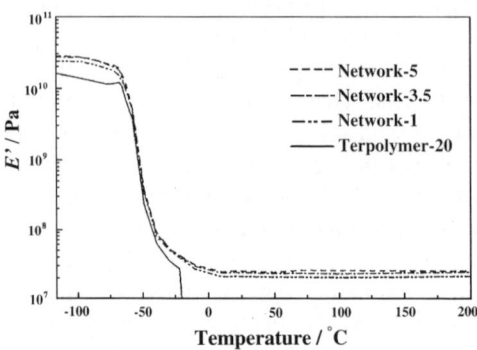

Fig. 6.2.8. Temperature dispersions of E' of Terpolymer-20 and the networks prepared from the terpolymers [16].

wide room temperature range, which suggests an excellent elastomeric behavior of the crosslinked samples.

References

1. D. E. Fenton, J. M. Parker and P. V. Wright, *Polymer*, **14** (1973) 589.
2. P. V. Wright, *Br. Polym. J.*, **7** (1975) 319.

3. M. B. Armand, J. M. Chabagno and M. J. Duclot, *Fast Ion Transport in Solids*, ed. P. Vashishta, North-Holland, Amsterdam (1979).
4. S. Kohjiya, T. Kawabata, K. Maeda, S. Yamashita and Y. Shibata, *Second Inter. Symp. on Polym. Electrolytes*, ed. B. Scrosati, Elsevier Appl. Sci., London p.187, (1990).
5. S. Kohjiya, T. Horiuchi, S. Yamashita, *Electochim. Acta*, **37** (1992) 1721.
6. S. Kohjiya, T. Horiuchi, K. Miura, M. Kitagawa, T. Sakashita, Y. Matoba and Y. Ikeda, *Polym. Int.*, **49** (2000) 197.
7. Y. Ikeda, H. Masui and Y. Matoba, *J. Appl. Polym. Sci.*, **95** (2005) 178.
8. Y. Ikeda, H. Masui, S. Syoji, T. Sakashita, Y. Matoba and S. Kohjiya, *Polym. Int.*, **43** (1997) 267.
9. A. Nishimoto, M. Watanabe, Y. Ikeda and S. Kohjiya, *Electrochim. Acta*, **43** (1998) 1177.
10. Y. Ikeda, Y. Wada, Y. Matoba, S. Murakami and S. Kohjiya, *Electrochim. Acta*, **45** (2000) 1167.
11. Y. Ikeda, Y. Wada, Y. Matoba, S. Murakami and S. Kohjiya, *Rubber Chem. Technol.*, **73** (2000) 720.
12. S. Murakami, K. Ueda, T. Kitade, Y. Ikeda and S. Kohjiya, *Solid State Ionics*, **154–155** (2002) 399.
13. Y. Takahashi and H. Tadokoro, *Macromolecules*, **6** (1973) 672.
14. M. L. Williams, R. F. Landel and J. D. Ferry, *J. Am. Chem. Soc.*, **77** (1955) 3701.
15. S. Kohjiya, H. Tsubata and K. Urayama, *Bull. Chem. Soc. Jpn.*, **71** (1998) 961.
16. Y. Matoba, Y. Ikeda and S. Kohjiya, *Solid State Ionics*, **147** (2002) 403.

6.3 Viscosity behaviors of branched poly(oxyethylene)s

For practical applications of highly conductive and amorphous poly(oxyethylene)s, processability of the polymers is the most important factor. Processability will determine a critical feasibility of polymeric materials to a specific product, because an excellent processability is one of the characteristics of polymeric materials in general. Here, for the evaluation of processability, viscosities of poly(oxyethylene)s are investigated.

Table 6.3.1 lists three samples subject to the measurements. PEO is a commercial product, the molar mass of which is near to the branched poly(oxyethylene)s *i.e.* BEC-05 and BEC-22. As described in Section 6.2.2, BEC-22 is completely amorphous at room temperature. PEO and BEC-05 are crystalline, though they also contain amorphous parts. For viscosity measurements, the polymers were dissolved in distilled water, and were subjected to measurements on MR-500 (Reologi Co., Kyoto) under the frequency range between 0.06 rad s^{-1} and 40 rad s^{-1} at room temperature.

Figure 6.3.1 shows the results of dependence of steady-state viscosity on shear rate. At a lower frequency region a constant viscosity is recognized from which zero-shear viscosity (η_0) is determined. However, at the other regions viscosity depends on shear rate *i.e.* the solutions are non-Newtonian. η_0 values indicate that the viscosity

Table 6.3.1. Characteristics of PEO, BEC-05 and BEC-22.

	PEO	BEC-05	BEC-22
m (mol%)	100	95	78
n (mol%)	0	5	22
M_w / g mol^{-1}	1.8×10^6	1.4×10^6	1.6×10^6
M_w/M_n	3.7	6.9	6.2
Number of monomer units	4.1×10^4	3.0×10^4	2.2×10^4

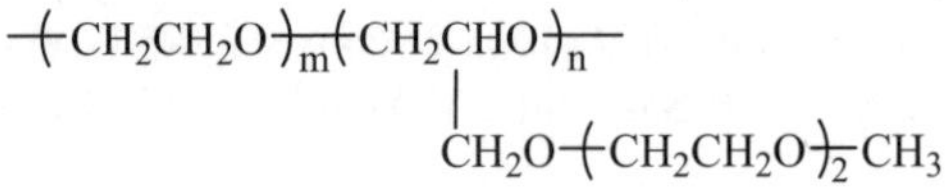

$$-(CH_2CH_2O)_m(CH_2CHO)_n-$$
$$|$$
$$CH_2O(CH_2CH_2O)_2CH_3$$

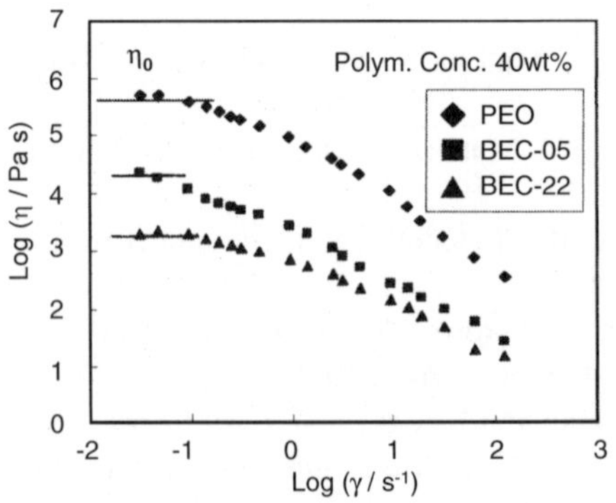

Fig. 6.3.1. Dependence of shear viscosity on shear rate.

has decreased by the introduction of branches to poly(oxyethylene). BEC-22 displays much lower viscosity than PEO and BEC-05, which suggests higher processability of BEC-22. This result suggests very favorable situations of BEC-22 for many applications, because it is not only amorphous but also of relatively low viscosity *i.e.* being a material of choice in terms of processing.

6.4 Composite electrolytes based on branched poly(oxyethylene)

6.4.1 Introduction: Composite-type polymer electrolytes

Generally, commercial rubber products are produced as composites with inorganic fillers. At the same time, such composites form a group of typical soft materials. Nowadays, soft composites are expanding the end uses of composite materials. For examples, pneumatic tires for automobiles and rubber bearings for a seismic isolation system are made from natural rubber in conjunction with suitable reinforcing filler and a sulfur-cure system for crosslinking. So far, carbon black has been widely used as a reinforcer. As described in Section 6.1, the rubbery state is actually a liquid from a rheological viewpoint, even though it is classified as a solid from its appearance [1–3]. Therefore, a lot of inorganic powders can be mixed with rubbers by a mechanical milling without any solvents.

Utilization of conventional milling technique in rubber technology has been focused on producing a novel organic/inorganic composite with excellent performance as a solid electrolyte. Up to now, several studies have been published on the preparation of composite-type solid electrolytes. For examples, ceramic powders such as TiO_2, SiO_2 and Al_2O_3 were mixed with poly(ethylene oxide) (PEO), and the mixing of these powders to the PEO matrix was found to decrease the crystallinity of PEO segment to result in the increase of ionic conductivity [4–7]. The lithium ion transference number in some composite-type electrolytes was also observed to be increased by the specific interaction between the surface of inorganic powder and the counter anion [8]. Some ceramic powders were found to be useful for preparation of solid electrolytes, but they were insulating materials themselves. Thus, in our study, an ion-conductive inorganic glass is employed to prepare composite-type electrolytes. Since there has been a question for the last decade that "Is it possible to prepare the amorphous polymer electrolyte having ionic conductivity higher than 10^{-4} S cm^{-1} ?" and the tentative answer has been negative from our results shown above, we have undertaken an alternative molecular design for elastomeric solid electrolyte, based on Equation 6.1.1 [9, 10] in Section 6.1. Namely, the first and second terms in this equation were combined in order to construct the excellent solid state electrolyte. The ionic conductivity on the composites prepared from ion-conductive inorganic glassy or crystalline materials and linear poly(oxyethylene)s were reported [11–15]. However, the matrixes used in these reports contain a crystalline region, leading to a low ionic conductivity at room temperature, because linear crystalline PEOs were used. Utilization of the low molar mass PEO may prevent the formation of crystalline regions in the composites, but mechanical properties of the composites are not good. We have developed the ion-conductive inorganic/organic composite using a high molar mass branched poly(oxyethylene), which is expected to be a good matrix for the high lithium ionic-conductive glass, because this polymer has been found to be less crystalline [16–20].

6.4.2 Hybrid solid electrolytes from oxysulfide glass and branched poly(oxyethylene)

Starting from the considerations in the previous section, we have designed a composite-type electrolyte using branched poly(oxyethylene)s as a matrix, because PEO is highly crystalline and it was difficult to mix with inorganic powders homogeneously. As an inorganic component, Li_2S-SiS_2-Li_4SiO_4 oxysulfide glass was chosen, because ionic conductivity of this glass was very high as reported to be 10^{-3} S cm^{-1} at room temperature [21–27]. The oxysulfide glass, $95(0.6Li_2S \cdot 0.4SiS_2) \cdot 5Li_4SiO_4$, was prepared from crystalline Li_4SiO_4 and reagent-grade Li_2S and SiS_2 using a twin-roller quenching technique [22, 23]. The obtained glass was ground for 1h by a planetary ball mill. The preparation of the glass was carried out in a glove box under a dry nitrogen atmosphere

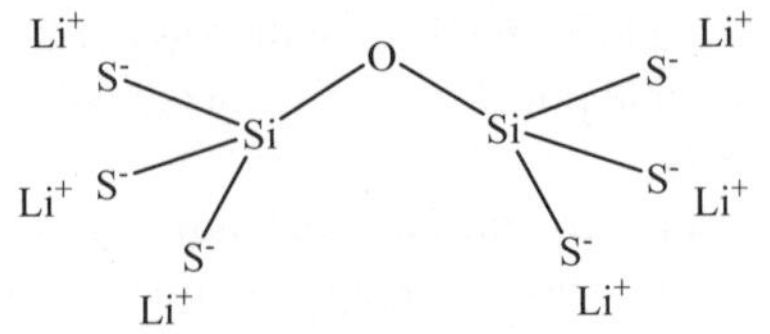

Fig. 6.4.1. Main structural unit in $95(0.6Li_2S \cdot 0.4SiS_2) \cdot 5Li_4SiO_4$ oxysulfide glass [28].

([H_2O] < 1ppm). The diameter of the powder glass was a few μm according to the scanning electron microscope. The main structural unit expected to be present in $95(0.6Li_2S \cdot 0.4SiS_2) \cdot 5Li_4SiO_4$ oxysulfide glass is illustrated in Fig. 6.4.1 [24, 25]. The weight-average molar mass (M_w) of branched poly(oxyethylene) (TEC) was 8×10^5 and its index of polydispersity (M_w/M_n) was 6.3, which were evaluated by a size exclusion chromatography. The content of side tri(oxyethylene) segments of TEC was 19 mol%. The glass powder and polymer were mixed at 80 °C for 2 h, and the composites with a large amount of glass were pressed under 5550 kg cm^{-2} in a dry glove box. As references, oligomeric PEOs were also used for the preparation of the composite-type electrolytes.

Figure 6.4.2 shows the ionic conductivity of the composites from the oxysulfide glass and several poly(oxyethylene)s, where oxysulfide glass and poly(oxyethylene)s were mixed in the weight ratio of Glass/PEO = 67/33 (*e.g.* in equal volume fraction: v/v = *ca.* 1/1) [28]. TEC displayed the comparable ionic conductivity (3.3×10^{-6} S cm^{-1} at 31°C and 2.5×10^{-5} S cm^{-1} at 75°C) with those from oligomeric linear PEOs, whose molar masses were 600 and 1000. High molar mass branched poly(oxyethylene) is preferable for the preparation of hybrid type electrolytes, where the ionic carrier is provided from the dispersant, *i.e.*, the oxysulfide glass in the present system. As shown in Fig. 6.4.3, the ionic conductivity of novel inorganic/organic composites (Glass/TEC) increased with the increase of the glass at the temperature under 150°C [29, 30]. This means that the high percentage of the glass in the composite was preferable for a higher ionic conduction, because TEC itself was simply a matrix. It was also found that the temperature dependence of ionic conductivity became smaller by increasing the glass content. For comparisons, the ionic conductivities of the bulk and the compressed powder pellet of $95(0.6Li_2S \cdot 0.4SiS_2) \cdot 5Li_4SiO_4$ oxysulfide glass [21, 22] are displayed in this figure. The ionic conductivity of the inorganic/organic composites was lower than those of the oxysulfide glasses. The ionic conductivity of Glass/TEC(99/1) was the highest among the composites and was in the orders of 10^{-5} S cm^{-1} at 30°C and 10^{-4} S cm^{-1} at 70 °C, which was comparable with those of single ion-conductive inorganic/organic hybrid polyelectrolytes [31] and lithium salt doped TECs [17–20] which are dual-ion conductors.

Figure 6.4.4 shows the current change with time after applying dc constant voltage, 0.1V, to Glass/TEC(99/1). When lithium metal plate (non-blocking electrodes) was used as electrodes, a constant current was observed. The value was about the same with the calculated current value from the conductivity measured

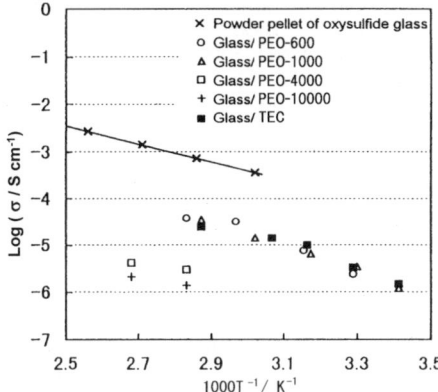

Fig. 6.4.2. The Arrhenius plots of ionic conductivity for the powder pellet of oxysulfide glass, oxysulfide glass/linear PEO composites and oxysulfide glass/TEC composite [28].

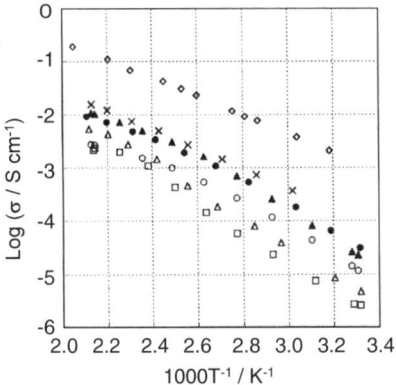

Fig. 6.4.3. The Arrhenius plots for ionic conductivity of the oxysulfide glasses, oxysulfide glass/TEC composites doped with and without LiClO$_4$. ◇: Bulk glass, ×: Pellet glass, ○: Glass/TEC (99/1), △: Glass/TEC (95/5), □: Glass/TEC (90/10); ●: Glass/TEC (99/1)-salt; ▲: Glass/TEC (95/5)-salt [29, 30].

by the ac method. In the case of stainless steel plates (blocking electrodes for lithium), a large decrease of current due to the polarization was initially observed, and the current became almost constant: the steady current was much lower than the value obtained using the lithium electrodes. These results suggest that electronic conductivity of Glass/TEC(99/1) was much (at least three orders of magnitude as seen from Fig. 6.4.4) lower than the ionic conductivity. Therefore, the lithium ion transference number of Glass/TEC(99/1) was calculated to be lager than 0.999. High molar mass branched poly(oxyethylene) was of use as a binder for the glass, which afforded a good adhesion between the electrolytes. This adhesion did not seem to inhibit the diffusion of lithium ion. But, further improvement of ionic conductivity

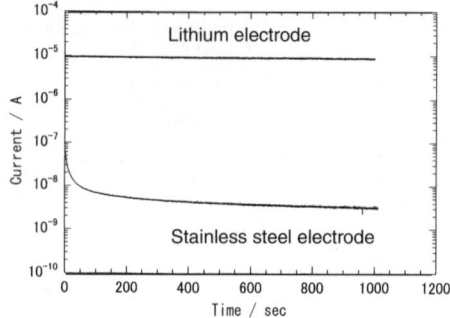

Fig. 6.4.4. Current change with time after applying dc constant voltage to Glass/TEC (99/1) [30].

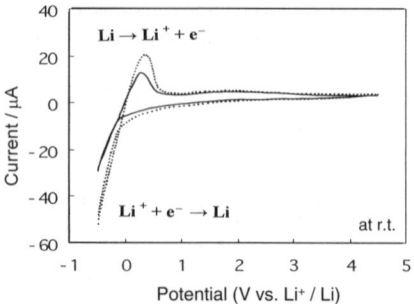

Fig. 6.4.5. Cyclic voltammograms of oxysulfide glass/TEC composites at room temperature [30]. —— : Glass/TEC (99/1), ······ : Glass/TEC (99/1)-salt.

of the composites is necessary for the practical application as a solid electrolyte. In the Arrhenius plots shown in Fig. 6.4.3, the temperature dependence of ionic conductivity for the bulk and powder pellet of the glass is linear, whereas the experimental points of ionic conductivity for the inorganic/organic composites seem to be convexly curved rather than linear. They are semi-quantitatively expressed by the Williams-Landel-Ferry (WLF) equation [32] or the Vogel-Tamman-Fulcher (VTF) [33–35] equation. This observation suggests that the segmental motion of oxyethylene units at least of those existing in the interface between the glass particles and matrix are surely involved in the transportation of lithium cation in the glass/TEC composites.

The electrochemical stability window up to $ca.$ 4 ~ 4.5V was obtained for the novel inorganic/organic composites as shown in the cyclic voltammograms of Fig. 6.4.5. The stability against lithium metal and mechanical properties of the composites were also good, $i.e.$, flexible and tough as a matrix of practical solid-state electrolyte. A clear increase of ionic conductivity was observed by adding a salt (LiClO$_4$) into the composites as shown in Fig.6.4.3. One reason of the increase is that the composites became dual-ion conductors, $i.e.$, both Li$^+$ and ClO$_4^-$ contributed to the conduction. The observed ionic conductivities at high temperature of the salt-doped composites were comparable with that of the powder pellet from $95(0.6Li_2S \cdot 0.4SiS_2) \cdot 5Li_4SiO_4$ oxysulfide glass. The ionic conductivity of Glass/TEC(99/1)-salt was 3×10^{-5} S cm^{-1} at 30°C and 1×10^{-3} S cm^{-1} at 100°C. The optimization of the particle size of the glass and the composition of the composite may increase the ionic conductivity of this new inorganic/organic solid electrolyte.

References

1. A. S. Lodge, *Elastic Liquid*, Academic Press, London (1960).
2. S. Kohjiya, *Proc. the 38th Japan Congress on Mater. Res.*, (1995) p.15.
3. S. Kohjiya and Y. Ikeda, *Recent Res. Dev. Electrochem.*, **4** (2001) 99.
4. F. Croce, G. B. Appetecchi, L. Persi and B. Scrosati, *Nature*, **394** (1998) 456.
5. G. B. Appetecchi, F. Croce, L. Persi, F. Ronci and B. Scrosati, *Electrochim. Acta*, **45** (2000) 1481.
6. A. S. Best, J. Adebahr, P. Jacobsson, D. R. MacFarlane and M. Forsyth, *Macromolecules*, **34** (2001) 4549.
7. M. Forsyth, D. R. MacFarlane, A. Best, J. Adebahr, P. Jacobsson and A. J. Hill, *Solid State Ionics*, **47** (2002) 203.
8. F. Croce, L. Persi, B. Scrosati, F. Serraino-Fiory, E. Plichta and M. A. Hendrickson, *Electrochim. Acta*, **46** (2001) 2457.
9. T. Miyamoto and K. Shibayama, *Kobunnshikagaku*, **28** (1971) 797.
10. T. Miyamoto and K. Shibayama, *J. Appl. Phys.*, **44** (1973) 5372.
11. S. Skaarup, K. West and B. Zachau-Christiansen, *Solid State Ionics*, **28–30**(1988) 975.
12. T. Sotomura, S. Itoh, S. Kondo and T. Iwaki, *Denki Kagaku*, **59** (1991) 129.
13. J. Cho and M. Liu, *Electrochim. Acta*, **42** (1997) 1481.
14. J. Cho, G. Kim, H. Lim and M. Liu, *J. Electrochem. Soc.*, **145** (1998) 1949.
15. Y. Ikeda, T. Hiraoka and S. Ohta, *Solid State Ionics*, **175** (2004) 261.
16. D. G. H. Ballard, P. Cheshire, T. S. Mann and J. E. Przeworski, *Macromolecules*, **23** (1990) 1256.
17. Y. Ikeda, H. Masui, S. Shoji, T. Sakashita, Y. Matoba and S. Kohjiya, *Polymer Inter.*, **43** (1997) 269.
18. A. Nishimoto, M. Watanabe, Y. Ikeda and S. Kohjiya, *Electrochim. Acta*, **43** (1998) 1177.
19. Y. Ikeda, Y. Matoba, S. Murakami and S. Kohjiya, *Electrochim. Acta*, **45** (2000) 1167.
20. Y. Ikeda, Y. Wada, Y. Matoba, S. Murakami and S. Kohjiya, *Rubber Chem. Technol.*, **73** (2000) 720.
21. M. Tatsumisago, K. Hirai, T. Minami, K. Takada and S. Kondo, *J. Ceram. Soc. Jpn.*, **101** (1993) 1315.
22. K. Hirai, M. Tatsumisago and T. Minami, *Solid State Ionics*, **78** (1995) 269.
23. M. Tatsumisago, K. Hirai, T. Hirata, M. Takahashi and T. Minami, *Solid State Ionics,* **86–88** (1996) 487.

24. A. Hayashi, M. Tatsumisago, T. Minami and Y. Miura, *J. Am. Ceram. Soc.*, **81** (1998) 1305.
25. A. Hayashi, M. Tatsumisago, T. Minami and Y. Miura, *Phys. Chem. Glasses*, **39** (1998) 145.
26. A. Hayashi, M. Tatsumisago and T. Minami, *J. Electrochem. Soc.*, **146** (1999) 3472.
27. M. Tatsumisago, H. Yamashita, A. Hayashi, H. Morimoto and T. Minami, *J.Non-Cryst. Solids*, **274** (2000) 30.
28. Y. Ikeda, T. Kitade, S. Kohjiya, A. Hayashi, A. Matsuda, M. Tatsumisago and T. Minami, *Polymer*, **42** (2001) 7225.
29. A. Hayashi, T. Kitade, Y. Ikeda, S. Kohjiya, A. Matsuda, M. Tatsumisago and T. Minami, *Chem. Lett.*, **2001**, 814.
30. S. Kohjiya, T. Kitade, Y. Ikeda, A. Hayashi, A. Matsuda, M. Tatsumisago and T. Minami, *Solid State Ionics*, **154–155** (2002) 1.
31. T. Fujinami, M. A. Metha, K. Sugie and K. Mori, *Electrochim. Acta*, **45** (2000) 1181.
32. M. L. Williams, R. F. Landel and J. D. Ferry, *J. Am. Chem. Soc.*, **77** (1955) 3701.
33. H. Vogel, *Phys. Z.*, **22** (1921) 645.
34. G. Tammann and W. Hesse, *Z. Anorg. Allg. Chem.*, **156** (1926) 245.
35. G. S. Fulcher, *J. Am. Ceram. Soc.*, **8** (1925) 339.

6.5 Effect of elongation of elastomer electrolytes on conductivity

We have reported the high ionic conductivities of the well-defined high molar mass branched poly(oxyethylene)s consisting of oxyethylene segments at both main and side chains [1, 2]. Introduction of side chains to PEO main chains afforded amorphous nature and rubber elasticity. The elastomers are a material of great deformability upon application of even a small stress. From this point of view, we study the effect of uniaxial stretching of the linear and branched poly(oxyethylene)s on ionic conductivity. The films with flexibility were prepared from linear PEO and the blends of linear PEO with BEC. The molar mass of linear PEO and that of BEC were $M_v \sim 4.0 \times 10^6$ and $M_w \sim 1.8 \times 10^6$, respectively. These films doped with LiClO$_4$ were stamped out with a dumbbell shaped die to make it possible to be stretched (Fig. 6.5.1) using a home-made tensile machine. Figure 6.5.2 shows the relationship between the lithium-ion conductivity along the stretching direction and the stretching ratio of films consisting of the linear PEO doped with varying lithium salt concentrations. The ionic conductivity increased by uniaxial stretching ratio at various lithium salt concentrations. Particularly, PEO-10 that was doped with 10 mol% of LiClO$_4$ gave the highest ionic conductivity, which was found 10^{-5} S cm^{-1} at 25 °C when the stretching ratio was about 3, while it had an ionic conductivity smaller than 10^{-7} S cm^{-1} at 25 °C before stretching. These results indicate that the ionic conductivity was increased to a two orders higher value than the original upon stretching. The structural changes of PEO during stretching were investigated by wide-angle X-ray diffraction (WAXD) experiments. The *in situ* measurements of

6.5 Effect of elongation of elastomer electrolytes on conductivity 207

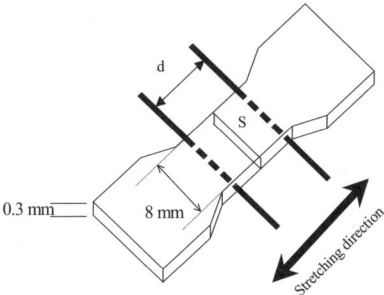

Fig. 6.5.1. Sample configuration for the conductivity measurements under stretching.

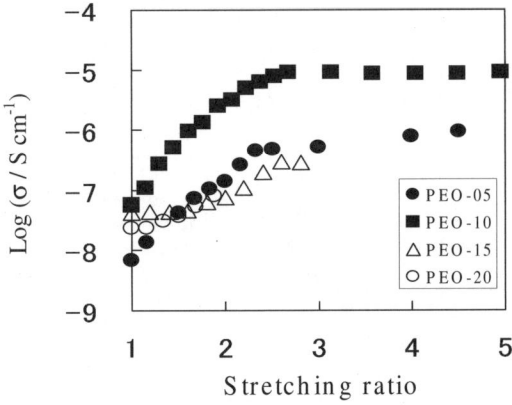

Fig. 6.5.2. Relationship between ionic conductivity and stretching ratio of PEO.

WAXD using a strong X-ray source from a synchrotron radiation was necessary in order to elucidate a change of crystalline structure of PEO. Figure 6.5.3 shows that the degree of orientation of the crystalline chain was increased continuously with an increase of the stretching ratio in the case of PEO-10.

The X-ray experiment revealed that no difraction ring was recognized in PEO-15, which doped with 15 mol% of lithium salt. This implies that it is completely amorphous because of the larger amount of the salt. Nevertheless, the continuous increase of ionic conductivity was also observed. From these results, it is estimated that the orientation of amorphous chains as well as crystalline chains play an important role on the increase of an ionic conductivity of PEO upon stretching.

In order to make use of amorphous character of BEC, polymer blends from PEO and BEC were prepared at various composition ratios. Figure 6.5.4 shows the relationship between ionic conductivities and uniaxial stretching ratio of the blend samples. The blend samples showed a higher ionic conductivity than the linear PEO before the elongation. The introduction of branched chains onto PEO increased its ionic conductivity, which was the case for PEO/BEC blends also. As the blend samples were uniaxially stretched, the ionic conductivity increased monotonously,

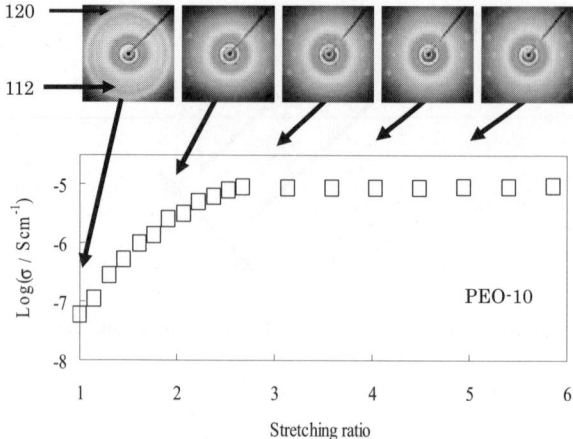

Fig. 6.5.3. Relationship between WAXD patterns and ionic conductivity of PEO-10.

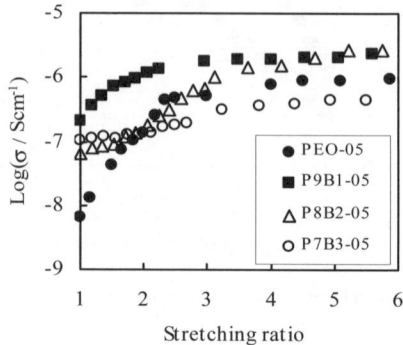

Fig. 6.5.4. Dependence of conductivity on stretching ratio of PEO and PEO/BEC blends.

and it became higher up to almost twice of the original when the stretching ratio was 3. Particularly, the blend sample P9B1-10 (PEO:BEC ratio was 9:1) attained a higher ionic conductivity than linear PEO after the uniaxial elongation. The WAXD measurement of the sample revealed that crystal orientation increased with an increase of the stretching ratio. By detailed examinations of WAXD patterns, the increase of crystal orientation in P9B1-10 was found to be small in comparison with that of an ionic conductivity, suggesting that the orientation of the amorphous chains contributes to the improvement of ionic conductivity.

We reported that the sample oriented by shearing at around the melt temperature was found to be a film with the stacked lamellar structure [3]. Branched PEO samples that were not completely amorphous exhibited the spherulites which had a lamellar structure. By shearing of it, the columnar texture appeared and the columns may be of use for ionic conduction. Taking this result into consideration, we have suggested that the lithium ion which can move freely in the blend system might find the easy path for conduction formed by the oriented chains resulted by the stretching.

References

1. Y. Ikeda, H. Masui, S. Syoji, T. Sakashita, Y. Matoba and S. Kohjiya, *Polym. Int.,* **43** (1997) 269.
2. A. Nishimoto, M. Watanabe, Y. Ikeda and S. Kohjiya, *Electrochimica Acta,* **43** (1998) 1177.
3. S. Murakami, K. Ueda, T. Kitade, Y. Ikeda and S. Kohjiya, *Solid State Ionics,* **154–155**, (2002) 39.

6.6 Ionene elastomers for polymer solid electrolytes

6.6.1 Introduction: Ionene elastomers

"Ionene" is defined as a polymer consisting of quaternary amines in the linear main chain, and the structure of ionene is shown in Fig. 6.6.1. The name was based on the abbreviation for "ionic amine" [1]. Ionene is also classified as a cationic polymer, and some with lower cationic contents are included in the category of ionomer. "Ionomer" is an ionic polymer and referred to as the amphiphilic polymer consisting of hydrophobic polymer segments and a small fraction of ionic sites. Generally, the ionic content of the ionomer is less than *ca.*15 mol%. Therefore, "ionene elastomer" is one type ionomer with low quaternary amino groups in the linear main chain. From the classification of chemical structure, the ionene is divided into three main classes, a random type ionene (quaternary amines randomly distributed in the main chain), a telechelic ionene (quaternary amine at the both ends of polymer) and a well-structurally defined ionene (quaternary amine is regularly repeated in the main chain). Alternatively, the ionene is classified into an aliphatic ionene, aromatic ionene and ionene consisting of heterocyclic structures, and so on. A most accepted classification of ionene elastomer is shown in Fig. 6.6.2. The introduction of ionic sites to polymer chains is known to drastically change the properties and morphology of mother polymers. Therefore, the physical properties of ionomers including ionene elastomers have been extensively researched as high performance materials over the last few decades both in the academic and industrial fields [2]. Some of them have attracted much attention as a physical polymer networks showing rubber elasticity.

The synthesis of ionenes was first reported in 1904 [3] and later in 1930s [4]. Most ionenes studied so far were prepared by the Menshutkin reaction [5] of aliphatic di-tertiary amines with dihalides. So far, wide varieties of starting materials have produced various ionenes. The synthesis and properties of ionenes have been reviewed by Tsutsui [6] and Meyer *et al.* [7]. On the other hand, the development of living ionic polymerization has enabled us to prepare the uniform-sized polymers with a suitable reactive end group [8] and to expand the variety of Menshutkin reaction to prepare novel ionenes. One of the examples is a living

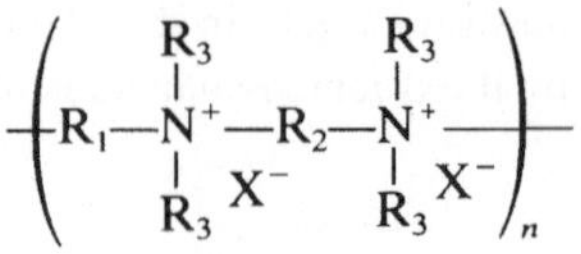

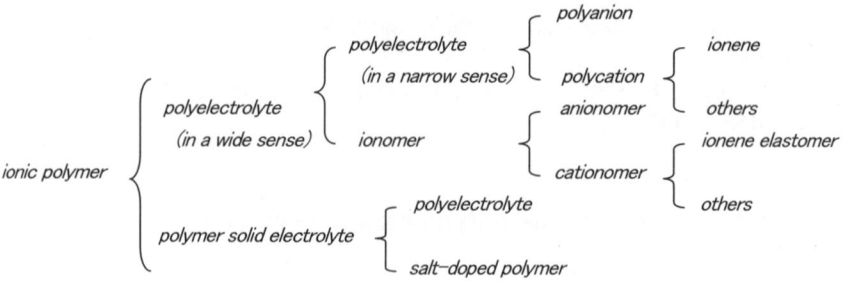

Fig. 6.6.1. Chemical structure of ionene.

Fig. 6.6.2. Classification of ionic polymers.

cationic polymerization of tetrahydrofuran (THF), which has been much utilized for the syntheses of tailored polymers such as block and graft polymers, star-shaped polymers and polymer networks [8–10]. The living poly(oxytetramethylene) (POTM) can be reacted with nucleophiles to give various kinds of macromolecule, because the oxonium group of the living POTM is stable. The living polymerization of THF and termination reaction using dimethylamine produced dimethylamino-terminated POTM, which was reacted with aliphatic or aromatic dihalide to give new ionenes [11–13]. The obtained ionenes were elastomeric and showed excellent mechanical properties. Aliphatic poly(isoprene) ionenes were also synthesized through the living anionic polymerization of isoprene [14].

The pioneer work on POTM ionene elastomers has been conducted by Kohjiya et al. They first synthesized viologen-type POTM ionenes by one-pot synthesis using a reaction between living POTM chains and 4,4'-bipyridine [15–22]. The viologen ionenes were high functional elastomers exhibiting thermochromism, photochromism and photomechanical effect [15–17, 20]. The synthesis of POTM ionenes with 2,2'-bipyridinium units [23, 24] and aliphatic quaternary ammonium units [25] were also reported. The structural analysis for these physically crosslinked elastomers were conducted based on the cascade model [26–29] in order to elucidate the relationship between the properties and higher-order structures [24, 30, 32]. The cascade model for randomly branched f-functional polycondensates was found to be useful for the small-angle X-ray scattering analysis of the ionene elastomers. Generally, unique characteristics of ionene elastomers arise from the ionic aggregation or clustering in the flexible polymer matrix. Therefore, not only the chemical structure but also higher-order structure of ionene elastomers becomes important for the molecular design of ionene elastomers.

For the applications to electronic devices, an ionene polymer was utilized as a material for electronic conductor by complexing with tetracyano-*p*-quinodimethane (TCNQ) [33, 34]. The ionene was expected to be useful for the column formation of TCNQ for the electronic conduction. However, a high electronic conductivity was not obtained due to the difficulty of the close stacking of TCNQ molecules. The study on the ionic conduction in the matrix of ionenes, on the other hand, has been carried out since 1988 [35]. For an example, polymer-salt complexes prepared from partially quaternized polyethyleneimine and sodium or lithium salts were reported to show the conductivities up to 10^{-3} S cm^{-1} at 150 °C [36, 37]. Conductivities of about 10^{-5} S cm^{-1} at room temperature of the ionenes have suggested us the possibility of ionenes as a new type of ionic conductor. However, the ionenes used in the most studies were those whose molar mass between the ionic sites was low, and many of them were glassy or crystalline polymers. Therefore, the use of ionene elastomers for the matrix of solid electrolytes may be preferable for the fast ionic conduction from the advantages of soft rubbery materials for solid state electrolytes as described in Section 6.1.

6.6.2 Polymer solid electrolytes prepared from poly(oxytetramethylene)s

Poly(ethylene oxide) (PEO) has been studied as a matrix of polymer solid electrolytes [38–44], because alkali metal salts are easily soluble in PEO to form a kind of solid solution and the mobility of alkaline metal ions in the amorphous PEO matrix is relatively high. For examples, amorphous poly(oxyethylene) solid electrolytes prepared from high molar mass branched poly(oxyethylene)s [45–51] and PEO networks with branched PEO chains [52, 53] were found to show very high ionic conductivities owing to their low crystallinities of oxyethylene segments described in Subsections 6.2.2 and 6.2.3. However, the lithium cation transference number of PEO electrolytes has been reported to be low due to the strong interaction between the ether dipole of oxyethylene units and lithium cation [38–41, 53]. POTM segment, on the other hand, may be more advantageous than PEO for the fast ionic conduction. It is because the glass-transition temperature (T_g) and melting temperature (T_m) of POTM are lower than those of PEO, which is preferable to make a good ion-conducting matrix in terms of temperature region of the rubbery state. In addition, the number of ether oxygen in a monomer unit of POTM is lower than that of PEO, when two polyethers (POTM and PEO) of the similar molar masses are compared. This means that the interaction between the ether oxygen and lithium cation is smaller in the POTM matrix than in the PEO matrix. Therefore, both higher mobility and larger transference number of alkaline metal cations in the former are expected than in the latter [54, 55].

Up to now, a few studies on the polyurethanes containing POTM soft segments were reported for the preparation of polymer solid electrolytes [54, 56]. However, the ionic conductivities of these polyurethanes doped with LiClO$_4$ were not high and in the order of 10^{-7} - 10^{-9} S cm^{-1} at room temperature. Abraham *et al.* [55] reported that the ionic conductivity of POTM oligomers doped with LiClO$_4$ was 10^{-6} S cm^{-1} at room temperature and its lithium ion transference number at 50 °C was 0.60. Silva

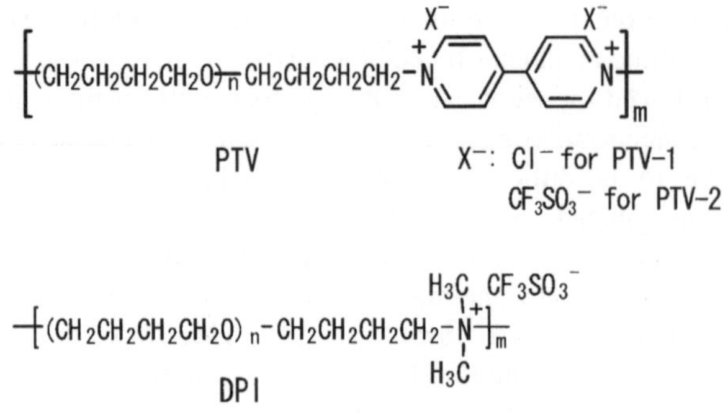

212   6 Polymer solid electrolytes for lithium-ion conduction

$$[(CH_2CH_2CH_2CH_2O)_{\overline{n}} CH_2CH_2CH_2CH_2-N^+ \underset{}{\diagup=\diagdown}-\underset{}{\diagdown=\diagup} N^+]_m$$

PTV X^-: Cl^- for PTV-1
             $CF_3SO_3^-$ for PTV-2

$$[(CH_2CH_2CH_2CH_2O)_n-CH_2CH_2CH_2CH_2-\underset{H_3C}{\overset{H_3C\ CF_3SO_3^-}{\overset{|+}{N}}}]_m$$

DPI

Fig. 6.6.3. Chemical structure of PTV and DPI.

et al. [57] investigated the effect of molar mass of oligomeric POTM electrolytes on the ionic conductivity and reported the highest ionic conductivity of 2.2×10^{-4} S cm^{-1} at room temperature for LiClO$_4$/oligomeric POTM whose molar mass was ca. 900. Based on the dynamic molecular simulations, Ferreria *et al.* [58] and Curtiss *et al.* [59] recently reported that a POTM containing system could be useful as an alternative polymer electrolyte and the migration barrier against lithium-ion conduction for POTM was likely to be lower than that of PEO. The good combination of POTM with an adequate lithium salt may bring about a high ionic conductivity for POTM-based solid electrolytes. However, oligomeric POTM was not effective for obtaining good mechanical properties. Therefore, an ionene composed of POTM segments will be one of the good choices for the preparation of novel polymer solid electrolytes. The ionic segments aggregate in the POTM matrix of ionene to form the ionic domains, which results in the formation of a physical network structure to give good mechanical properties [22–25, 30–32]. The introduction of ionic units to the POTM backbone chain decreases the crystallinity of POTM, which is also preferable for the fast ionic conduction. Furthermore, obtaining an optimal higher-order structure of ionene elastomers may produce the most excellent matrix for the fast ionic conduction, even when the chemical structure (first-order structure) of the ionene molecule is the same. In this section, the results of POTM ionenes for the application of polymer solid electrolytes for a lithium ion rechargeable battery are described. Two ionene-type elastomeric polymer solid electrolytes, one prepared from viologen-type POTM ionenes (PTV) and the other from aliphatic POTM ionene (DPI) are investigated, the chemical structures of which are illustrated in Fig. 6.6.3.

6.6.3 Viologen-type poly(oxytetramethylene) ionene elastomers

PTVs were synthesized by living cationic ring-opening polymerizations of THF followed by chain-extension reactions using 4,4'-bipyridine in accord with the reported method [15, 22, 23]. The molar mass of the POTM segment between the viologen units of PTV-1 was 8700, and the intrinsic viscosity of PTV-1 in THF was 0.35 dl g^{-1}, which showed that the overall molar mass of PTV-

1 was not high, but it was enough for a mechanically good film formation. Lithium bis(trifluoromethyl sulfonyl)imide ($LiN(SO_2CF_3)_2$), lithium perchlorate ($LiClO_4$) and lithium trifluoromethanesulfonate (CF_3SO_3Li) were doped into PTV-1 respectively, by casting the THF solutions of samples on Teflon® molds. The concentration of lithium salt was changed in the range of [Li]/[-O-] = 0.05 - 0.175, where "-O-" stands for the ether oxygen in POTM segments.

The ionic conductivity of PTV-1 was significantly influenced by the salt, as shown in Fig. 6.6.4. $LiN(SO_2CF_3)_2$ doped PTV film gave the highest ionic conductivity (σ) among the films doped with $LiN(SO_2CF_3)_2$, $LiClO_4$ and CF_3SO_3Li at the concentration of [Li]/[-O-] = 0.05 [60]. The ionic conductivity was determined by an alternating current impedance measurement in a thermostat oven at temperatures increasing from 25 °C to 80 °C and followed by temperatures decreasing from 80 °C to -20 °C. The difference of ionic conductivity by the salts was ascribable to the solubility or compatibility of the salts with the PTV-1 matrix, which was investigated by a differential scanning calorimetry (DSC) analysis. Since the morphology of PTV-1 film is significantly influenced by the dissociation of salts in the matrix, the DSC measurement is useful to elucidate the relationship between morphology and ionic conductivity of the salt-doped PTV-1 films. Figure 6.6.5 shows DSC curves of the samples doped with and without salts, which were measured during heating from *ca.* -150 °C. T_g of the POTM amorphous phase, T_m of the POTM crystalline phase and phase transition temperature of the ionic domains (T_i) were detected in all the samples. These results suggest that the films were composed of three phases at room temperature, *i.e.*, POTM amorphous phase, POTM crystalline phase and ionic domains. Comparison of these transition temperatures among the samples indicates that T_m and T_i of PTV-1 doped with $LiN(SO_2CF_3)_2$ were shifted to lower temperatures compared with those of PTV-1 film doped with $LiClO_4$ or CF_3SO_3Li. Additionally, the T_g of POTM amorphous phase of PTV-1 doped with $LiN(SO_2CF_3)_2$ was the highest among the samples. These findings clearly suggest that $LiN(SO_2CF_3)_2$ was the most dissolved in the PTV-1 matrix among the salts, resulting in the strongest interaction of polyether oxygens with lithium cation in the PTV-1 matrix. $LiN(SO_2CF_3)_2$ was reported to be easily dissociated in PEO matrix, and the anionic charge in $(CF_3SO_2)_2N^-$ was estimated to be delocalized [61, 62]. These characteristics of $LiN(SO_2CF_3)_2$ are considered to have brought about a high ionic conductivity to viologen-type POTM ionene similarly with PEO electrolytes.

In the case of PTV-1 doped with CF_3SO_3Li, on the other hand, the high T_i of ionic domains and two melting peaks of POTM crystalline phase were detected, which suggests the predominant dissolution of CF_3SO_3Li into the ionic domains. Namely, an appreciable amount of CF_3SO_3Li was dissolved in the ionic domains and the ionic domains became more rigid to result in the higher T_i and larger heat of fusion of T_i (ΔH_i). The rigid ionic domains prevented the expansion of POTM crystalline phases, which caused two melting peaks of POTM segments to appear as shown in Fig. 6.6.5. The low T_g of PTV-1 film doped with CF_3SO_3Li also suggests a low dissociation of the salt in the POTM amorphous phase. Number of the carrier in this case seems to be the lowest, which brought about the lowest ionic conductivity among the samples. PTV-1 doped with $LiClO_4$ showed the intermediate thermal properties (T_g

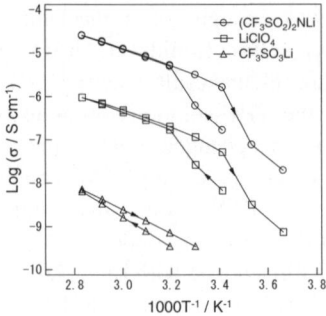

Fig. 6.6.4. The Arrhenius plots of ionic conductivity of PTV-1 films doped with LiN(SO$_2$CF$_3$)$_2$, LiClO$_4$ or CF$_3$SO$_3$Li at the concentration of [Li]/[-O-] = 0.05 [60].

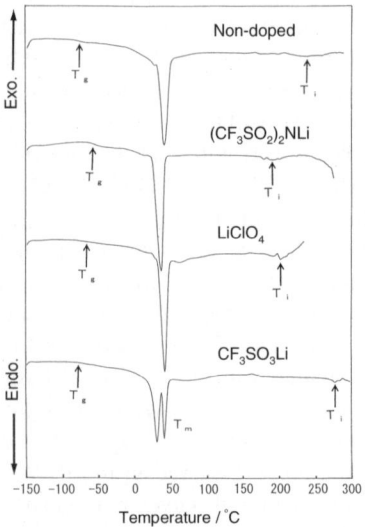

Fig. 6.6.5. DSC curves of PTV-1 films non-doped and doped with LiN(SO$_2$CF$_3$)$_2$, LiClO$_4$ or CF$_3$SO$_3$Li at the concentration of [Li]/[-O-] = 0.05 [60].

and T_i) between the PTV-1 samples doped with LiN(SO$_2$CF$_3$)$_2$ and LiClO$_4$. Thus, the ionic conductivity of this system was also intermediate between them as shown in Fig. 6.6.4. Above 20 °C, all the Arrhenius plots of ionic conductivity of these samples exhibit convexly curved profiles. The plots were analyzed by the Williams-Landel-Ferry (WLF) equation [63] (see Equation 6.2.2), which reasonably explains the observed behavior on ionic conductivity of PTVs doped with the salts. The good fitting to the WLF equation above 20 °C implies that the ionic conduction in these ionene electrolytes was governed by the diffusion of ionic species in the amorphous rubbery matrix, and the free volume was a determining factor for the ionic conduction above 20 °C.

The maximum ionic conductivity was observed in PTV-1 doped with $LiN(SO_2CF_3)_2$ at the concentration of [Li]/[-O-] = 0.125, and it was 1.6×10^{-5} S cm^{-1} at 30 °C and 8.1×10^{-5} S cm^{-1} at 60 °C under the cooling process of the complex impedance measurement. These results were among the highest ionic conductivity of POTM-based polymer electrolytes reported. Interestingly, the temperature dependence above 20 °C of the electrolyte was small compared with that of PEO polymer solid electrolytes. The PTV-1 electrolyte films became softer with the increase of $LiN(SO_2CF_3)_2$ content, but were still mechanically tough even when the salt was doped up to the concentration of [Li]/[-O-] = 0.125. The presence of maximum ionic conductivity of PTV-1 against the salt concentration was explained by Equation 6.1.2 in Section 6.1. With the increase of the salt concentration, the carrier number increased. However, T_g shifted to higher temperatures with the increase of the salt, which gave rise to the smaller mobility of ion. These two effects were opposite for increasing the ionic conductivity, and resulted in the appearance of the maximum ionic conductivity in the plots as shown in Fig. 6.6.6. The ionic conductivity of PTV ionene electrolytes seems to be also influenced by a morphological effect, because the PTV electrolytes were composed of three phases, and the POTM crystalline phase became smaller and T_i significantly lowered to ca. 165 °C with increasing the salt concentration as shown in Fig. 6.6.6. In addition, ΔH_i showed a maximum at the concentration of [Li]/[-O-] = 0.10 and T_g did not linearly change with the salt concentration. The values of T_g were more or less similar with each other among the PTV-1 samples whose salt concentrations were within the range of [Li]/[-O-]= 0.05 - 0.125. Since these phenomena were different from the monophasic polymer solid electrolytes, the selective dissolution of salt in the multiphase matrix of POTM ionene should be taken into account for the consideration on the ionic conduction. The adequate transportation of ionic species in the microphase-separated structure may occur in the POTM ionene matrix.

PTV-1 exhibited the good ionic conductivity at temperatures higher than room temperature. However, it contained the crystalline phase of POTM segments, which decreased the ionic conductivity at room temperature and low temperature region. Thus, an alternative PTV electrolyte (PTV-2) without a crystalline phase of POTM at ambient temperature was prepared by cationic polymerization of THF. The molar mass between the ionic units and overall molar mass of PTV-2 were 3030 and 40500, respectively. The counter-anion of PTV-2 was $CF_3SO_3^-$. PTV-2 formed a microphase-separated structure, which was composed of amorphous POTM phase and the ionic aggregated domains at room temperature. The shear dynamic modulus (G') at 30 °C of PTV-2 was ca. 7 MPa and its rubbery plateau region expanded to ca. 140 °C as shown in Fig. 6.6.7. PTV-2 film was elastic and very tough. Figure 6.6.8 shows the concentration dependence of salt for the ionic conductivities at 30 °C and 80 °C of PTV-2 films doped with $LiN(SO_2CF_3)_2$. The ionic conductivity increased with the increase of salt concentration up to [Li]/[-O-] = 0.125. Over this concentration, the ionic conductivity was almost equal, although T_g of samples increased with the increase of salt concentration, except for the sample whose salt concentration was [Li]/[-O-] = 0.025. This observation shows that the ionic conductivity of PTV-2 was influenced by the morphology effect. However, the ionic conductivity of this

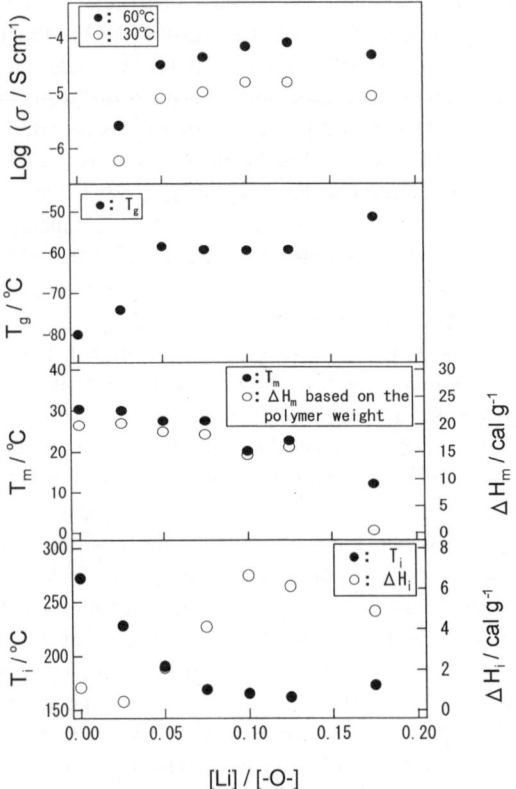

Fig. 6.6.6. Effect of $LiN(SO_2CF_3)_2$ concentration on the ionic conductivity and thermal properties of PTV-1 films [60].

elastomeric viologen-type POTM ionene was not so high probably due to the strong interaction of ionic segments, which means that rigid ionic domains worked as strong crosslinking sites to result in the decrease of ionic conductivity. Based on the results of PTV films described above, a new molecular design for elastomeric polymer solid electrolyte with fast ionic conduction was conducted, and the results are shown in the next subsection.

6.6.4 Aliphatic poly(oxytetramethylene) ionene elastomer

Generally, unique characteristics of ionene elastomers arise from the ionic aggregation in the flexible polymer matrix. Therefore, the chemical structure of ionic segment becomes important for the molecular design of functionality ionene elastomers. The first order structure of ionic segment, such as mono-cation or di-cations type ionenes and aliphatic or aromatic segments, is recognized to significantly influence the properties of ionenes. Up to now, many studies on

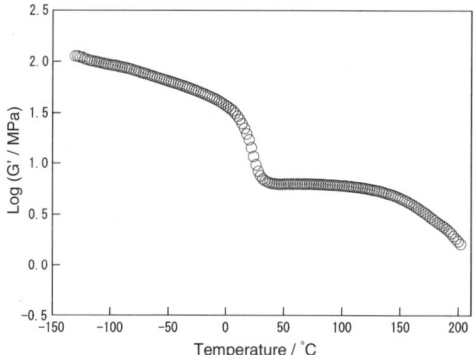

Fig. 6.6.7. Temperature dispersion of G' of PTV-2 film.

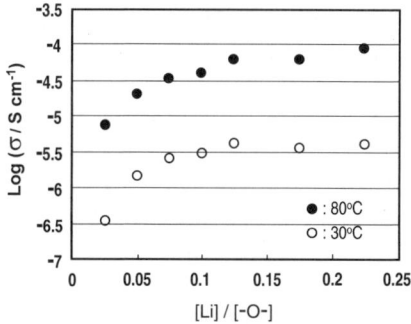

Fig. 6.6.8. Concentration dependence of salt on the ionic conductivities at 30°C and 80°C for PTV-2 films doped with $LiN(SO_2CF_3)_2$.

the aliphatic or aromatic ionene elastomers were reported, however, the most of them were classified as di-cations type ionenes. Recently, we found a new synthetic method to produce an ionene elastomer alternatively consisting of one dimethylammonium group and POTM segment in the repeating unit [25]. This aliphatic POTM ionene (DPI) is expected to be a good matrix of polymer solid electrolyte, because the ionic aggregation of dimethylammonium groups of DPI must be weaker than that of PTV whose number of ionic site was two in the repeating unit. Additionally, the morphological effect of DPI may be expected to increase the ionic conductivity, because DPI also forms a microphase-separated structure.

DPI with $CF_3SO_3^-$ was prepared by a cationic polymerization of THF followed by the chain extension reaction of living POTM chain with N,N-dimethylaminotrimethylsilane [25]. The molar mass between the ionic sites of DPI was 3300. The DPI film possessed the mechanical properties similar to those of conventional raw synthetic rubbers as shown Fig. 6.6.9. The ionic conductivity of DPI doped with $LiN(SO_2CF_3)_2$ is shown in Fig. 6.6.10 with those of POTM homopolymer and PTVs doped with $LiN(SO_2CF_3)_2$ at the same salt concentration

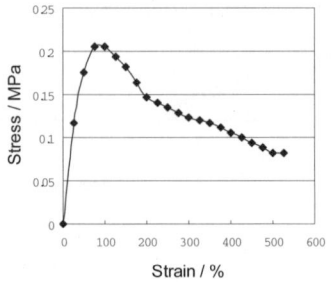

Fig. 6.6.9. Stress-strain curve of DPI film at room temperature.

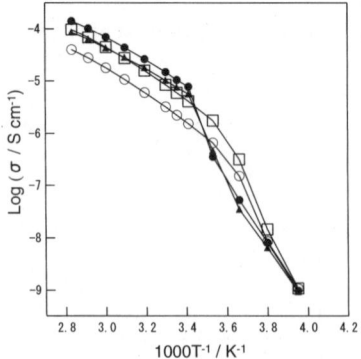

Fig. 6.6.10. The Arrhenius plots of ionic conductivity of DPI (□), PTV-1 (●), PTV-2 (○) and POTM homopolymer (▲) doped with LiN(SO$_2$CF$_3$)$_2$ at the concentration of [Li]/[-O-] = 0.10.

of [Li]/[-O-] = 0.10. The molar mass of POTM homopolymer was 19000 and it was a semi-crystalline polymer at room temperature. Among the samples, DPI shows the highest ionic conductivity below room temperature. The ionic conductivity above 20 °C of DPI was comparable with that of POTM homopolymer doped with LiN(SO$_2$CF$_3$)$_2$. The low molar mass of POTM segment between the ionic sites and the small number of ionic sites of DPI seem to expand the amorphous phase and to increase the mobility of POTM chains, which may result in the good ionic conductivity of DPI electrolyte in a wide temperature range. The ionic conductivity of DPI/LiN(SO$_2$CF$_3$)$_2$ was not changed so much even after the storage under dry argon atmosphere at room temperature for 50 days.

It is noted that the lithium ion transference number (T_+) of the DPI doped with LiN(SO$_2$CF$_3$)$_2$ at the concentration [Li]/[-O-] = 0.10 was *ca.* 0.54 at 60 °C, which was calculated using the Evans equation [64] based on the direct current polarization measurement.

$$T_+ = I_s(\Delta V - I_0 R_{CT,0})/I_0(\Delta V - I_s R_{CT,s}) \quad (6.6.1)$$

where R_{CT} is the charge transfer resistance, I is the current, ΔV is the applied potential and 0 and s refer to the initial and steady-states, respectively. The bulk electrolyte sandwiched between non-blocking lithium electrodes and interfacial characteristics of the cell were probed before and after the direct current polarization at 10 mV by alternating current impedance measurement. $LiN(SO_2CF_3)_2$ has been well utilized for the preparation of good PEO solid electrolytes, because $LiN(SO_2CF_3)_2$ is more easily dissolved in polyether matrix in general and brings about a high ionic conductivity. However, the lithium ion transference number of these PEO/$LiN(SO_2CF_3)_2$ electrolytes was very low due to the fast migration of $(CF_3SO_2)_2N^-$ in the PEO matrix [53]. Therefore, the observed lithium transference number of DPI is concluded to be much better than those of PEO/$LiN(SO_2CF_3)_2$ electrolytes. Since DPI is a physical gel (a physically crosslinked polymer network) and soluble in solvents such as anhydrous THF and methanol, no chemical crosslinking reaction is necessary for the processing. This characteristic of DPI is also preferable for the preparation of solid state electrolytes. For the development of solid state ionics, the results on DPI/$LiN(SO_2CF_3)_2$ electrolyte will be valuable. Especially, the use of POTM segment and control of highly ordered structure for multi-phased elastomers will become more and more important for preparing the excellent polymer solid electrolytes with fast ionic conduction.

References

1. A. Rembaum, *J. Macromol. Sci., Chem.*, **3** (1969) 87.
2. S. Yano and E. Hirasawa ed., *Ionomers-Ionic Polymers*, CMC Publisher, Tokyo (2003).
3. I. Knorr, *Berichte*, **37** (1904) 3507.
4. E. R. Littmann and C. S. Marvel, *J. Am. Chem. Soc.*, **52** (1930) 287.
5. N. Menschutkin, *Z. Phys. Chem.*, **5** (1890) 589.
6. T. Tsutsui, *Developments in Ionic Polymers - 2*, eds. A. D. Wilson and H. J. Prosser, Elsevier Applied Science, London, Chapter 4 (1986).
7. W. H. Meyer and L. Dominingez, *Polymer Electrolyte Reviews-2*, eds. J. R. MacCallum and C. A. Vincent, Elsevier, London, p.191 (1989).
8. E. J. Goethals, ed, *Telechelic Polymers: Synthesis and Applications*, CRC Press, Boca Raton, Florida (1989).
9. E. J. Goethals, P. V. Caeter, J. M. Geeraert and F. E. Du Prez, *Angew. Makromol. Chem.*, **223** (1994) 1.
10. Y. Tezuka, T. Shida, T. Shiomi and K. Imai, *Macromolecules*, **26** (1993) 575.
11. S. Kohjiya, T. Ohtuki and S. Yamashita, *Makromol. Chem. Rapid Commun.*, **2** (1981) 417.
12. S. Kohjiya, T. Ohtuki and S. Yamashita, *Makromol. Chem.*, **191** (1990) 397.
13. S. Kohjiya, T. Ohtuki, S. Yamashita, M. Taniguchi, T. Hashimoto, *Bull. Chem. Soc. Jpn.*, **63** (1990) 2089.
14. P. Charlier, R. Jerome, P. Teyssie and L. A. Utracki, *Macromolecules*, **25** (1992) 617.
15. S. Kohjiya, T. Hashimoto, S. Yamashita and M. Irie, *Chem. Lett.*, **1985**, 1497.
16. S. Kohjiya, T. Hashimoto and S. Yamashta, *Makromol. Chem. Rapid Commun.*, **10** (1989) 9.

17. S. Kohjiya, Y. Ikeda, N. Moriya, T. Hashimoto, S. Yamashta and Y. Shibata, *MRS Intl Mtg. on Adv. Mats* ., **12** (1989) 255.
18. T. Hashimoto, S. Kohjiya, S. Yamashita and M. Irie, *J. Polym. Sci. Part A: Polym. Chem. Ed.*, **29** (1991) 651.
19. S. Kohjiya and S. Yamashita, *Kautschuk Gummi Kunst.*, **44** (1991) 1128.
20. S. Kohjiya, T. Hashimoto and S. Yamashita, *J. Appl. Polym. Sci.*, **44** (1992) 555.
21. T. Hashimoto, S. Sakurai, M. Morimoto, S. Nomura, S. Kohjiya and T. Kodaira, *Polymer*, **35** (1994) 2672.
22. T. Murakami, Y. Ikeda, H. Urakawa, K. Kajiwara and S. Kohjiya, *Rubber Chem. Technol.*, **73** (2000) 864.
23. T. Murakami, and Y. Ikeda, *Kobnnshi Ronbunshu*, **55** (1998) 637.
24. Y. Ikeda, T. Murakami, Y. Yuguchi and K. Kajiwara, *Macromolecules*, **31** (1998) 1246.
25. Y. Ikeda, T. Murakami, H. Urakawa, S. Kohjiya, R. Grottenmuller and M. Schmidt, *Polymer*, **43** (2002) 3483.
26. M. Gordon, *Proc. Roy. Soc.* (London), **A268** (1961) 240.
27. K. Kajiwara, W. Burchard, and M. Gordon, *Br. Polymer J.* , **2** (1970) 110.
28. K. Kajiwara, *J. Chem. Phy.*, **54** (1971) 296.
29. H. Urakawa, Y. Ikeda, Y. Yuguchi, K. Kajiwara, Y. Hirata and S. Kohjiya, *Polymer Gels and Networks*, eds. Y. Osada and A. R. Khokhlov, Marcel Dekker, New York, p.1 (2001).
30. Y. Ikeda, T. Murakami and K. Kajiwara, *J. Macromol. Sci. -Phys.*, **B40** (2001) 171.
31. Y. Ikeda, Y. Yuguchi, K. Kajiwara, T. Murakami and S. Kohjiya, *Abstract of German Rubber Conference 2000*, Nurnberg, p.199 (2000).
32. Y. Ikeda, J. Yamato, T. Murakami and K. Kajiwara, *Polymer*, **45** (2004) 8367.
33. J. M. Lupinski, K. D. Kopple and J. J. Hertz, *J. Polym. Sci., Part C*, **16** (1967) 1561.
34. A. Rembaun, W, Baumgartner and A. Eisenberg, *J. Polym. Sci., Polym. Lett. Ed.*, **6** (1968) 159.
35. L. Dominguez and W. H. Mayer, *Solid State Ionics*, **28/38** (1988) 941.
36. C. K. Chiang, G. T. Davis, C. A. Harding and T. Takahashi, *Solid State Ionics*, **18/19** (1986) 300.
37. T. Takahashi, G. T Davis, C. K. Chiang and C. A. Harding, *Solid State Ionics*, **18/19** (1986) 321.
38. J. R. MacCallum and C. A. Vincent, eds, *Polymer Electrolyte Reviews 1*, Elsevier Applied Science, London (1987).
39. J. R. MacCallum and C. A. Vincent, eds, *Polymer Electrolyte Reviews 2*, Elsevier Applied Science, London (1989).
40. F. M. Gray, Polymer Solid Electrolytes: Fundamentals and Technological Applications, VCH Publishers, New York (1991).
41. B. Scrosati, ed., *Second International Symposium on Polymer Electrolytes*, Elsevier Applied Science, London & New York (1990).
42. S. Kohjiya and Y. Ikeda, *Mater. Sci. Res. Inter.*, **4** (1998) 73.
43. Y. Ikeda, *Kobunshi Ronbunshu*, **57** (2000) 761.

44. Y. Ikeda, *J. Appl. Polym. Sci.*, **78** (2000) 1530.
45. Y. Ikeda, H. Masui, S. Shoji, T. Sakashita, Y. Matoba and S. Kohjiya, *Polym. Inter.*, **43** (1997) 269.
46. A. Nishimoto, M. Watanabe, Y. Ikeda and S. Kohjiya, *Electrochim. Acta*, **43** (1998) 1177.
47. Y. Ikeda, Y. Wada, Y. Matoba, S. Murakami and S. Kohjiya, *Electrochim. Acta*, **45** (2000) 1167.
48. Y. Ikeda, Y. Wada, Y. Matoba, S. Murakami and S. Kohjiya, *Rubber Chem. Technol.*, **73** (2000) 720.
49. Y. Ikeda, H. Masui and Y. Matoba, *J. Appl. Polym. Sci.*, **95** (2005) 178.
50. Y. Matoba, S. Shoji and Y. Ikeda, *J. Appl. Polym. Sci.*, in press.
51. S. Murakami, K. Ueda, T. Kitade, Y. Ikeda and S. Kohjiya, *Solid State Ionics*, **154–155** (2002) 399.
52. Y. Matoba, Y. Ikeda and S. Kohjiya, *Solid State Ionics*, **147** (2002) 403.
53. A. Nishimoto, K. Agehara, N. Furuya, T. Watanabe and M. Watanabe, *Macromolecules*, **32** (1999) 1541.
54. S. Kohjiya, S. Takesako, Y. Ikeda and S. Yamashita, *Polym. Bull.*, **23** (1990) 299.
55. M. Alamgir, R. D. Moulton and K. M. Abraham, *Electrochim. Acta*, **36** (1991) 773.
56. M. Watanabe, K. Nagaoka, K. Katsuro, M. Kanba and I. Shinohara, *Polym. J.*, **14** (1982) 877.
57. C. A. Furtado, G. G. Silva, M. A. Pimenta and J. C. Machado, *Electrochim. Acta*, **43** (1998) 1477.
58. B. A. Ferreira, F. Muller-Plathe, A. T. Bernardes and W. B. De Almeira, *Solid State Ionics*, **147** (2002) 361.
59. P. C. Redfern and L. A. Curtiss, *J. Power Sources*, **110** (2002) 401.
60. Y. Ikeda, M. Ikeda and F. Ito, *Solid State Ionics*, **169** (2004) 35.
61. A. Vallee, S. Besner and M. Prud'homme, *Electrochim. Acta*, **37** (1992) 1579.
62. F. Alloin, J.-Y. Sanchez and M. Armand, *Solid State Ionics*, **60** (1993) 3.
63. M. L. Williams, R. F. Landel and J. D. Ferry, *J. Am. Chem. Soc.*, **77** (1955) 3701.
64. J. Evans, C. A. Vincent and P. G. Bruce, *Polymer*, **28** (1987) 2324.

6.7 Further usefulness of rubbery matrix

Other than ion-conducting functionality, lots of possibilities may be suggested on rubbery matrix. Essentially, flexible and soft amorphous matrix can be afforded using rubbers or polymers by some modifications, both chemical and physical. In Table 6.1.1, possible functionality by using elastomeric materials is listed. Developments of functionality elastomers are especially to be emphasized now when nanotechnology is in need [1–4]. Practically the same branched poly(oxyethylene)s described here as BEC-18 was subject to gas-permeability measurements [5]. CO_2 permeability P (in 10^{-10} cm^3(STP) cm (cm^{2}s cm Hg)$^{-1}$) was found to be 770 and that of poly(dimethyl siloxane) (PDMS) was 2700. The $P(CO_2)$ of PDMS was larger, but in term of selectivity by $P(CO_2)/P(N_2)$ the value of BEC-18 was 46 *vs.* 11 of

PDMS, which suggests much higher selectivity of the branched PEO over PDMS. The introduction of oxyethylene side chains was effective not only to make the solid matrix amorphous, but also to give rise to a selective gas permeability.

We have already demonstrated the further usefulness of rubbery matrix for elastomeric functionality materials; antithrombogenic poly(urethaneurea) from poly(oxyethylene)-poly(oxytetramethylene)-poly(oxyethylene) triblock copolymer as a prepolymer which showed very good blood-compatibility as well as excellent mechanical properties [6-8], amphiphilic ionene-type elastomers [9, 10], viologen-type ionenes showing photochromism and photomechanical property [11-13] and their higher-order structures [14, 15]. Using a poly(urethaneurea) containing PEO segments, two functionalities *i.e.* biocompatibility and Li^+ ion conduction were obtained [16]. For the details refer to these literatures.

References

1. K. Eric Drexler, Engines of Creation: *The Coming Era of Nanotechnology*, Doubleday, New York (1986).
2. A C&EN Special Report on Nanotechnology, *Chem. & Eng. News*, Oct. 16 (2000).
3. D. Mulhall, *Our Molecular Future*, Prometheus Books, New York (2002).
4. S. Kohjiya, Paper presented at International Conference and Exhibition on Rubber and Allied Materials, New Delhi, India, Nov. 28–30 (2002).
5. M. Yoshino, H. Kita, K. Okamoto, M. Tabuchi and T. Sakai, *Polymer Preprints, Japan*, **51** (2002) 2620.
6. S. Kohjiya, Y. Ikeda and S. Yamashita, *Polyurethanes in Biomedical Engineering II*, eds. H. Plank, I. Syre, M. Dauner and G. Egbers, Elsevier, Amsterdam (1987).
7. Y. Ikeda, S. Kohjiya, S. Takesako and S. Yamashita, *Biomaterials*, **11** (1990) 553.
8. Y. Ikeda, S. Kohjiya, S. Yamashita and H. Fukumura, *J. Mater. Sci.: Mater. Medicine*, **2** (1991) 110.
9. S. Kohjiya, T. Ohtsuki and S. Yamashita, *Makromol. Chem., Rapid Commun.*, **2** (1981) 417.
10. S. Kohjiya, T. Ohtsuki, S. Yamashita, M. Taniguchi and T. Hashimoto, *Bull. Chem. Soc. Jpn.*, **63** (1990) 2089.
11. S. Kohjiya, T. Hashimoto, S. Yamashita and M. Irie, *Chem. Lett.*, **1985**, 1497.
12. S. Kohjiya, T. Hashimoto, S. Yamashita and M. Irie, *Makromol. Chem., Rapid Commun.*, **10** (1989) 9.
13. T. Hashimoto, S. Kohjiya, S. Yamashita and M. Irie, *J. Polym. Sci.: Part A: Polym. Chem.*, **29** (1991) 651.
14. T. Hashimoto, S. Sakurai, M. Morimoto, S. Nomura, S. Kohjiya and T. Kodaira, *Polymer*, **35** (1994) 2672.
15. Y. Ikeda, T. Murakami, Y. Yuguchi and K. Kajiwara, *Macromolecules*, **31** (1998) 1246.
16. S. Kohjiya, S. Takesako, Y. Ikeda and S. Yamashita, *Polym. Bull.*, **23** (1980) 299.

6.8 Concluding remarks

Among soft materials, elastomers are expected to play the most important role. Elastomers contain crosslinked rubbers or rubber vulcanizates and thermoplastic elastomers (TPE). The former are chemically crosslinked, and quite often mixed with inorganic fillers, while in the latter network structures are due to physical interaction not by chemical bonds. In both, some kinds of network structure are assumed, which are needed in order to display rubber elasticity. Therefore, elastomers are expected to be of use where rubber elasticity is needed. From this point of view, it has already been emphasized that a compromise may be necessary between rubber elasticity and a targeted high functionality.

In this chapter, elastomers have been subjected to the developmental studies of lithium-ion conducting polymers by taking advantage of the characteristics of rubbery state, which is possible to be realized even without crosslinking if the molar mass of a polymer is high and its T_g is well below room temperature. Polyethers, *i.e.*, poly(oxyethylene)s and poly(oxytetramethylene)s, are among promising candidates to be developed as elastomers. Poly(oxyethylene)s were found to show promising behaviors for lithium-ion conduction when doped with a suitable lithium salt. They may be mixed with ion-conducting inorganic filler for composite-type solid electrolytes, or may be crosslinked for better mechanical performances. Ionene-type elastomers having oxytetramethylene units are other promising materials, the performance of which should be developed more in a future.

7

First principles calculations of lithium battery materials

7.1 Introduction

A number of materials with the ability of insertion and/or extraction of lithium have been investigated for application to electrode materials of primary and/or secondary lithium batteries. In 1970s, transition-metal chalcogenides, e.g., TiS_2 and MoS_2, attracted attention as positive electrode materials [1–3]. In 1980, Mizushima et al. first reported $LiCoO_2$ as the positive electrode material [4]. Its voltage is as high as approximately 4 V against metallic lithium, though the voltages of the many chalcogenides are approximately 2 V. Since then, numerous investigations on the $3d$ transition-metal oxides as the electrode materials have been performed [5–8]. $3d$ transition-metal atoms are lighter and smaller than $4d$ and $5d$ ones. Oxygen atom is also lighter and smaller than chalcogens. Therefore, $3d$ transition-metal oxides are more attractive than $4d/5d$ transition-metal oxides and chalcogenides from the viewpoint of larger electric capacities both per weight and volume. Then, the $3d$ transition-metal oxides using a solid-state redox reaction between +3 and +4 valences of transition-metal ion associated with Li insertion/extraction have been widely investigated. Some of them, e.g., $LiCoO_2$, $LiNiO_2$ and $LiMn_2O_4$, show high voltages of approximately 4 V against metallic lithium. On the basis of the systematic studies on the voltages of $3d$, $4d$ and $5d$ transition-metal oxides, it is found that $3d$ transition-metal oxides tend to show higher voltages than $4d$ and $5d$ transition-metal oxides [5]. Thus, $3d$ transition-meal oxides are also preferable from the viewpoint of higher battery voltage and larger energy density.

Layered $LiMO_2$ (M = $3d$ transition-metal element) with α-$NaFeO_2$-structural type is a series that has been widely investigated [7]. For it, the redox reaction of M(III)/M(IV) can be used. Its crystal structure is of an ordered rock-salt type such that Li and M occupy alternate (111) layers. Among the layered oxides, $LiCoO_2$ was first investigated as electrode material and has been used as the positive electrode material of commercial lithium-ion batteries. It has good electrochemical properties [4, 9–11], namely, high operating voltage of approximately 4 V against metallic lithium, large reversible capacity, high energy density and excellent cyclic durability. Three other oxides, $LiVO_2$, $LiCrO_2$ and $LiNiO_2$, also have the isomorphic crystal structure as $LiCoO_2$. $LiNiO_2$ has somehow lower voltage and larger reversible capacity than $LiCoO_2$ [12, 13]. In contrast, $LiVO_2$ and $LiCrO_2$ have less reversible capacities [14–16]. Crystal structures of $LiTiO_2$, $LiMnO_2$ and $LiFeO_2$ are related to the rock-salt type, but they are not really isostructural to α-$NaFeO_2$ [17–19]. Although $LiFeO_2$ and $LiMnO_2$ would obviously be of great interest, the structures conventionally prepared do not allow Li to be intercalated reversibly to any significant degree. Several different approaches [20–24] have been attempted to prepare $LiMnO_2$ and $LiFeO_2$ with the α-$NaFeO_2$-type structure. However, those with good reversibility of insertion and extraction of Li have not been prepared yet.

Spinel-type lithium transition-metal oxides are another series of popular oxides for electrode materials [6–8]. Among them, $LiMn_2O_4$ has been investigated widely [25–29]. Its capacity is a little smaller than that of $LiCoO_2$ and fading of the capacity during charge-discharge cycles is not negligible for application to commercial use. It, however, has some advantages, e.g., low cost, low toxicity and safety in overcharged states. Thus it is believed as a hopeful candidate to replace $LiCoO_2$. Elaborate studies to improve the capacity fading have been performed with control of stoichiometry [27–29] and doping of cations [30–33]. Some alloyed oxides of $LiMn_{2-x}M_xO_4$ (M = Cr, Fe, Co, Ni and Cu) [34–41] are known to have additional plateaus of approximately 5 V in charge/discharge curves. Therefore, they have been investigated with great interests from the viewpoints of both application for high voltage electrode material and understanding solid-state chemistry.

Recently, other types of transition-metal compounds, i.e., ordered olivine-type $LiMXO_4$ (M = transition-metal element, and X = P and V) [42–44] and NASICON-type $Li_xM_2(XO_4)_3$ (M = transition-metal element, and X = S, P and Mo) [45–47], are investigated. Among them, $LiFePO_4$ and $Fe_2(SO_4)_3$ are of great interest from the viewpoint of low cost and low toxicity, despite their low voltages of 3.4 V and 3.6 V, respectively. Because original NASICON, $Na_{1+3x}Zr_2(P_{1-x}Si_xO_4)_3$, is a superionic conductor of Na ion, some NASICON-type compounds, e.g., $Li_{1+x}Ti_{2-x}Al_x(PO_4)_3$, are investigated as Li ion conductors [48, 49], which will be applied for solid-state electrolytes of advanced lithium batteries.

For electrode materials of rechargeable lithium batteries, examination for many properties such as operating voltage, reversible capacity, cyclic durability, chemical stability and mobility of Li ion, is very important. While many studies on the electrochemical and physical properties of the compounds have been performed, the mechanism on Li intercalation process has not been clear. For example, removal of Li from the layered oxides, e.g., $LiCoO_2$, does not seem desirable because of increase in electrostatic repulsion between oxygen layers. It is in fact, however, possible to extract almost all Li from $LiCoO_2$ and $LiNiO_2$ without destruction of the layered structure [10, 11, 13]. Conversely, it is impossible in the cases of $LiVO_2$ and $LiCrO_2$ [14–16]. The reason for such a difference has not been clarified. Some properties, e.g., operating voltages and conductivity of Li ion, have been examined in detail by experiments. Roughly speaking, the voltage increases with rising atomic number of the transition-metal element in each period in the Periodic table. Anion with high polarizability is believed to be preferable for higher Li conductivity. The factors determining these properties, however, have not been clear. Almost all the properties

7.1 Introduction

of compounds are influenced by defects. Occasionally the defects play decisive roles of the properties. Some compounds, e.g., $LiNiO_2$ and $LiMn_2O_4$, are known to have nonstoichiometry easily [13, 27–29]. Although effects of the nonstoichiometry on the electrochemical properties are studied, our knowledge on the defects associated with the nonstoichiometry is rather poor. In some cases, it is not established even what type of defect is introduced by the nonstoichiometry. In order to design and to develop advanced electrode materials rationally, the electronic mechanism behind the intercalation process should be thoroughly understood. Theoretical calculations are indispensable for such a purpose.

Recently some theoretical studies have been performed. Czyżyk *et al.* have first applied the first principles calculation to the electrode material of the lithium battery [50]. They reported electronic structure of $LiCoO_2$ using the localized spherical waves (LSW) method. Miura *et al.* were the first who used the first principles calculation to explain the voltage of $LiMn_2O_4$ using the discrete variational (DV)-$X\alpha$ method on model clusters [51]. Wolverton and Zunger [52–54] studied Li/Co ordering in $LiCoO_2$ with several crystal structures. Li/vacancy ordering associated with the Li intercalation process in layered $LiCoO_2$ was also investigated using a combination of the first principles total energies by the full-potential linearized augmented plane wave (FLAPW) method, a cluster expansion technique and Monte Carlo simulations. They succeeded in finding a finite temperature order-disorder transition as well as in reproducing of the stable ground state configuration. Its voltage was also discussed in detail. Ceder and his coworkers [55–61] conducted detailed and systematic studies on the intercalation process in layered and spinel-type lithium transition-metal oxides using the first principles pseudopotential method, a cluster expansion technique and Monte Carlo simulations with special emphasis on the voltages and phase diagrams.

Despite these elaborate studies and successes, the important question from the viewpoint of the design and development of electrode materials has not been answered: What is the factor determining the electrochemical properties, e.g., operating voltage, reversible capacity, cyclic durability and mobility of Li ion? In the present thesis, the first principles calculations are performed in order to understand electronic mechanisms which determine the important electrochemical properties.

In Section 7.2, the focus is directed to understanding of the difference of intercalation process in α-$NaFeO_2$-structural $LiMO_2$ (M = V, Cr, Co and Ni) from the viewpoint of the change in chemical bondings associated with the insertion/extraction of Li. The first principles molecular orbital calculation by the DV-$X\alpha$ method using model clusters is employed. The difference of the reversible capacities is discussed from the viewpoint of the stability of layered structure of Li-extracted MO_2. A part of this work was published in ref [62]. In Section 7.3, the results by the molecular orbital calculation are reexamined in order to discuss the factor determining the voltage of layered $LiMO_2$ (M = Ti - Ni). This is followed by the first principles band-structure calculation using the FLAPW method in order to make quantitative discussion of the voltage [63]. In Section 7.4, on the basis of the discussion on the factors determining voltage, new cobalt fluorides are proposed as positive electrode materials for high voltage lithium batteries. The proposal is quantitatively examined through first principles band-structure calculations. They are predicted to have high voltages of approximately 6 V against metallic lithium despite the use of the same redox reaction of Co(III)/Co(IV) as that in $LiCoO_2$. A part of results was published in ref [64].

In Section 7.5, formation energies of defects in $LiMn_2O_4$ are calculated. Even though it is well known that exactly stoichiometric $LiMn_2O_4$ is difficult to be prepared and that the nonstoichiometry influences its performance as the electrode material, we have little information about the defects introduced by the nonstoichiometry. The preferable defect species in oxygen deficient $LiMn_2O_4$ is determined by the first principles pseudopotential method using plane-wave basis. Local electronic structure around the preferable defect is also investigated and discussed. The work was partially described in ref [65].

References

1. A. H. Thompson, *J. Electrochem. Soc.*, **126** (1979) 608.
2. T. Jacobsen, K. West and S. Atlung, *J. Electrochem. Soc.*, **126** (1979) 2169.
3. K. West, T. Jacobsen, B. Zachau-Christiansen and S. Atlung, *Electrochim. Acta*, **28** (1983) 97.
4. K. Mizushima, P. C. Jones, P. J. Wiseman and J. B. Goodenough, *Mater. Res. Bull.*, **15** (1980) 783.
5. T. Ohzuku and A. Ueda, *Solid State Ionics*, **69** (1994) 201.
6. E. Ferg, R. J. Gummow, A. de Kock and M. M. Thackeray, *J. Electrochem. Soc.*, **141** (1994) L147.
7. R. Koksbang, J. Barker, H. Shi and M. Y. Saïdi, *Solid State Ionics*, **84** (1996) 1.
8. M. M. Thackeray, *J. Am. Ceram. Soc.*, **82** (1999) 3347.
9. J. N. Reimers and J. R. Dahn, *J. Electrochem. Soc.*, **139** (1992) 2091.
10. T. Ohzuku and A. Ueda, *J. Electrochem. Soc.*, **141** (1994) 2972.
11. G. G. Amatucci, J. M. Tarascon and L. C. Klein, *J. Electrochem. Soc.*, **143** (1996) 1159.
12. T. Ohzuku, A. Ueda, M. Nagayama, Y. Iwakoshi and H. Komori, *Electrochim. Acta*, **38** (1993) 1159.
13. A. Hirano, R. Kanno, Y. Kawamoto, Y. Takeda, K. Yamaura, M. Takano, K. Ohyama, M. Ohashi and Y. Yamaguchi, *Solid State Ionics*, **78** (1995) 123.
14. L. A. de Picciotto, M. M. Thackeray, W. I. F. David, P. G. Bruce and J. B. Goodenough, *Mater. Res. Bull.*, **19** (1984) 1497.
15. L. A. de Picciotto and M. M. Thackeray, *Mater. Res. Bull.*, **20** (1985) 1409.
16. C. D. W. Jones, E. Rossen and J. R. Dahn, *Solid State Ionics*, **68** (1994) 65.
17. T. A. Hewston and B. L. Chamberland, *J. Phys. Chem. Solids*, **48** (1987) 97.
18. J. N. Reimers, E. W. Fuller, E. Rossen and J. R. Dahn, *J. Electrochem. Soc.*, **140** (1993) 3396.
19. I. Koetschau, M. N. Richard, J. R. Dahn, J. B. Soupart and J. C. Rousche, *J. Electrochem. Soc.*, **142** (1995) 2906.
20. F. Capitaine, P. Gravereau and C. Delmas, *Solid State Ionics*, **89** (1996) 197.

21. Y.-I. Jang, B. Huang, Y.-M. Chiang and D. R. Sadoway, *Electrochem. Solid-State Lett.*, **1** (1998) 13.
22. B. Ammundsen, J. Desilvestro, T. Groutso, D. Hassell, J. B. Metson, E. Regan, R. Steiner and P. J. Pickering, *J. Electrochem. Soc.*, **147** (2000) 4078.
23. K. Ado, M. Tabuchi, H. Kobayashi, H. Kageyama, O. Nakamura, Y. Inaba, R. Kanno, M. Takagi and Y. Takeda, *J. Electrochem. Soc.*, **144** (1997) L177.
24. R. Kanno, T. Shirane, Y. Inaba and Y. Kawamoto, *J. Power Sources*, **68** (1997) 145.
25. M. M. Thackeray, P. J. Johnson, L. A. de Picciotto, P. G. Bruece and J. B. Goodenough, *Mater. Res. Bull.*, **19** (1984) 179.
26. J. M. Tarascon, E. Wang, F. K. Shokoohi, W. R. McKinnon and S. Colson, *J. Electrochem. Soc.*, **138** (1991) 2859.
27. M. M. Thackeray, A. de Kock, M. H. Rossouw, D. Eiles, R. Bittihn and D. Hoge, *J. Electrochem. Soc.*, **139** (1992) 363.
28. Y. Gao and J. R. Dahn, *J. Electrochem. Soc.*, **143** (1996) 100.
29. Y. Xia, T. Sakai, T. Fujieda, X. Q. Yang, X. Sun, Z. F. Ma, M. McBreen and M. Yoshio, *J. Electrochem. Soc.*, **148** (2001) A723.
30. R. J. Gummow, A. de Kock and M. M. Thackeray, *Solid State Ionics*, **69** (1994) 59.
31. L. Guohua, H. Ikuta, T. Uchida and M. Wakihara, *J. Electrochem. Soc.*, **143** (1996) 178.
32. G. Pistoia, A. Antonini, R. Rosati, C. Bellitto and G. M. Ingo, *Chem. Mater.*, **9** (1997) 1443.
33. J. H. Lee, J. K. Hong, D. H. Jang, Y.-K. Sun and S. M. Oh, *J. Power Sources*, **89** (2000) 7.
34. C. Sigala, D. Guyomard, A. Verbaere, Y. Piffard and M. Tournoux, *Solid State Ionics*, **81** (1995) 167.
35. Y. Gao, K. Myrtle, M. Zhang, J. N. Reimers and J. R. Dahn, *Phys. Rev. B*, **54** (1996) 16670.
36. Q. Zhong, A. Bonakdarpour, M. Zhang, Y. Gao and J. R. Dahn, *J. Electrochem. Soc.*, **144** (1997) 205.
37. Y. Ein-Eli and W. F. Howard, Jr., *J. Electrochem. Soc.*, **144** (1997) L205.
38. M. N. Obrovac, Y. Gao and J. R. Dahn, *Phys. Rev. B*, **57** (1998) 5728.
39. H. Kawai, M. Nagata, H. Tukamoto and A. R. West, *J. Mater. Chem.*, **8** (1998) 837.
40. T. Ohzuku, S. Takeda and M. Iwanaga, *J. Power Sources*, **81–82** (1999) 90.
41. M. Okada, Y.-S. Lee and M. Yoshio, *J. Power Sources*, **90** (2000) 196.
42. A. K. Padhi, K. S. Nanjundaswamy and J. B. Goodenough, *J. Electrochem. Soc.*, **144** (1997) 1188.
43. K. Amine, H. Yasuda and M. Yamachi, *Electrochem. Solid-State Lett.*, **3** (2000) 178.
44. A. Yamada, Y. Kudo and K.-Y. Liu, *J. Electrochem. Soc.*, **148** (2001) A747.
45. C. Delmas, A. Nadiri and J. L. Soubeyroux, *Solid State Ionics*, **28–30** (1988) 419.
46. K. S. Nanjundaswamy, A. K. Padhi, J. B. Goodenough, S. Okada, H. Ohtsuka, H. Arai and J. Yamaki, *Solid State Ionics*, **92** (1996) 1.

47. A. K. Padhi, K. S. Nanjundaswamy, C. Masquelier and J. B. Goodenough, *J. Electrochem. Soc.*, **144** (1997) 2581.
48. H. Aono, E. Sugimoto, Y. Sadaoka, N. Imanaka and G. Adachi, *J. Electrochem. Soc.*, **136** (1989) 590.
49. S. Wong, P. J. Newman, A. S. Best, K. M. Nairn, D. R. MacFarlane and M. Forsyth, *J. Mater. Chem.*, **8** (1998) 2199.
50. M. T. Czyżyk, R. Potze and G. A. Sawatzky, *Phys. Rev. B*, **46** (1992) 3729.
51. K. Miura, A. Yamada and M. Tanaka, *Electrochim. Acta*, **41** (1996) 249.
52. C. Wolverton and A. Zunger, *Phys. Rev. B*, **57** (1998) 2242.
53. C. Wolverton and A. Zunger, *J. Electrochem. Soc.*, **145** (1998) 2424.
54. C. Wolverton and A. Zunger, *Phys. Rev. Lett.*, **20** (1998) 606.
55. M. K. Aydinol, A. F. Kohan, G. Ceder, K. Cho and J. Joannopoulos, *Phys. Rev. B*, **56** (1997) 1354.
56. G. Ceder, M. K. Aydinol and A. F. Kohan, *Comp. Mater. Sci.*, **8** (1997) 161.
57. M. K. Aydinol and G. Ceder, *J. Electrochem. Soc.*, **144** (1997) 3832.
58. A. Van der Ven, M. K. Aydinol, G. Ceder, G. Kresse and J. Hafner, *Phys. Rev. B*, **58** (1998) 2975.
59. A. Van der Ven, M. K. Aydinol and G. Ceder, *J. Electrochem. Soc.*, **145** (1998) 2149.
60. S. K. Mishra and G. Ceder, *Phys. Rev. B*, **59** (1999) 6120.
61. A. Van der Ven, C. Marianetti, D. Morgan and G. Ceder, *Solid State Ionics*, **135** (2000) 21.
62. Y. Koyama, Y-S. Kim, I. Tanaka and H. Adachi, *Jpn. J. Appl. Phys.*, **38** (1999) 2024.
63. Y. Koyama, Y-S. Kim, I. Tanaka, S.R. Nishitani and H. Adachi, *Jpn. J. Appl. Phys.*, **38** (1999) 4804.
64. Y. Koyama, I. Tanaka and H. Adachi, *J. Electrochem. Soc.*, **147** (2000) 3633.
65. Y. Koyama, I. Tanaka, H. Adachi, Y. Uchimoto and M. Wakihara, *J. Electrochem. Soc.*, **150** (2003) A63.

7.2 Changes in chemical bondings by lithium insertion/extraction in $LiMO_2$ (M = V, Cr, Co and Ni)

7.2.1 Introduction

Lithium transition-metal oxides have been extensively studied for application to electrode materials of lithium batteries [1–4]. Among them, layered $LiCoO_2$ has favorable attributes such as high voltage, large capacity, high energy density and excellent cyclic durability [5–8]. Its crystal structure is an ordered rock-salt structure such that Li and Co occupy alternate (111) layers, as shown in Fig. 7.2.1. While many studies of the electrochemical properties have been conducted, there still exist some unanswered questions. For example, the extraction of Li from layered $LiCoO_2$ results in an increase in electrostatic repulsion between bare oxygen layers. Thus it seems undesirable to extract much Li from layered $LiCoO_2$. However, we

7.2 Changes in chemical bondings by lithium insertion/extraction in LiMO$_2$

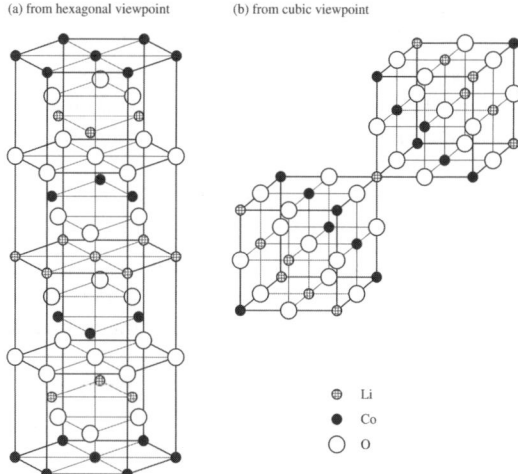

Fig. 7.2.1. Crystal structure of α-NaFeO$_2$-structural type LiCoO$_2$ (a) as hexagonal view and (b) cubic view. Li, Co and O atoms are denoted by crossed-patterned, filled and open circles, respectively. Other layered LiMO$_2$ are isomorphic.

can actually extract almost all Li from LiCoO$_2$ without destruction of the layered structure. Three other compounds, LiVO$_2$, LiCrO$_2$ and LiNiO$_2$ have an isomorphic crystal structure. A large fraction of Li can be extracted from LiNiO$_2$ [9, 10] but not from LiVO$_2$ [11, 12] and LiCrO$_2$ [13]. The reason for such a difference has not been clarified quantitatively. In order to design and develop rationally advanced electrode materials, the electronic mechanism behind the intercalation process should be thoroughly understood. First principles electronic structure calculations are indispensable for such a purpose.

Czyżyk et al. first reported electronic structure of LiCoO$_2$ using the localized spherical waves (LSW) method [14]. Aydinol and coworkers [15–18] conducted a more detailed and systematic study on the intercalation process of LiMO$_2$ (M = Ti, V, Mn, Co, Ni, Cu, Zn and Al) by the first principles pseudopotential method with special emphasis on their voltages. Their initial success was reinforced by including a cluster expansion technique and Monte Carlo simulations for predicting the phase diagram of Li$_x$CoO$_2$. Wolverton and Zunger [19 21] studied Li/Co and Li-vacancy/Co orderings associated with the intercalation process of LiCoO$_2$ using a combination of first principles total energies by the full-potential linearized augmented plane wave (FLAPW) method, a cluster expansion technique and Monte Carlo simulations. They succeeded in finding a finite temperature order-disorder transition as well as in reproducing of the stable ground state configuration. Its voltage was also discussed in detail.

In spite of these extensive studies, the naive questions remain unanswered: Why can we extract almost all Li from LiCoO$_2$ and LiNiO$_2$ despite electrostatic repulsion between bare oxygen layers? Why is it not the case for LiVO$_2$ and LiCrO$_2$? In the present section, the focus is directed to understanding the difference among these isomorphic oxides from the viewpoint of the change in chemical bondings associated with the intercalation of Li. A first principles molecular orbital calculation by the discrete variational (DV)-Xα method using model clusters was employed. It is true that the use of the cluster method does not allow us to predict the voltage quantitatively because of the ambiguity associated with the presence of the cluster surface. However, electronic structures obtained by cluster calculations have been found to be reliable at least qualitatively in a number of metal oxides when the cluster size is sufficiently large. Molecular orbital calculations using a minimal number of atomic orbitals as basis functions are very useful for understanding the phenomena from the viewpoint of chemical bondings. This is the greatest advantage in exchange for the lack of precision in the present type of calculations.

7.2.2 Computational procedure

All calculations were performed by means of a nonrelativistic first principles molecular orbital (MO) method using model clusters. The computer code called SCAT [22], which is a modified version of the original DV-Xα program [23, 24], was employed. Spin polarization was taken into account. Numerical atomic orbitals (NAOs) were used as basis functions. They were generated flexibly by solving the radial part of the Schrödinger equation for a given environment. Minimal basis sets were used in order to clarify the simple relationship between spectral features and chemical bondings. Basis sets were $1s$, $2s$ and $2p$ for Li and O, and $1s$, $2s$, $2p$, $3s$, $3p$, $3d$, $4s$ and $4p$ for $3d$ transition-metal elements. Integrations to obtain energy eigenvalues and eigenfunctions were made numerically.

Population analyses were made in the standard Mulliken's manner [25]. The overlap population between the i-th atomic orbital and the j-th atomic orbital at the l-th MO is given by

$$Q^l_{ij} = C_{il} C_{jl} S_{ij}, \qquad (7.2.1)$$

where S_{ij} is the overlap integral given by

$$\int \chi_i^*(\mathbf{r}) \chi_j(\mathbf{r}) d\mathbf{r} = S_{ij}. \qquad (7.2.2)$$

χ_i and C_{il} are the i-th atomic orbital and its coefficient for the l-th MO. The overlap population between atoms A and B at the l-th MO is given by

$$Q^l_{AB} = \sum_{i \in A} \sum_{j \in B} Q^l_{ij}. \qquad (7.2.3)$$

Overlap population diagrams were made by broadening Q^l_{AB} at individual MOs using Gaussian function with the full width at half maximum (FWHM) of 1.0 eV. The sum of Q^l_{AB} over occupied orbitals is called bond-overlap population, Q_{AB}, which is defined by

7.2 Changes in chemical bondings by lithium insertion/extraction in LiMO$_2$ 233

$$Q_{AB} = \sum_{l} f_l Q^l_{AB}, \quad (7.2.4)$$

where f_l is the occupation number of the l-th MO. The orbital population of the i-th orbital is given by

$$Q_i = \sum_{l} f_l \sum_{j} Q^l_{ij}. \quad (7.2.5)$$

The net charge ΔQ_A of each atom is obtained by

$$\Delta Q_A = Z_A - \sum_{i \in A} Q_i, \quad (7.2.6)$$

where Z_A is the atomic number of atom A.

Two kinds of model clusters have been chosen for LiMO$_2$ and MO$_2$, as shown in Fig.7.2.2. A set of smaller clusters has a transition-metal ion at its center. Li ions are located at the surface of the small cluster. On the other hand, Li ions are not located at the surface of the larger clusters. As will be seen later, results from the clusters of two sizes are the same from the viewpoint of qualitative chemical bondings.

Four elements, V, Cr, Co and Ni were selected as M, since experimental crystal structures were established only for these four LiMO$_2$ oxides. Lattice parameters were adopted from the experimental data of LiMO$_2$. The same lattice parameters were used for the calculations of MO$_2$. In other words, lattice relaxation associated with the extraction of Li was neglected. The model clusters were embedded in electrostatic potential generated by approximately 10,000 point charges of formal values. The values of Li, M and O in LiMO$_2$ were +1, +3 and –2, respectively. In MO$_2$, they were +4 and –2 for M and O, respectively. Convergence of the electrostatic potential with respect to dipole and quadrupole sums [26] was established within an accuracy of 0.1 %.

7.2.3 Electronic and bonding states of LiCoO$_2$ and CoO$_2$

Total density of states (DOS) and partial DOS of LiCoO$_2$ are shown in Fig.7.2.3 together with energy level diagram. DOS by the two clusters resemble each other in the valence band except for an energy shift of 1.5 eV. The filled band located from –7 to 0 eV is mainly composed of O-$2p$ orbitals. The partially filled band located around 3 eV is mainly composed of Co-$3d$ orbitals. An unoccupied band located above 10 eV is made up of Li-$2s$, $2p$ and Co-$4s$, $4p$ orbitals. The shape of the unoccupied band exhibits small dependence on the cluster size. Regardless of the cluster size, however, Li-$2s$ and $2p$ states have very small contributions to the valence band. Li in LiCoO$_2$ is therefore nearly completely ionized. On the other hand, significant amounts of Co-$3d$, $4s$ and $4p$ states in the O-$2p$ band are found. A mixture of O-$2p$ states in the Co-$3d$ band is also notable. Strong covalent interaction between Co and O is likewise noted. This can be ascribed to the fact that Li is virtually in the state of Li$^+$ ion and make slight covalent interaction with the surrounding atoms. The location of Li therefore does not affect the electronic states of the clusters significantly. The

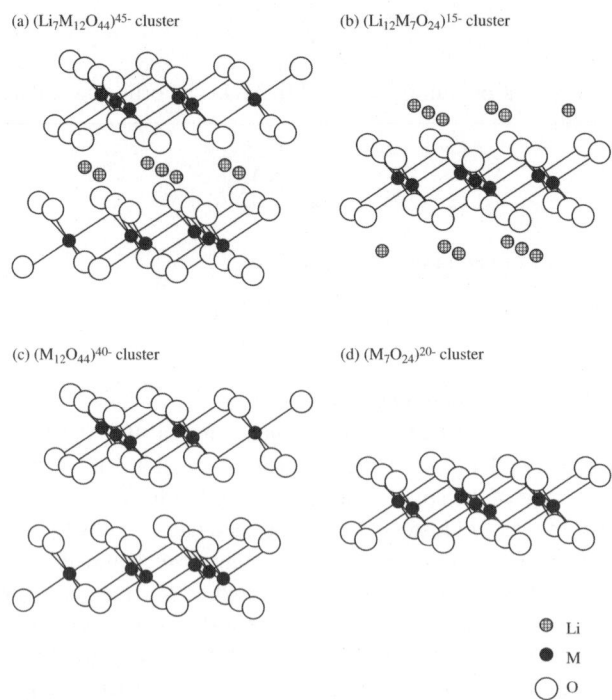

Fig. 7.2.2. Model clusters used in the present study, (a) $(Li_7M_{12}O_{44})^{45-}$ and (b) $(Li_{12}M_7O_{24})^{15-}$ clusters for $LiMO_2$, and (c) $(M_{12}O_{44})^{40-}$ and (d) $(M_7O_{24})^{20-}$ clusters for MO_2 (M = V, Cr, Co and Ni). Notations of atoms are the same as in Fig. 7.2.1.

results of the larger clusters will be used hereafter. However, the same arguments can be made when the small ones are used. In order to discuss the covalent interactions in more detail, overlap population diagrams for Co-O, Li-O, Co-Co and Li-Li bonds are plotted, as shown in Fig.7.2.4. The value given in each diagram is the bond overlap population that is given by the integration of the diagram with respect to the occupied bands. As qualitatively discussed, strong covalent bonding can be noted between Co and O. On the other hand, the bond overlap population between Li and O is small, but not zero. Cation-cation bond overlap populations are very small, except for the higher energy region of the conduction band. Energy level structures, total and partial DOS of CoO_2 by two kinds of clusters are shown in Fig.7.2.5. Compared with those of $LiCoO_2$, the removal of Li results in O-$2p$ and Co-$3d$ bands being closer in energy. Components of Co-$3d$, $4s$ and $4p$ in the O-$2p$ band significantly increase, resulting in stronger Co-O covalent bonding. Overlap population diagrams for Co-O and Co-Co bonds are shown in Fig.7.2.6. Since an electron is removed from the Co-$3d$ band, in which the Co-O interaction is remarkably antibonding, the bond overlap population of Co-O is increased from 0.184 to 0.263 by the removal of Li. It has been conventionally assumed that the removal of Li from $LiCoO_2$ transforms the

7.2 Changes in chemical bondings by lithium insertion/extraction in LiMO$_2$

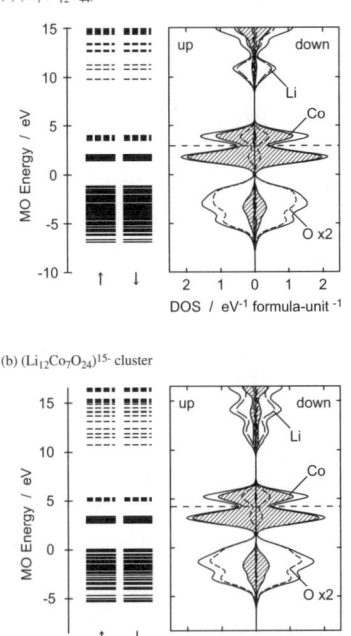

Fig. 7.2.3. Energy level diagrams and total and partial density of states of LiCoO$_2$ obtained (a) by (Li$_7$Co$_{12}$O$_{44}$)$^{45-}$ cluster and (b) by (Li$_{12}$Co$_7$O$_{24}$)$^{15-}$ cluster.

electronic state of Co from Co(III) to Co(IV). However, because of the significant admixture of O-2p and Co-3d due to the lowering of the Co-3d band, Co cannot be "oxidized" during the extraction of Li. The net charges of Co according to Eq. 7.2.6 are +1.47 in LiCoO$_2$ and +1.30 in CoO$_2$ (see Fig.7.2.9). Contrary to the formal redox notation, Co is virtually "reduced" by the extraction of Li. In exchange for the "reduction", oxygen ions are "oxidized". As a matter of fact, the net charge of O is changed from −1.00 to −0.48 by the removal of Li. The result might appear puzzling for some readers. However, this is quite normal for 3d transition-metal ions having a high formal valence. The formal valence of oxygen ion is always fixed at −2 in conventional (formal) textbooks, which may be the source of the puzzle. Electronic state of oxygen in CoO$_2$ is very close to a neutral atom. The "oxidation" of oxygen can also be seen in the spatial charge distribution. Figure 7.2.7 shows the difference in charge density before and after the removal of Li. Darker areas in the figure correspond to a greater decrease in charge density by the removal. The charge density is significantly decreased around oxygen. It is also decreased from one of the d-orbitals of Co. Conversely, it is increased in the region between Co and O. As a result, the net charge of Co decreases by 0.17. The same quantitative results were reported by Aydinol et al. [15, 16] who applied the first principles pseudopotential method using plane-wave basis functions. Although first principles methods are

236   7 First principles calculations of lithium battery materials

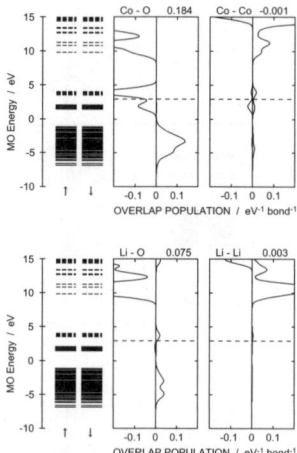

Fig. 7.2.4. Overlap population diagrams for Co-O, Co-Co, Li-O and Li-Li bonds in LiCoO$_2$ obtained by (Li$_7$Co$_{12}$O$_{44}$)$^{45-}$ cluster.

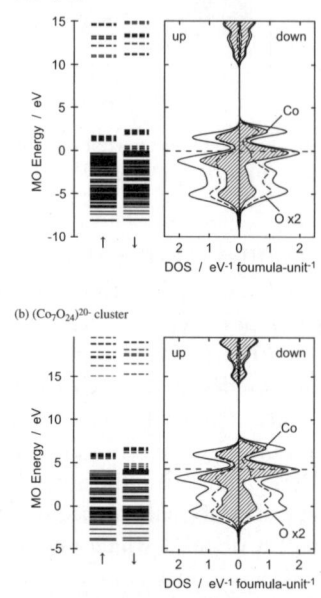

Fig. 7.2.5. Energy level diagrams and total and partial density of states of CoO$_2$ obtained (a) by (Co$_{12}$O$_{44}$)$^{40-}$ cluster and (b) by (Co$_7$O$_{24}$)$^{20-}$ cluster.

7.2 Changes in chemical bondings by lithium insertion/extraction in LiMO$_2$ 237

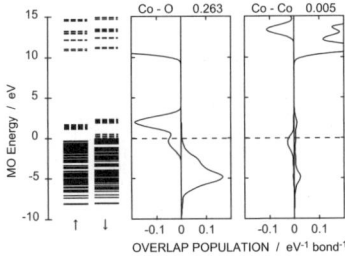

Fig. 7.2.6. Overlap population diagrams for Co-O and Co-Co bonds in CoO$_2$ obtained by (Co$_{12}$O$_{44}$)$^{40-}$ cluster.

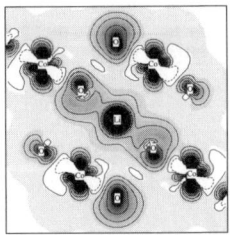

Fig. 7.2.7. Difference in charge density on the (11$\bar{2}$0) plane before and after the removal of Li in LiCoO$_2$. Darker areas indicate a greater decrease in charge density by the removal of Li. Contour maps of −0.02, −0.01, −0.005, 0 and +0.005 are overlaid.

applied, results using different kinds of basis functions are contradictory in general. Agreement between the result obtained by them and the present one confirms the idea that the "oxidation" takes place not on Co but on O.

7.2.4 Differences in bonding states among LiMO$_2$/MO$_2$

Figure 7.2.8 shows energy level diagrams for model clusters of LiMO$_2$ and MO$_2$ (M = V, Cr, Co and Ni). The four LiMO$_2$ results differ mainly in the energy and the magnitude of spin polarization of the M-3d band. The response of the electronic structure due to the removal of Li is qualitatively the same among the four compounds. As can be seen in Fig.7.2.9, the net charge of Li in the model clusters for LiMO$_2$ is approximately +0.7, which is almost independent of the transition-metal species. The net charge of M in LiMO$_2$ gradually decreases with increasing atomic number of M, because of the greater overlapping of O-2p and M-3d orbitals.

The "oxidation" associated with the extraction of Li takes place not on M but on O in all four LiMO$_2$, as shown in Fig.7.2.9. As a result, the net charges of O in CoO$_2$ and NiO$_2$ are closer to zero compared with those in VO$_2$ and CrO$_2$. As pointed out above, the extraction of Li from LiMO$_2$ brings about a strong electrostatic repulsion between bare oxygen layers from a viewpoint on formal charges. However, the real electrostatic repulsion is much smaller than that expected for oxygen having a formal charge of −2. It should be even weaker for CoO$_2$ and NiO$_2$. This seems to be the

238   7 First principles calculations of lithium battery materials

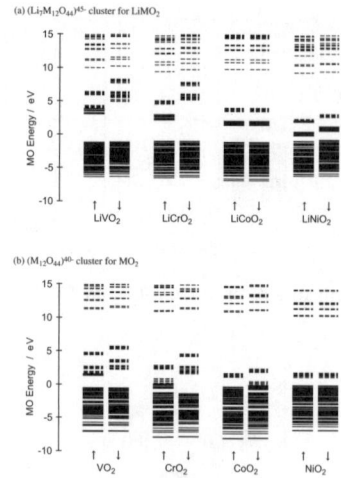

Fig. 7.2.8. Energy level diagrams (a) of $(Li_7M_{12}O_{44})^{40-}$ clusters for $LiMO_2$, and (b) of $(M_{12}O_{44})^{40-}$ clusters for MO_2 (M = V, Cr, Co and Ni).

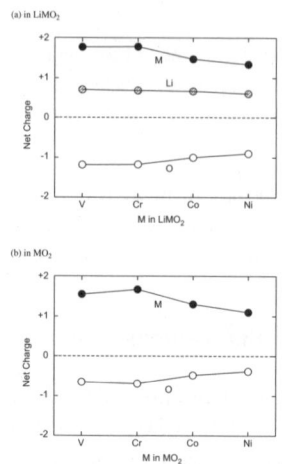

Fig. 7.2.9. (a) Net charges of Li, M and O in model clusters of $(Li_7M_{12}O_{44})^{40-}$ for $LiMO_2$ (M = V, Cr, Co and Ni). (b) Those of M and O in model clusters of $(M_{12}O_{44})^{40-}$ for MO_2.

electronic mechanism behind the large capacity of intercalation and the excellent cycle life of $LiCoO_2$ and $LiNiO_2$ compared to those of $LiVO_2$ and $LiCrO_2$.

7.2.5 Conclusion

The electronic and bonding states of $LiMO_2$ and MO_2 (M = V, Cr, Co and Ni) are investigated by the first principles molecular orbital calculation employing the DV-Xα method using model clusters. Special attention is paid to understanding the changes in chemical bondings by the extraction of Li. Li is found to be nearly completely ionized in $LiMO_2$. Strong covalent interaction between M and O is noted. The extraction of Li significantly increases the interaction between M and O. This results in the "oxidation" associated with the extraction of Li not of M but of O. The formal redox notation for the intercalation, i.e., M(III)/M(IV), is thus far from reality. The difference in the intercalation capability among four $LiMO_2$ oxides is ascribed to the difference in the magnitude of the electrostatic repulsion between the oxygen layers in MO_2. The net charges of oxygen in CoO_2 and NiO_2 are found to be close to zero, resulting in weak electrostatic repulsion and stabilization of the MO_2 structure.

References

1. T. Ohzuku and A. Ueda, *Solid State Ionics*, **69** (1994) 201.
2. E. Ferg, R. J. Gummow, A. de Kock and M. M. Thackeray, *J. Electrochem. Soc.*, **141** (1994) L147.
3. R. Koksbang, J. Barker, H. Shi and M. Y. Saïdi, *Solid State Ionics*, **84** (1996) 1.
4. M. M. Thackeray, *J. Am. Ceram. Soc.*, **82** (1999) 3347.
5. K. Mizushima, P. C. Jones, P. J. Wiseman and J. B. Goodenough, *Mater. Res. Bull.*, **15** (1980) 783.
6. J. N. Reimers and J. R. Dahn, *J. Electrochem. Soc.*, **139** (1992) 2091.
7. T. Ohzuku and A. Ueda, *J. Electrochem. Soc.*, **141** (1994) 2972.
8. G. G. Amatucci, J. M. Tarascon and L. C. Klein, *J. Electrochem. Soc.*, **143** (1996) 1114.
9. T. Ohzuku, A. Ueda, M. Nagayama, Y. Iwakoshi and H. Komori, *Electrochim. Acta*, **38** (1993) 1159.
10. A. Hirano, R. Kanno, Y. Kawamoto, Y. Takeda, K. Yamaura, M. Takano, K. Ohyama, M. Ohashi and Y. Yamaguchi, *Solid State Ionics*, **78** (1995) 123.
11. L. A. de Picciotto, M. M. Thackeray, W. I. F. David, P. G. Bruce and J. B. Goodenough, *Mater. Res. Bull.*, **19** (1984) 1497.
12. L. A. de Picciotto and M. M. Thackeray, *Mater. Res. Bull.*, **20** (1985) 1409.
13. C. D. W. Jones, E. Rossen and J. R. Dahn, *Solid State Ionics*, **68** (1994) 65.
14. M. T. Czyżyk, R. Potze and G. A. Sawatzky, *Phys. Rev. B*, **46** (1992) 3729.
15. M. K. Aydinol, A. F. Kohan, G. Ceder, K. Cho and J. Joannopoulos, *Phys. Rev. B*, **56** (1997) 1354.
16. G. Ceder, M. K. Aydinol and A. F. Kohan, *Comp. Mater. Sci.*, **8** (1997) 161.
17. A. Van der Ven, M. K. Aydinol, G. Ceder, G. Kresse and J. Hafner, *Phys. Rev. B*, **58** (1998) 2975.
18. A. Van der Ven, M. K. Aydinol and G. Ceder, *J. Electrochem. Soc.*, **145** (1998) 2149.
19. C. Wolverton and A. Zunger, *Phys. Rev. B*, **57** (1998) 2242.
20. C. Wolverton and A. Zunger, *J. Electrochem. Soc.*, **145** (1998) 2424.
21. C. Wolverton and A. Zunger, *Phys. Rev. Lett.*, **20** (1998) 606.
22. H. Adachi, M. Tsukada and C. Satoko, *J. Phys. Soc. Jpn.*, **45** (1978) 875.
23. F. W. Averill and D. E. Ellis, *J. Chem. Phys.*, **59** (1973) 6412.
24. D. E. Ellis, H. Adachi and F. W. Averill, *Surf. Sci.*, **58** (1976) 497.

25. R. S. Mulliken, *J. Chem. Phys.*, **23** (1955) 1833.
26. H. Coker, *J. Phys. Chem.*, **87** (1983) 2512.

7.3 First principles study on factors determining voltages of layered LiMO$_2$ (M = Ti - Ni)

7.3.1 Introduction

Lithium can be repeatedly extracted from and inserted into some transition-metal oxides, which have been applied as electrode materials in rechargeable lithium batteries [1–4]. Among them, LiCoO$_2$ has been widely investigated [5–8] because of its high voltage, large capacity, high energy density and excellent cyclic durability. LiCoO$_2$ and other LiMO$_2$ (M = V, Cr and Ni) exhibit a layered structure that is isostructural to α-NaFeO$_2$. In Section 7.2, results of first principles molecular orbital (MO) calculations for LiMO$_2$ and MO$_2$ were reported with special focus on the changes of chemical bondings by the removal of Li. The electronic mechanism that determines the voltages is the prime interest in the present section.

The operating voltages of lithium transition-metal oxides are examined in detail by experiments [1]. Roughly speaking, it increases with increasing atomic number of transition-metal element. For example, LiVO$_2$ shows a voltage of approximately 3 V against metallic lithium although only about one-third of Li can be extracted from it [9, 10]. On the other hand, LiCoO$_2$ and LiNiO$_2$ can be used at approximately 4 V [5–8, 11, 12]. The voltage should be related to sum of energy required to remove a Li ion from LiMO$_2$ and that to remove an electron. The former is related to the electrostatic potential at the Li position. The latter is analogous to the work function of the system, although these two factors are dependent on each other.

Some theoretical studies showed the voltages by first principles calculations. Miura *et al.* were the first to attempt to correlate the voltage of spinel-type Li$_x$Mn$_2$O$_4$ ($0 \leq x \leq 2$) using a first principles calculation by the discrete variational (DV)-Xα method on model clusters [13]. They discussed the voltage from the viewpoint of the difference in work function, but they did not take account of the chemical potential of Li ion. Moreover, they performed calculations only for LiMn$_2$O$_4$. Aydinol *et al.* made a systematic theoretical study [14, 15] on the voltages of LiMO$_2$ (M = Ti, V, Mn, Co, Ni, Cu, Zn and Al) using the first principles pseudopotential plane-wave method. Although the work truly contributed to gaining an insight into the problem, it cannot be fully sure of their results. The reason is that they did not include spin polarization into their calculations, which may be essential for the discussion of certain transition-metal oxides. Wolverton and Zunger [16–18] studied Li/Co and Li-vacancy/Co orderings associated with the intercalation process of LiCoO$_2$ using a combination of first principles total energies by the full-potential linearized augmented plane wave (FLAPW) method, a cluster expansion technique and Monte Carlo simulations. Its voltage was also discussed in detail. Additional study may be worthwhile.

7.3 First principles study on factors determining voltages of layered LiMO$_2$ 241

The present section is composed of two parts. In the first part, the results by molecular orbital calculations are reexamined in order to discuss the electronic mechanism that determines the voltage of layered LiMO$_2$. This is followed by first principles band-structure calculations using the FLAPW method in order to allow for a quantitative discussion of the voltages.

7.3.2 Computational procedure

Molecular orbital calculations were performed by means of a nonrelativistic first principles method using model clusters. The computer code SCAT [19], a modified version of an original DV-Xα program [20, 21], was employed. Exchange and correlation term by Slater with $\alpha = 0.7$ was used. Details of the computational procedure are the same as in Section 7.2. Two kinds of model clusters as shown in Fig.7.3.1 were chosen. One is centered by a transition-metal ion. Li ions are located on the surface of the clusters. On the other hand, Li ions are located at the center of the other set of clusters. As will be seen later, the results of the two kinds of clusters are the same except for a small difference in the absolute energy level. Four elements, V, Cr, Co and Ni were selected as M in Section 7.2. Although LiTiO$_2$ LiMnO$_2$ and LiFeO$_2$ with α-NaFeO$_2$-type structure are metastable or not yet synthesized, additional calculations were made in the present study. Metastable LiFeO$_2$ with α-NaFeO$_2$-type structure was synthesized by hydrothermal and ion exchange method [22, 23]. Lattice parameters were adopted from the experimental data for layered LiFeO$_2$ reported by Ado et al. [23]. The structure of layered LiMnO$_2$ [24, 25] was a distorted α-NaFeO$_2$-type due to the strong cooperative Jahn-Teller effect of high-spin Mn^{3+} ion. In the present study, the distortion was ignored and the structure was considered as a regular α-NaFeO$_2$-type, keeping the same volume and interlayer distances as the experimental data reported by Capitaine et al. [24]. Experimental synthesis of LiTiO$_2$ that is isostructural to α-NaFeO$_2$-type has not been reported yet. The lattice parameters reported by Aydinol et al. [14], which were optimized using the first principles pseudopotential plane-wave method, were adopted. Spin polarization was allowed in all of the present calculations. Low-spin states were used for LiCoO$_2$ and LiNiO$_2$ as initial states of the calculation, whereas high-spin states for LiTiO$_2$ to LiFeO$_2$. In the present DV-Xα calculations, the spin state was not constrained during self-consistent calculations. In other words, the resultant spin configuration is an energetically favorable one, at least in the sense of local minimum. As for the magnetic structure, all of the compounds were assumed to exhibit ferromagnetic ordering, since detailed magnetic structures were unknown. Singh reported that the difference in the total energy obtained by first principles band-structure calculations for ferromagnetic and antiferromagnetic LiMnO$_2$ is 0.173 eV [26]. The assumption of ferromagnetic ordering does not significantly affect the discussion in the present study.

FLAPW band-structure calculations were made using the program code WIEN97 [27], developed by Blaha et al. The sphere radius was chosen to be 0.85 Å for all atoms. The exchange and correlation term of Perdew et al. [28] was employed. The number of basis functions was determined by an energy cutoff of 340 eV. Li-1s and

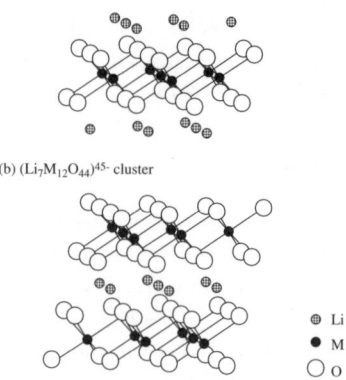

Fig. 7.3.1. Model clusters used in the present study, (a) $(Li_{12}M_7O_{24})^{15-}$ and (b) $(Li_7M_{12}O_{44})^{45-}$ clusters for $LiMO_2$ (M = Ti - Ni). Li, M and O atoms are denoted by crossed-patterned, filled and open circles, respectively.

M-3s and 3p orbitals were treated as semicore states. Reciprocal space integration was carried out using 5×5×5 mesh points in the first Brillouin zone. Convergence tests of the energy differences with respect to the basis-function cutoff, the number of mesh-points for the integration in reciprocal space and the sphere radii were carefully checked. The primitive cell of α-NaFeO$_2$-type structure is a rhombohedral one that contains only one chemical formula unit. The lattice parameters were the same as those used in cluster calculations. Spin polarization was taken into account, assuming a ferromagnetic structure.

7.3.3 Molecular orbital calculations using model clusters

Figure 7.3.2 shows MO energy level diagrams of all LiMO$_2$ by cluster calculations using the M-centered clusters, $(Li_{12}M_7O_{24})^{15-}$. With the increase in atomic number of M from Ti to Fe, the M-3d levels of majority spin (up spin) decrease in energy and the energy separation between majority and minority spins increases. This behavior is quite natural for transition-metal oxides, leading to a high-spin electronic configuration of Mn^{3+} and Fe^{3+}. On the other hand, LiCoO$_2$ and LiNiO$_2$ exhibit a low-spin configuration. This agrees well with experimental data [29]. The results of a series of Li-centered clusters, $(Li_7M_{12}O_{44})^{45-}$ are not greatly different except for absolute energy levels, which will be explained later.

The open-circuit voltage $V_{OCV}(x)$ depends on the difference in the chemical potential of Li in the positive electrode of Li$_x$MO$_2$, and that in negative electrode of metallic lithium.

$$V_{OCV}(x) = -\frac{\mu_{Li}^{positive}(x) - \mu_{Li}^{negative}}{zF} \quad . \quad (7.3.1)$$

7.3 First principles study on factors determining voltages of layered LiMO$_2$ 243

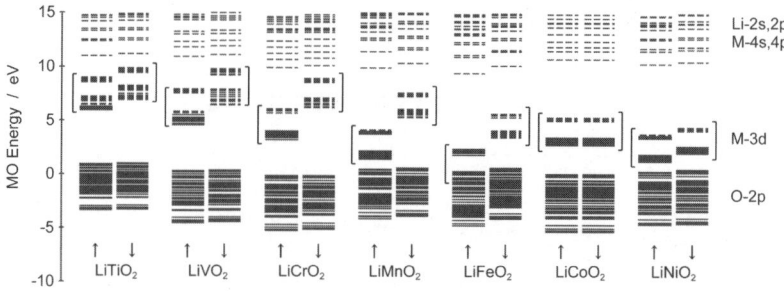

Fig. 7.3.2. Energy level diagrams of the M-centered clusters, $(Li_{12}M_7O_{24})^{15-}$, by DV-Xα calculation. ↑ and ↓ denote majority and minority spin states, respectively.

F is the Faraday constant and z is the charge transported by Li in the electrolyte. In most nonelectrically conducting electrolytes, $z = 1$. $\mu_{Li}^{positive}(x)$ and $\mu_{Li}^{negative}$ are chemical potential of Li in the positive and negative electrode, respectively. $\mu_{Li}^{positive}(x)$ may be related to sum of energy required to remove a Li ion from Li$_x$MO$_2$ and that to remove an electron. The former is related to the electrostatic potential at the Li position. The latter is analogous to the work function of the system. These two factors will be evaluated on the basis of cluster calculations. Figure 7.3.3 shows net charges of ions obtained by Mulliken's population analysis [30] at the central part of the M-centered clusters. Regarding charges of Li, two values are shown. One is the net charge of Li, Q_{Li}, directly obtained by Mulliken's population analysis and the other is the modified net charge of Li, Q'_{Li}, determined by

$$Q'_{Li} = -(Q_M + 2Q_O) \quad . \tag{7.3.2}$$

Because of the use of clusters that are different from the formal stoichiometry of the corresponding compounds, charge neutrality is not strictly satisfied. In order to discuss the Madelung potential, a modified value, as given by Eq. 7.3.2, is practically useful. Q_{Li} may be better evaluated by the Li-centered cluster. It is approximately +0.7, which is not significantly different from Q'_{Li} in the M-centered cluster, which is approximately +0.75. On the other hand, net charges of M and O are almost half of formal charges. This provides the evidence for the strong covalent bonding between M and O, as discussed in Section 7.2. The values and their dependence on the atomic number of M are almost the same when a series of Li-centered clusters are used.

The electrostatic potential (Madelung potential) at the Li position is then computed using the net charges. Calculations were made by the Ewald summation technique employing the program code GULP [31]. In this program, the acceleration parameter, η, was optimized in order to have an optimum convergence with respect to distance. The values obtained by a set of formal charges are compared with those by net charges in Fig.7.3.4. The potentials deduced from the net charges are smaller by approximately 7 V, because of their smaller ionicity. Besides the absolute values, the dependence on the atomic number of M is different. It exhibits a maximum at M = Cr when the formal charges are used, whereas it increases monotonically when the

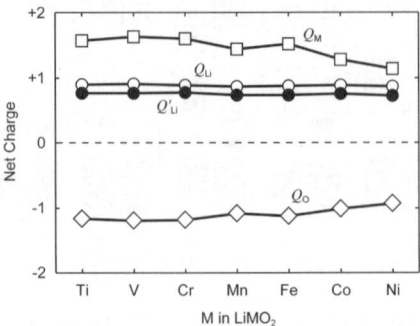

Fig. 7.3.3. Net charges of Li, M and O, and the modified charge of Li given by Eq. 7.3.2 in the M-centered clusters obtained by DV-Xα calculation.

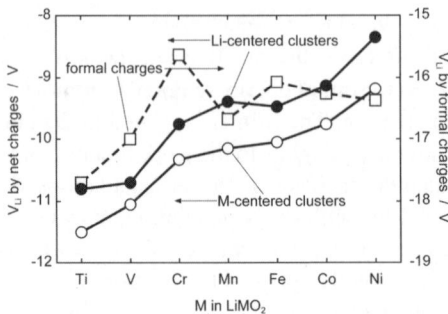

Fig. 7.3.4. Electrostatic potential at the Li position deduced by the formal and net charges.

net charges are used. This could be explained by the trade-off between interatomic distances and net charges. The potential by formal charges is overly sensitive to the variation of the interatomic distance. The difference of about 1 V in the effective potentials evaluated using the two types of clusters is found. The dependence on M of the potentials by the two types of clusters is almost the same.

The energy required to remove an electron from the system may be correlated with the energy level of the highest occupied molecular orbital (HOMO). It is true that some ambiguity is always present for absolute energy levels. However, relative values may be reliable among a series of isostructural compounds. Figure 7.3.5 displays the energies of the HOMO. Results of the two types of clusters are almost the same except for the difference in absolute energy of about 1 eV. It is interesting that the decrease in the HOMO energy is not monotonic with increasing atomic number of M; it shows a minimum at M = Fe and then increases. In the case of isolated transition-metal ions of a fixed valence, the energy of $3d$ orbital decreases with the increase in the atomic number when spin polarization is not allowed, because of the increase in the nuclear potential. In the present case, however,

7.3 First principles study on factors determining voltages of layered LiMO$_2$

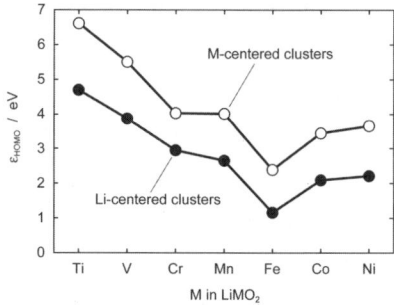

Fig. 7.3.5. Energy of the HOMO for two kinds of model clusters obtained by DV-Xα calculation.

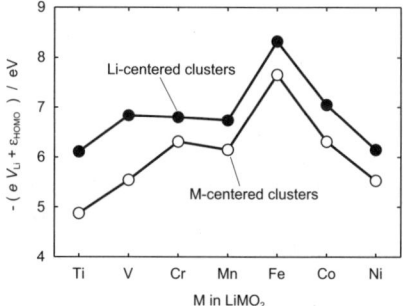

Fig. 7.3.6. Sum of energy required to remove a Li ion from LiMO$_2$ and that to remove an electron given by Eq. 7.3.3.

spin configuration is changed from high-spin state to low-spin one at M = Co. This leads to the minimum of the HOMO energy at M = Fe, as shown in Fig.7.3.5.

The sum of energy required to remove a Li ion from LiMO$_2$ and that to remove an electron can be expressed as

$$-(eV_{Li}^{electrostatic} + \varepsilon_{HOME}) \quad , \tag{7.3.3}$$

which is shown in Fig.7.3.6. Comparison of these quantities with experimental voltages will be made in the following section.

7.3.4 FLAPW band-structure calculations for LiMO$_2$ and MO$_2$

Band-structures of LiMO$_2$ and MO$_2$ have been systematically reported by Aydinol et al. [14] using the pseudopotential plane-wave method. As described above, they did not include spin polarization. Figure 7.3.7 compares density of states (DOS) of LiCoO$_2$ obtained 1) by the present FLAPW method, 2) by the present cluster calculation using Co-centered clusters, 3) by Aydinol et al. using the pseudopotential method, and 4) by Czyżyk et al. [32] using the localized spherical waves (LSW)

246   7 First principles calculations of lithium battery materials

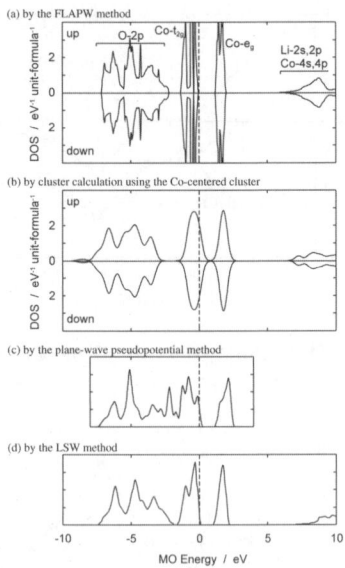

Fig. 7.3.7. Density of states of LiCoO$_2$ obtained (a) by the FLAPW method, (b) by cluster calculation using the Co-centered cluster, (c) by the plane-wave pseudopotential method (Ref. 14), and (d) by the LSW method (Ref. 32).

method. They are aligned so as to make the highest occupied energy zero. The cluster result was broadened by Gaussian functions with a full width at half maximum (FWHM) of 0.5 eV. As can be seen in Fig.7.3.7, the DOS of up and down spins are the same. In other words, the net magnetic moment is zero. Therefore, inclusion of spin polarization does not change the electronic structure in this case. The DOS of the present FLAPW calculation using the tetrahedron method looks almost the same as that by Czyżyk et al. The O-2p band by Czyżyk et al. is composed of three major peaks. A gap can be seen between the O-2p and the Co-3d (t$_{2g}$-like) bands. The result of the present cluster calculation agrees well with those of the band-structure calculations except for the larger band gap. Contrary to them, the DOS by the pseudopotential calculation did not reproduce these features, particularly around the band gap. Aydinol et al. [14] pointed out that the choice of computational methods did not affect the voltages. However, the claim may not always be valid. The DOS of the compounds by the FLAPW calculations are compared in Fig.7.3.8. No contradictory results are found with the cluster results, which confirm the validity of each computational result.

The averaged voltage of LiMO$_2$/MO$_2$ is given by

$$V_{\text{AVE}} = \int_0^1 V_{\text{OCV}}(x)dx \quad , \qquad (7.3.4)$$

using $V_{\text{OCV}}(x)$ defined by Eq.7.3.1. The following quantity V_{AVE}^* is often used [14–17, 33, 34] as an approximated V_{AVE} by the band-structure calculation.

7.3 First principles study on factors determining voltages of layered LiMO₂ 247

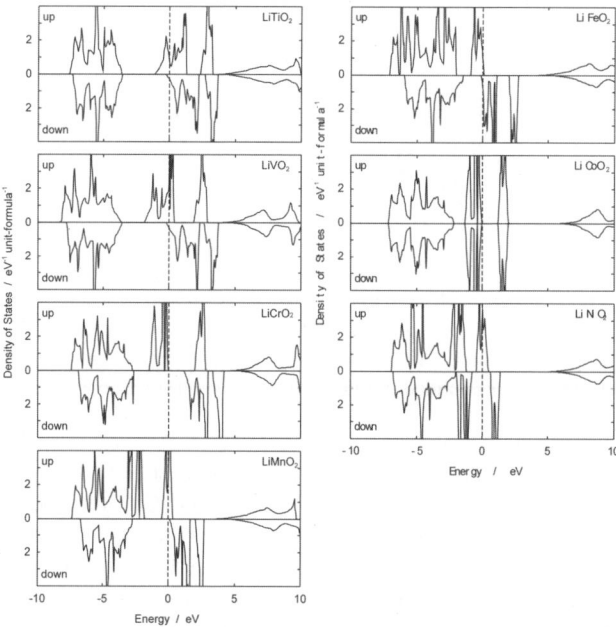

Fig. 7.3.8. Density of states of LiMO₂ (M = Ti - Ni) obtained by the FLAPW method. They are aligned so as to make the highest occupied energy zero.

$$V_{AVE}^* = -\frac{1}{F}\{E_T[LiMO_2] - E_T[MO_2] - E_T[Li]\} \quad , \tag{7.3.5}$$

where E_T is the total energy of the system. V_{AVE}^* is equal to V_{AVE} when both entropy and volume-change terms are negligible in free-energy difference between LiMO₂ and MO₂. The latter term is known to be on the order of 10^{-5} eV [14]. In the present study, the total energies per unit formula for LiMO₂, MO₂ and metallic lithium with the body centered cubic (BCC) structure were computed independently, and V_{AVE}^* was obtained. The same lattice parameters of LiMO₂ as those for cluster calculations were used. As experimental data of MO₂ are not known for most of the cases, the same lattice parameters as those for LiMO₂ were used for MO₂. In other words, structural optimization was not carried out. Wolverton and Zunger [16] reported that lattice constants are reduced by 1.1% for a and 10.8% for c in the extraction of Li from LiCoO₂ by FLAPW calculation. However, their V_{AVE}^* for LiCoO₂/CoO₂ is 3.78 V, which is only 0.32 V larger than the present FLAPW result. The present V_{AVE}^* values are displayed in Fig.7.3.9, together with those reported by the pseudopotential calculations [14]. Experimental voltages are shown for comparison. The values of V_{AVE}^* calculated by the two methods are not significantly different. This is a surprising result, at least for the authors, because the magnitude of voltages is only about 10^{-4} of the total energies of LiMO₂/MO₂.

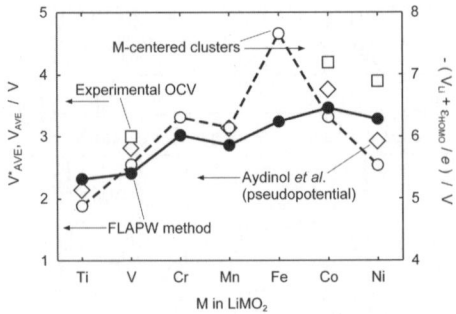

Fig. 7.3.9. V^*_{AVE} by the present FLAPW calculation and that by the plane-wave pseudopotential calculation (Ref. 14). Values defined in Fig.7.3.6 are shown for comparison.

The tendency of V^*_{AVE} by the FLAPW calculation and the quantity shown in Fig.7.3.6 agree well for M = Ti, V, Cr, Mn and Co. The FLAPW result is approximately 1 V smaller for $LiFeO_2$, and is approximately 1 V larger for $LiNiO_2$ when values in Figs.7.3.6 and 7.3.9 are aligned to meet the best agreement. Since two calculations were conducted for the same pair of $LiMO_2/MO_2$ structures, excluding the lattice relaxation associated with the extraction of Li, the discrepancy cannot be attributed to the relaxation. However, the value shown in Fig.7.3.9 includes both the total energies of $LiMO_2$ and MO_2, whereas only the response of $LiMO_2$ to the deintercalation is taken into account in the case of Fig.7.3.6. The differences of $LiFeO_2$ and $LiNiO_2$ may be explained by the differences in the relative stability of MO_2.

Unfortunately, experimental data for $LiMO_2$ (M = Ti, Cr, Mn and Fe) are not available. In addition, experimental data for $LiVO_2$ is obtained in the range of x from 0 to 1/3 in $Li_{1-x}VO_2$, whereas a whole range of x from 0 to 1 is used to obtain the experimental voltages for $LiCoO_2$ and $LiNiO_2$. They may not be comparable. Nevertheless, the experimental voltages agree with the calculated V^*_{AVE} within an error of 1 V.

7.3.5 Conclusion

First principles MO calculations using model clusters by the DV-Xα method and band-structure calculations by the FLAPW method are combined in order to understand the electronic mechanisms that determine the voltages of $LiMO_2/MO_2$. The major results can be summarized as follows:

1. Electronic structures determined by the two methods agree well. However, the present results differ from those of the pseudopotential plane-wave method [14]. The origin of the discrepancy may be twofold: 1) inclusion of the spin polarization in the present work, and 2) validity of their pseudopotential for the $3d$ orbital.

2. Mulliken's population analysis of MO has found that Li is 70% ionic. On the other hand, strong covalent bonding between M and O can be seen. Their net charges are almost half that of the formal charges.
3. Electrostatic potential (Madelung potential) at the Li position was evaluated using the net charges by the Ewald summation technique. The effective potential should be related to the energy required to remove a Li ion from $LiMO_2$. It is smaller by approximately 7 V than the potential deduced by the formal charges because of the smaller ionicity. Besides the absolute values, the dependence on the atomic number of transition-metal element is different. The effective potential monotonically increases with the atomic number of transition-metal elements. The potential by formal charges should be overly sensitive to the variation of the interatomic distances.
4. The energy required to remove an electron from the system was estimated from the HOMO energy. It shows a minimum value at $LiFeO_2$, which can be explained by the balance of two factors: 1) the decrease in energy of the M-$3d$ bands with an atomic number of M, and 2) the spin configuration of the M-$3d$ band.
5. The sum of the two factors as described above is found to explain the dependence of the voltage on the atomic number of M.
6. The value V^*_{AVE}, which may be a good approximation of theoretical voltage, reproduces experimental voltage within an error of 1 V. The sum of the two factors by MO calculations agrees well with the V^*_{AVE} except for $LiFeO_2$ and $LiNiO_2$ which show discrepancies of about 1 V.

References

1. T. Ohzuku and A. Ueda, *Solid State Ionics*, **69** (1994) 201.
2. E. Ferg, R. J. Gummow, A. de Kock and M. M. Thackeray, *J. Electrochem. Soc.*, **141** (1994) L147.
3. R. Koksbang, J. Barker, H. Shi and M. Y. Saïdi, *Solid State Ionics*, **84** (1996) 1.
4. M. M. Thackeray, *J. Am. Ceram. Soc.*, **82** (1999) 3347.
5. K. Mizushima, P. C. Jones, P. J. Wiseman and J. B. Goodenough, *Mater. Res. Bull.*, **15** (1980) 783.
6. J. N. Reimers and J. R. Dahn, *J. Electrochem. Soc.*, **139** (1992) 2091.
7. T. Ohzuku and A. Ueda, *J. Electrochem. Soc.*, **141** (1994) 2972.
8. G. G. Amatucci, J. M. Tarascon and L. C. Klein, *J. Electrochem. Soc.*, **143** (1996) 1114.
9. L. A. de Picciotto, M. M. Thackeray, W. I. F. David, P. G. Bruce and J. B. Goodenough, *Mater. Res. Bull.*, **19** (1984) 1497.
10. L. A. de Picciotto and M. M. Thackeray, *Mater. Res. Bull.*, **20** (1985) 1409.
11. T. Ohzuku, A. Ueda, M. Nagayama, Y. Iwakoshi and H. Komori, *Electrochim. Acta*, **38** (1993) 1159.
12. A. Hirano, R. Kanno, Y. Kawamoto, Y. Takeda, K. Yamaura, M. Takano, K. Ohyama, M. Ohashi and Y. Yamaguchi, *Solid State Ionics*, **78** (1995) 123.
13. K. Miura, A. Yamada and M. Tanaka, *Electrochim. Acta*, **41** (1996) 249.

14. M. K. Aydinol, A. F. Kohan, G. Ceder, K. Cho and J. Joannopoulos, *Phys. Rev. B*, **56** (1997) 1354.
15. G. Ceder, M. K. Aydinol and A. F. Kohan, *Comp. Mater. Sci.*, **8** (1997) 161.
16. C. Wolverton and A. Zunger, *Phys. Rev. B*, **57** (1998) 2242.
17. C. Wolverton and A. Zunger, *J. Electrochem. Soc.*, **145** (1998) 2424.
18. C. Wolverton and A. Zunger, *Phys. Rev. Lett.*, **20** (1998) 606.
19. H. Adachi, M. Tsukada and C. Satoko, *J. Phys. Soc. Jpn.*, **45** (1978) 875.
20. F. W. Averill and D. E. Ellis, *J. Chem. Phys.*, **59** (1973) 6412.
21. D. E. Ellis, H. Adachi and F. W. Averill, *Surf. Sci.*, **58** (1976) 497.
22. B. Ammundsen, J. Desilvestro, T. Groutso, D. Hassell, J. B. Metson, E. Regan, R. Steiner and P. J. Pickering, *J. Electrochem. Soc.*, **147** (2000) 4078.
23. K. Ado, M. Tabuchi, H. Kobayashi, H. Kageyama, O. Nakamura, Y. Inaba, R. Kanno, M. Takagi and Y. Takeda, *J. Electrochem. Soc.*, **144** (1997) L177.
24. F. Capitaine, P. Gravereau and C. Delmas, *Solid State Ionics*, **89** (1996) 197.
25. Y.-I. Jang, B. Huang, Y.-M. Chiang and D. R. Sadoway, *Electrochem. Solid-State Lett.*, **1** (1998) 13.
26. D. J. Singh, *Phys. Rev. B*, **55** (1997) 309.
27. P. Blaha, K. Schwarz and J. Luitz, WIEN97, Vienna University of Technology 1997. Improved and updated Unix version of the original copyrighted WIEN code, which was published by P. Blaha, K. Schwarz, P. Sorantin and S. B. Trickey, in *Comput. Phys. Commun.*, **59** (1990) 399.
28. J. P. Perdew, K. Burke and M. Ernzerhof, *Phys. Rev. Lett.*, **77** (1996) 3865.
29. Experiments for $LiCoO_2$ can be found in J. van Elp, J. L. Wieland, H. Eskes, P. Kuiper, G. A. Sawatzky, F. M. F. de Groot and T. S. Turner, *Phys. Rev. B*, **44** (1991) 6090.
30. R. S. Mulliken, *J. Chem. Phys.*, **23** (1955) 1833.
31. J. D. Gale, *J. Chem. Soc. Faraday Trans.*, **93** (1997) 629.
32. M. T. Czyżyk, R. Potze and G. A. Sawatzky, *Phys. Rev. B*, **46** (1992) 3729.
33. E. Deiss, A. Wokaun, J. L. Barras, C. Daul and P. Dufek, *J. Electrochem. Soc.*, **144** (1997) 3877.
34. S. K. Mishra and G. Ceder, *Phys. Rev. B*, **59** (1999) 6120.

7.4 New fluorides electrode materials for advanced lithium batteries

7.4.1 Introduction

Redox reaction between +3 and +4 states of transition-metal ions associated with intercalation of lithium into oxides has been successfully utilized for rechargeable lithium batteries [1–4]. Among them, $LiCoO_2$ is most widely used [5–8]. Extensive researches are in progress for spinel-type $LiMn_2O_4$ [9–13]. Both of them show voltages of approximately 4 V when graphitic carbon or lithium based metals are used as negative electrode materials. Some of alloyed oxides show the voltages close to 5 V [14–21]. The increase in battery voltage should be very advantageous for applications that require high voltages without increase of the weight or volume of the cells. Currently, the use of liquid electrolytes usually limits the battery voltages at approximately 4.5 V, where the liquid electrolytes start to dissociate. The bottleneck may be broken when good solid-state electrolytes become available. For example, Hayashi et al. [22] reported that some Si-Li-oxysulfide glasses exhibit a wide electrochemical window of more than 10 V. Kanno and Murayama [23] reported that some kinds of thio-LISICON crystals show high Li conductivity far exceeding that in oxide crystals. All solid-state batteries using this type of solid-state electrolytes may expand the limit for the height of the battery voltages.

In Section 7.3, factors determining the voltages of layered $LiMO_2$ (M = Ti - Ni) were discussed on the basis of two kinds of first principles calculations. The voltages were quantitatively evaluated using the full potential linearized augmented plane wave (FLAPW) method. The electronic structures were then analyzed using first principles molecular orbital calculations. Two factors that determine the voltages were pointed out: 1) the effective electrostatic potential at the Li position, and 2) the energy of the highest occupied orbital of $LiMO_2$. They are related to the energies required to remove a Li ion and an electron, respectively. The sum of these two terms reproduces qualitatively well the theoretical voltages by the FLAPW method. The idea may be applicable to other compounds. If this is really the case, we should be able to manipulate the voltages through optimization of chemistry and atomic arrangement of the electrode materials.

The energy required to remove an electron from an isolated atom or ion is called ionization energy. It is determined by experiments to be 43.27 eV and 51.30 eV for Ti(III)/Ti(IV) and Co(III)/Co(IV), respectively [24]. The difference of 8.03 eV may be the ultimate difference in the voltages between Ti(III)/Ti(IV) and Co(III)/Co(IV) based electrode materials under the hypothesis that the effective electrostatic potential at the Li site is constant. In reality, the voltages of titanium oxides are 1.5 - 1.8 V [25, 26] and that of $LiCoO_2$ is 4.2 V [5–8]. The difference between voltages of titanium oxides and cobalt oxides is smaller than 3 V. This is significantly smaller than the ultimate value, i.e., 8.03V. The smaller difference can be ascribed to significant "oxidation" of oxygen associated with the extraction of Li from late transition-metal oxides, as described in Section 7.2. It is therefore quite natural to expect higher voltage by using more electronegative anion in the electrode material. Theoretical study of fluoride systems was thus motivated since fluorine is the only atomic species showing higher electronegativity than oxygen. Provided that the effective electrostatic potential at the Li position is almost constant, the voltage can be determined by the energy required to remove an electron. Presence of anions with high electronegativity should therefore result in the increase of the voltage. At the same time, the magnitude of the "oxidation" of anion associated with the extraction of Li can be reduced when more electronegative anion is used instead of oxide ion. The idea is indeed very naive. However, the validity of the idea can be quantitatively examined through first principles calculations.

The only fluorides that have been reported as electrode materials of lithium batteries are MF_3 (M = Ti, V, Fe, Mn and Co) [27]. However, they use the redox reaction of M(II)/M(III) and the voltage of FeF_3 is as low as 3.4 V. The initial voltage of CoF_3 was reported to be 4.5 V, but it did not work as the electrode because of the decomposition of the electrolyte. In order to obtain higher operating voltages, the redox reaction of Co(III)/Co(IV) is favorable, which is the same as $LiCoO_2$. As compared with oxides, however, the knowledge on fluorides with respect to their structures and properties is rather poor. Among them, a series of $LiAMF_6$ (A = Mg, Ca, Sr, Ba, Ni, Cu, Zn, Cd, and M = Ti, V, Cr, Mn, Fe, Co and Ni) are known to be present. A and M are expected to have formal charges of +2 and +3, respectively. A part of these compounds has been systematically synthesized by Fleischer and Hoppe with special interests on their magnetic properties [28]. But, neither experimental nor theoretical studies on such compounds as electrode materials have thus far been reported. A part of the reasons would be that ionic compounds, such as fluorides, are generally more soluble in a polar solvent than covalent compounds. Experimental study on fluorides as electrode materials is more difficult than that on oxides in general.

In the present Section, the voltages of CoF_3, $LiCoF_4$, $LiCaCoF_6$ and $LiCdCoF_6$ are investigated using the FLAPW method and compared with that of $LiCoO_2$. Since the magnitude of lattice relaxation associated with the intercalation of Li is unknown by experiments, the plane-wave pseudopotential (PW-PP) method was employed in order to examine the crystal structure theoretically.

7.4.2 PW-PP calculation for structural optimization

The crystal structures of CoF_3, $LiCoF_4$ and $LiACoF_6$ (A = Ca and Cd) are shown in Fig.7.4.1. In these structures, cations occupy octahedral positions in a sublattice of fluorine. The structures of CoF_3 [29] and $LiACoF_6$ [28] crystallize in the space group of $R\bar{3}c$ (No. 167) and $P\bar{3}_1c$ (No. 163), respectively. In the two structures, fluorine constitutes distorted hexagonal close packed (HCP) sublattice. The structure of $LiCoF_4$ [30] is an ordered rutile-type structure such that Li and Co occupy the octahedral positions of the alternate (101) layers. It crystallizes in the space group of $P2_1/c$ (No. 14).

Since these fluorides show complicated crystal structures, the PW-PP method using a program code CASTEP [31] was employed to optimize all the parameters under the given symmetries. Ultrasoft pseudopotentials included in the program package [32] with a plane-wave cut-off of 380 eV were used. The local density approximation (LDA) was applied. However, calculated volume of CoF_3 under the LDA was significantly smaller than the experimental value by 22 %. This may be the result of the Co^{3+} ion with non-spin state. Thus the local spin density approximation (LSDA) was applied only for the calculations of Li_xCoF_3. Theoretical volume of CoF_3 with high-spin Co^{3+} ion was then 9 % smaller than the experimental value. Geometry optimization was made by the BFGS approach [33]. The optimization was terminated when all the residual forces on atoms became less than 0.25 eV/Å.

7.4.3 FLAPW calculation for electronic structure and voltage

FLAPW calculations were made for the optimized structures using a program code WIEN97, developed by Blaha *et al.* [34] within the LDA except for the case of

7.4 New fluorides electrode materials for advanced lithium batteries

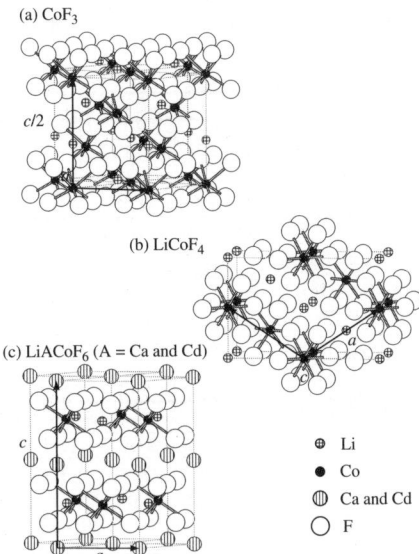

Fig. 7.4.1. Crystal structure of (a) CoF$_3$, (b) LiCoF$_4$ and (c) LiACoF$_6$ (A = Ca and Cd). Li, Co, A and F atoms are denoted by crossed-patterned, filled, striped-patterned and open circles, respectively. In the case of CoF$_3$, crossed-patterned circle denotes the position where inserted Li occupies.

Li$_x$CoF$_3$, where the LSDA was used. The sphere radii were chosen to be 0.95 Å for Li, Co, Ca and Cd and 0.85 Å for F. The number of basis functions was determined by an energy cutoff of 340 eV. Li-1s, Co-3sp, Ca-3sp and Cd-4sp orbitals were treated as semicore states.

The theoretical procedure to calculate average voltage through first principles calculations has already been established [35, 36]. The average voltage, V_{AVE}, is given by

$$V_{AVE} = -\Delta G/F, \qquad (7.4.1)$$

where ΔG is the change in Gibbs free energy for the intercalation reaction, and F is the Faraday constant. Assuming that the changes in volume and entropy associated with the intercalation are negligibly small, the change in Gibbs free energy ΔG can be approximated by the internal energy term, ΔE at 0 K. ΔE is given by the difference in total energies calculated for Li inserted and extracted compounds and metallic lithium.

The first discharge capacity of FeF$_3$ is experimentally reported to be about a half of the theoretical capacity. Thus, the average voltage of Li$_x$CoF$_3$ was calculated in the two ranges of x separately; one is from CoF$_3$ to Li$_{1/2}$CoF$_3$, and the other is from Li$_{1/2}$CoF$_3$ to LiCoF$_3$.

Table 7.4.1. Calculated and experimental cell constants and volume of unit cell. Z denotes the number of unit formula per unit cell.

Compound		Cell constants and volume of unit cell
LiCoF$_3$	theo.	$a = 4.97$ Å, $c = 13.7$ Å, V(Z=6) = 292.1 Å^3
Li$_{1/2}$CoF$_3$	theo.	$a = 4.95$ Å, $c = 13.5$ Å, V(Z=6) = 286.7 Å^3
CoF$_3$	theo.	$a = 4.85$ Å, $c = 13.2$ Å, V(Z=6) = 269.6 Å^3
	exp. (Ref. 15)	$a = 5.071$ Å, $c = 13.29$ Å, V(Z=6) = 296.0 Å^3
LiCoF$_4$	theo.	$a = 5.37$ Å, $b = 4.38$ Å, $c = 5.22$ Å, = 113.4°, V(Z=2) = 112.5 Å^3
	exp. (Ref. 14)	$a = 5.540$ Å, $b = 4.665$ Å, $c = 5.447$ Å, = 114.25°, V(Z=2) = 128.4 Å^3
CoF$_4$	theo.	$a = 5.24$ Å, $b = 4.48$ Å, $c = 4.99$ Å, = 115.6°, V(Z=2) = 105.8 Å^3
LiCaCoF$_6$	theo.	$a = 4.95$ Å, $c = 9.70$ Å, V(Z=2) = 205.6 Å^3
	exp. (Ref. 13)	$a = 5.1024$ Å, $c = 9.783$ Å, V(Z=2) = 220.6 Å^3
CaCoF$_6$	theo.	$a = 4.97$ Å, $c = 9.66$ Å, V(Z=2) = 206.4 Å^3
LiCdCoF$_6$	theo.	$a = 4.96$ Å, $c = 9.50$ Å, V(Z=2) = 202.6 Å^3
	exp. (Ref. 13)	$a = 5.0860$ Å, $c = 9.5181$ Å, V(Z=2) = 213.2 Å^3
CdCoF$_6$	theo.	$a = 4.91$ Å, $c = 9.44$ Å, V(Z=2) = 197.6 Å^3
LiCoO$_2$	theo.	$a = 2.85$ Å, $c = 13.7$ Å, V(Z=3) = 96.2 Å^3
	exp.	$a = 2.815$ Å, $c = 14.05$ Å, V(Z=3) = 96.4 Å^3
CoO$_2$	theo.	$a = 2.85$ Å, $c = 12.6$ Å, V(Z=3) = 88.9 Å^3

7.4.4 Lattice parameters and relaxation by delithiation

The calculated lattice parameters and volume of unit cell of the Li inserted and extracted compounds are listed in Table 7.4.1, together with the experimental values. The calculated volume is 5 - 12 % smaller than the experimental values. The smaller volume may be ascribed to use of the LDA or the LSDA.

It is interesting that the changes in volume due to the Li extraction of LiCaCoF$_6$ and LiCdCoF$_6$ are very small. The changes are only +0.4 % and –2.5 % in LiCaCoF$_6$ and LiCdCoF$_6$, respectively. The changes in volume of LiCoF$_4$ and LiCoO$_2$ are –6.0 % and –7.6 %, respectively. It seems that the degree of change in volume due to the extraction of Li depends on the density of Li. However, the change in volume from Li$_{1/2}$CoF$_3$ to CoF$_3$ is –6.0 %. The change is much greater than that of LiCaCoF$_6$ and LiCdCoF$_6$ even though the density of Li is the same. In contrast to it, the change in volume from LiCoF$_3$ to Li$_{1/2}$CoF$_3$ is only –1.8 %. Generally speaking, interatomic distance between Co and anion decreases and that between Li and anion increases by the extraction. The former is explained as the oxidation of Co, and the latter can be ascribed to the loss of the attraction between Li and anion. The change in volume depends on the balance of these interactions in opposite sign together with the arrangement of Li and Co.

7.4.5 Electronic structure

Figure 7.4.2 shows density of states (DOS) of LiCoF$_4$, LiACoF$_6$, LiCoO$_2$ and Li extracted ones computed by the FLAPW method. They are aligned so as to make the energy of the highest occupied states to be zero. Co is coordinated by six F ions

7.4 New fluorides electrode materials for advanced lithium batteries

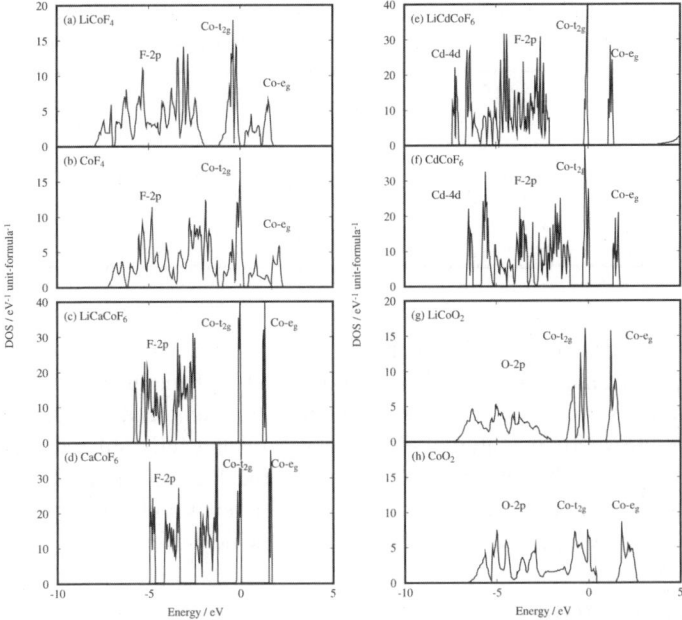

Fig. 7.4.2. Density of states of (a) LiCoO$_4$, (b) CoF$_4$, (c) LiCaCoF$_6$, (d) CaCoF$_6$, (e) LiCdCoF$_6$, (f) CdCoF$_6$, (g) LiCoO$_2$ and (h) CoO$_2$. They are aligned so as to make the energy of the highest occupied states to be zero.

in each compound, which brings about typical t$_{2g}$-e$_g$ splitting of the Co-3d band. Because of Co^{3+} has d^6 configuration in a formal sense, t$_{2g}$ band is fully occupied. After the removal of Li, Co^{4+} (d^5) is made and an empty state is introduced at the top of the t$_{2g}$ band. Two important differences should be noted: 1) Narrowing of the Co-3d band, especially in the case of LiACoF$_6$. This can be explained by smaller covalency in fluorides. Greater nearest-neighbor Co-Co distance in LiACoF$_6$, i.e., 5.1 Å as compared with 2.8 Å in LiCoO$_2$ or 3.4 Å in LiCoF$_4$ is also the reason for much smaller dispersion of the Co-3d bands. 2) The Co-t$_{2g}$ band in CoO$_2$ is significantly admixed with the top of the O-2p band. As a result, the shapes of the top of the valence band, i.e., O-2p and Co-3d bands, are remarkably changed from that of LiCoO$_2$. On the other hand in the fluorides, there is still a margin between the top of the F-2p band and the Co-t$_{2g}$ band. The t$_{2g}$ and e$_g$ bands are virtually translated within the rigid gap of the F-2p band. The shapes of the F-2p and Co-3d bands are not sensibly altered by the removal of Li. Although the shape of DOS between LiCaCoF$_6$ and LiCdCoF$_6$ looks different, the electronic states around Co and F are basically the same. The shape of the F-2p band in LiCdCoF$_6$ is modified because of the presence of fully occupied Cd-4d band near the bottom of the F-2p band. However, the interaction between Cd-4d and F-2p has only subtle contribution on the Co-F interaction.

Table 7.4.2. Calculated average intercalation voltages of cobalt fluorides and $LiCoO_2$ by the FLAPW method. Relative values to the voltage of $LiCoO_2$ are shown in parentheses.

Compound	Average Voltage (V)
CoO_2 / $LiCoO_2$	3.77
CoF_4 / $LiCoF_4$	5.99 (+2.22)
$CaCoF_6$ / $LiCaCoF_6$	5.80 (+2.03)
$CdCoF_6$ / $LiCdCoF_6$	5.71 (+1.94)
CoF_3 / $Li_{1/2}CoF_3$	4.70 (+0.93)
$Li_{1/2}CoF_3$ / $LiCoF_3$	3.76 (−0.01)

As described in Section 7.2, not only Co but also O is "oxidized" by the extraction of Li because of the admixture of the top of the O-$2p$ band and Co-$3d$ band in the CoO_2. In other words, Co^{3+} in $LiCoO_2$ cannot be fully oxidized to be Co^{4+} in CoO_2. As a result, the redox potential of Co(III)/Co(IV) is not fully utilized. On the other hand in the present fluorides, the admixture of the F-$2p$ band and Co-$3d$ band is much smaller even after the extraction. The difference should be important for the increase in the voltages, which will be shown in the next subsection.

7.4.6 Average voltage

The voltages calculated by Eq. 7.4.1 are summarized in Table 7.4.2. The theoretical voltage for $LiCoO_2$ using the LDA has been reported by several groups using different kind of computational methods to be 3.75 V [35] and 3.78 V [36]. These values are in good agreement with the present result, though it is about 0.4 V lower than the experimental value. The theoretical voltages for $LiCoF_4$, $LiCaCoF_6$ and $LiCdCoF_6$ are higher than the voltage of $LiCoO_2$ by about 2 V, although these values were obtained using the same redox reaction of Co(III)/Co(IV) as that in $LiCoO_2$. Although the reason for the underestimation of voltage has not been clearly described yet, the underestimation by less than 1 V seems to be a quite common phenomenon for many oxides when computed using Eq. 7.4.1. Provided that the theoretical voltages for the fluorides are underestimated by the similar amount, real voltages of greater than 6 V can be expected when these fluorides are used as positive electrode materials of lithium batteries.

The drop in voltage by about 1 V at $x = 1/2$ in Li_xCoF_3 is found. This behavior is similar to that of spinel-type $Li_xMn_2O_4$ at $x = 1$. The average voltage of Li_xCoF_3 from $x = 0$ to $x = 1$, i.e., in the whole range of Co(II)/Co(III) redox reaction, is expected to be somewhat higher than that of $LiCoO_2$ even though the redox reaction of Co(II)/Co(III) is used in the former instead of Co(III)/Co(IV) in the latter.

7.4.7 Conclusion

Even when the same redox reaction of Co(III)/Co(IV) is used, higher voltages by 2 V in the fluorides than in $LiCoO_2$ can be expected. Higher voltages of fluorides may be simply ascribed to the greater electronegativity of fluorine than oxygen. Several cobalt fluorides are known to be present. Some of them exhibit similar local atomic arrangement around Li and Co to $LiCoF_4$ or $LiACoF_6$. The subtle difference in the atomic arrangements may not change the electronic structures and therefore not change drastically the voltages of these compounds, because the electronic structures are mainly determined by their chemical composition and local atomic coordination. Many other properties including the mobility of Li, cyclic durability and chemical stability are also very important for electrode materials. Chemical composition and crystal structure can be optimized according to other properties without changing much of their voltages. Thus, it may be worthwhile to develop fluorides as new generation electrode materials for high voltage lithium batteries.

References

1. T. Ohzuku and A. Ueda, *Solid State Ionics*, **69** (1994) 201.
2. E. Ferg, R. J. Gummow, A. de Kock and M. M. Thackeray, *J. Electrochem. Soc.*, **141** (1994) L147.
3. R. Koksbang, J. Barker, H. Shi and M. Y. Saïdi, *Solid State Ionics*, **84** (1996) 1.
4. M. M. Thackeray, *J. Am. Ceram. Soc.*, **82** (1999) 3347.
5. K. Mizushima, P. C. Jones, P. J. Wiseman and J. B. Goodenough, *Mater. Res. Bull.*, **15** (1980) 783.
6. J. N. Reimers and J. R. Dahn, *J. Electrochem. Soc.*, **139** (1992) 2091.
7. T. Ohzuku and A. Ueda, *J. Electrochem. Soc.*, **141** (1994) 2972.
8. G. G. Amatucci, J. M. Tarascon and L. C. Klein, *J. Electrochem. Soc.*, **143** (1996) 1114.
9. M. M. Thackeray, P. J. Johnson, L. A. de Picciotto, P. G. Bruece and J. B. Goodenough, *Mater. Res. Bull.*, **19** (1984) 179.
10. J. M. Tarascon, E. Wang, F. K. Shokoohi, W. R. McKinnon and S. Colson, *J. Electrochem. Soc.*, **138** (1991) 2859.
11. M. M. Thackeray, A. de Kock, M. H. Rossouw, D. Eiles, R. Bittihn and D. Hoge, *J. Electrochem. Soc.*, **139** (1992) 363.
12. Y. Gao and J. R. Dahn, *J. Electrochem. Soc.*, **143** (1996) 100.
13. Y. Xia, T. Sakai, T. Fujieda, X. Q. Yang, X. Sun, Z. F. Ma, M. McBreen and M. Yoshio, *J. Electrochem. Soc.*, **148** (2001) A723.
14. C. Sigala, D. Guyomard, A. Verbaere, Y. Piffard and M. Tournoux, *Solid State Ionics*, **81** (1995) 167.
15. Y. Gao, K. Myrtle, M. Zhang, J. N. Reimers and J. R. Dahn, *Phys. Rev. B*, **54** (1996) 16670.
16. Q. Zhong, A. Bonakdarpour, M. Zhang, Y. Gao and J. R. Dahn, *J. Electrochem. Soc.*, **144** (1997) 205.
17. Y. Ein-Eli and W. F. Howard, Jr., *J. Electrochem. Soc.*, **144** (1997) L205.
18. M. N. Obrovac, Y. Gao and J. R. Dahn, *Phys. Rev. B*, **57** (1998) 5728.
19. H. Kawai, M. Nagata, H. Tukamoto and A. R. West, *J. Mater. Chem.*, **8** (1998) 837.
20. T. Ohzuku, S. Takeda and M. Iwanaga, *J. Power Sources*, **81–82** (1999) 90.
21. M. Okada, Y.-S. Lee and M. Yoshio, *J. Power Sources*, **90** (2000) 196.

22. A. Hayashi, M. Tassumisago and T. Minami, *J. Electrochem. Soc.*, **146** (1999) 3472.
23. R. Kanno and M. Murayama, *J. Electrochem. Soc.*, **148** (2001) A742.
24. E. C. Moore, *Atomic Energy Levels, Vol. III*, Circular of the National Bureau of Standards, (1958) 467.
25. T. Ohzuku, A. Ueda and N. Yamamoto, *J. Electrochem. Soc.*, **142** (1995) 1431.
26. L. Kavan, M. Grätzel, J. Rathouský and A. Zukal, *J. Electrochem. Soc.*, **143** (1996) 394.
27. H. Arai, S. Okada, Y. Sakurai and J. Yamaki, *J. Power Sources*, **68** (1997) 716.
28. T. Fleischer and R. Hoppe, *Z. Naturforsch.*, **37b** (1982) 988.
29. K. H. Jack and V. Gutmann, *Acta Cryst.*, **4** (1951) 245.
30. T. Fleischer and R. Hoppe, *Z. Naturforsch.*, **37b** (1982) 1132.
31. V. Milman, B. Winkler, J. A. White, C. J. Pickard, M. C. Payne, E. V. Akhmatskaya and R. H. Nobes, *Int. J. Quant. Chem.*, **77** (2000) 895.
32. CASTEP Program on Cerius 4.2 (Molecular Simulations Inc., San Diego, CA).
33. W. H. Press, S. A. Teukolsky, W. T. Vetterling and B. P. Flannery, *Numerical Recipes*, 2nd ed., Cambridge University Press, Cambridge (1992). ISBN 0-521-43064-X.
34. P. Blaha, K. Schwarz and J. Luitz, WIEN97, *A Full Potential Linearized Augmented Plane Wave Package for Calculating Crystal Properties*, (Karlheinz Schwarz, Techn. Univ. Wien, Vienna 1999). ISBN 3-9501031-0-4.
35. M. K. Aydinol, A. F. Kohan, G. Ceder, K. Cho and J. Joannopoulos, *Phys. Rev. B*, **56** (1997) 1354.
36. C. Wolverton and A. Zunger, *Phys. Rev. B*, **57** (1998) 2242.

7.5 First principles calculations of formation energies and electronic structures of defects in oxygen-deficient $LiMn_2O_4$

7.5.1 Introduction

Spinel type $LiMn_2O_4$ [1–5] and its derived materials [6–13] have been widely investigated as positive electrode materials of rechargeable lithium batteries. Because of their advantages such as low cost, low toxicity and safety in overcharged states, they are candidates to replace $LiCoO_2$ that is used as the positive electrode material of commercial lithium-ion batteries. It is known that exactly stoichiometric $LiMn_2O_4$ is difficult to obtain. Generally speaking, nonstoichiometry always leads some defects, e.g., vacancies, interstitial atoms and substituted atoms. Almost all the properties of compounds are influenced by such defects. Occasionally the defects play decisive roles of the properties. $LiMn_2O_4$ is not an exception. Effects associated with the nonstoichiometry have been experimentally reported. Phase transformation between cubic and orthorhombic structures is reported in nonstoichiometric $LiMn_2O_4$ nearly at room temperature [14–19]. Extra electrochemical reactions at 3.3 and 4.5 V are found in oxygen deficient $LiMn_2O_4$ [4, 20, 21]. Lithium excess improves the capacity fading during charge-discharge cycles [21–24] that

7.5 First principles calculations of formation energies and electronic structures of defects

is the main problem for wider use of LiMn$_2$O$_4$. Recently, Xia *et al.* reported that oxygen deficiency affects the capacity fading on the basis of systematic studies of nonstoichiometric LiMn$_2$O$_4$ [5]. In spite of the importance of the knowledge of defects, there has been little information about the defects in LiMn$_2$O$_4$ because of the difficulty to observe them directly by experiments.

In the present study, the focus is directed to the nonstoichiometric defects in oxygen deficient LiMn$_2$O$_{4-\delta}$. In this compound, oxygen deficient defects, e.g., oxygen vacancy, and/or metal excess defects, e.g., interstitial lithium and manganese atoms must exist for its nonstoichiometry. However, it has not been established which type of defects occurs in LiMn$_2$O$_{4-\delta}$. By usual chemical analysis, we can get only the elemental composition. It is hard to know what type of defects exists. Extra efforts are necessary to get the information on defects. Sugiyama *et al.* measured the relationship in LiMn$_2$O$_{4-\delta}$ between partial pressure of O$_2$ gas in environment and oxygen deficiency, δ [25]. In order to explain the experimental results, they proposed a cluster defect model composed of two oxygen vacancies and localized Mn^{3+} ion. The defect model was consistent with the experimental results. However, they did not consider metal-excess type defects. Kanno *et al.* reported that they observed not interstitial cations but a small amount of oxygen vacancy with interstitial oxygen in LiMn$_2$O$_{4-\delta}$ using neutron diffraction measurements [26]. Based on their result, they proposed another defect model composed of oxygen vacancies and interstitial oxygen. Contrary to them, Hosoya *et al.* reported that the density of LiMn$_2$O$_{4-\delta}$ increased linearly with the degree of oxygen deficiency, δ [27]. They concluded that not oxygen-vacancy type defects but metal-excess defects exist in oxygen deficient LiMn$_2$O$_{4-\delta}$. The present study aims to determine the preference of defect species in oxygen deficient LiMn$_2$O$_{4-\delta}$ by theoretical calculations. The formation energies of several kinds of defects in LiMn$_2$O$_{4-\delta}$, both of oxygen-deficient and metal-excess defects, are calculated by first principles calculations with a supercell method. Local electronic structures around the desirable defects are also investigated.

7.5.2 Computational procedure

In compound systems, formation energies of nonstoichiometric defects depend on the chemical potential of each element. The formation energy is defined as

$$E^F = E_T[\text{defective}] - \sum_X n_X \mu_X \quad , \qquad (7.5.1)$$

where $E_T[\text{defective}]$ is the energy of the system with the defect that is composed of n_X atoms of each element X. μ_X is the chemical potential in the perfect system without defects. The energy of the perfect system is described as

$$E_T[\text{perfect}] - \sum_X n_X(\text{perfect})\mu_X \quad . \qquad (7.5.2)$$

In the present study, several kinds of defects in oxygen deficient LiMn$_2$O$_{4-\delta}$ are considered. Because of the variety of the defects, their formation energies are compared per one oxygen deficiency. Therefore, the formation energy of the defect with m oxygen deficiency can be given by

$$E^F = 1/m\{E_T[Li_N Mn_{2N} O_{4N-m}] - NE_T[LiMn_2O_4]\} + \mu_O \quad , \tag{7.5.3}$$

where N is the number of chemical formula in the system. The formation energy of the defect with m excess Li and $2m$ excess Mn atoms can be evaluated as

$$E^F = 1/4m\{E_T[Li_{N+m} Mn_{2N+2m} O_{4N}] - (N+m)E_T[LiMn_2O_4]\} + \mu_O \quad . \tag{7.5.4}$$

The energies with the defects were calculated by the supercell method. A supercell composed of 56 atoms, e.g., [$Li_8Mn_{16}O_{32}$]-cell, was employed as the original supercell to calculate the perfect crystal. Defects were introduced in the original supercell. Since the attention was paid to the defects to make oxygen deficiency, only neutral defects were considered. Their total energies and electronic structures were calculated by the plane-wave pseudopotential method using a program code CASTEP [28]. Ultrasoft pseudopotentials [29] were used with a plane-wave cutoff of 500 eV. Convergence of the formation energy of oxygen vacancy with respect to the plane-wave cutoff was found to be smaller than 0.01 eV up to 800 eV. The local spin density approximation (LSDA) [30] was applied. All Mn ions were fixed to exhibit high spin states. Reciprocal space integration was carried out using 2×2×2 mesh points in the first Brillouin zone. Further increase in the number of mesh-points up to 4×4×4 does not change the formation energy of oxygen vacancy by more than 0.02 eV. The cell parameter, a, and internal parameter, u, of the perfect $LiMn_2O_4$ crystal (space-group $Fd\bar{3}m$, No. 227) were calculated to be 8.22 Å and 0.386, respectively. Both of them are in good agreement with experimental values, i.e., a = 8.240 Å and u = 0.390. Atomic arrangements around the defects were optimized allowing structural relaxation of atoms within the distance from the defects smaller than a half of the lattice constant. The structural relaxation to break the space-group symmetry of the initial defective structure was not allowed. The optimization was performed so that all the residual forces for the unconstrained atoms were less than 0.25 eV/Å under a constant lattice condition. All defect calculations were made using the same cell parameter, a, that was optimized for the perfect crystal, in order to model a system with dilute defect concentration.

As described above, the formation energies of defects depend on the chemical potential of oxygen. To evaluate the formation energies, the chemical potential of oxygen at the equilibrium condition of Mn_2O_3 and MnO_2 is chosen, i.e.,

$$\mu_O = 2E_T[MnO_2] - E_T[Mn_2O_3] \quad . \tag{7.5.5}$$

Both $E_T[MnO_2]$ and $E_T[Mn_2O_3]$ were obtained by separate calculations in the present study using the same method. It should be noted that the relative value of formation energies of each defect does not depend on the choice of chemical potential of oxygen. Only the absolute value changes.

7.5.3 Defects of oxygen-vacancy type

In $LiMn_2O_4$ with cubic spinel-type structure, Li, Mn and O atoms occupy the 8a, 16d and 32e positions in Wyckoff notation, respectively. Oxygen sublattice is a distorted

7.5 First principles calculations of formation energies and electronic structures of defects 261

Table 7.5.1. Formation energies of oxygen-vacancy type defects, together with their relative values to the simple oxygen mono-vacancy model.

Defect model	Formation energy (eV)	Relative formation energy (eV)
Fig. 7.5.1-a	2.17	0
Fig. 7.5.1-b	4.09	+1.92
Fig. 7.5.1-c	4.15	+1.98
Fig. 7.5.1-d	3.14	+0.97
Fig. 7.5.1-e	2.82	+0.65

cubic close packed (CCP) structure. Li and Mn occupy the tetrahedral and octahedral positions of the oxygen sublattice, respectively. In the spinel-type structure, another octahedral position, $16c$, still remains unoccupied, which may play an important role in defect process.

As defects of oxygen-vacancy type, five kinds of models were considered. Schematic structures of all models are shown in Fig.7.5.1. The first is a simple mono-vacancy of oxygen (Fig.7.5.1-a). For this model, a supercell of $[(Li_8)^{8a} (Mn_{16})^{16d} (O_{31})^{32e}]$ was employed and its formation energy was calculated using Eq. 7.5.3 with $N = 8$ and $m = 1$. The second to fourth are "modified" oxygen mono-vacancy models (Fig.7.5.1-b, c and d): An oxygen atom neighboring to the vacancy was moved toward the vacancy and put at the middle point of the two oxygen positions. In other words, two oxygen atoms were removed from the normal $32e$ positions and one oxygen atom was inserted in between these vacancies. In the spinel-type structure, each oxygen atom has three kinds of oxygen neighbors. One constitutes an edge of a [LiO$_4$] tetrahedron. The second constitutes an edge of a [MnO$_6$] octahedron and the edge is shared by two [MnO$_6$] octahedrons. The last constitutes an unshared edge of a [MnO$_6$] octahedron. The last model of the oxygen-vacancy type is a modified tri-vacancy of oxygen (Fig.7.5.1-e) proposed by Kanno et al [26]. Four oxygen atoms were removed from the [Mn$_4$O$_4$] cube in the spinel-type structure. Then one oxygen atom was put into not the original oxygen position but the central position of the [Mn$_4$O$_4$] cube, i.e., the $8b$ position. For this model, a supercell of $[(Li_8)^{8a} (Mn_{16})^{16d} (O_{28})^{32e} (O_1)^{8b}]$ was employed and its formation energy was calculated using Eq. 7.5.3 with $N = 8$ and $m = 3$.

The formation energies of the oxygen-vacancy models are listed in Table 7.5.1. The formation energy of the simple oxygen mono-vacancy is calculated to be 2.17 eV. Modification of the model simply results in an increase of the formation energy. Although the atomic arrangement around the vacancy was optimized, the "modified" mono-vacancy models give much higher formation energies. This can be ascribed to the location of the interstitial oxygen at a confined interstitial position. The modified tri-vacancy model shows the second smallest formation energy. This should be related to the presence of the relatively larger open space for the $8b$ position than the other interstitial positions. Although only a small number of models of oxygen-vacancy type were investigated, complexity of the atomic arrangement around a vacancy seems to make its energy higher. The simple oxygen mono-vacancy can be energetically most favorable defect among this type.

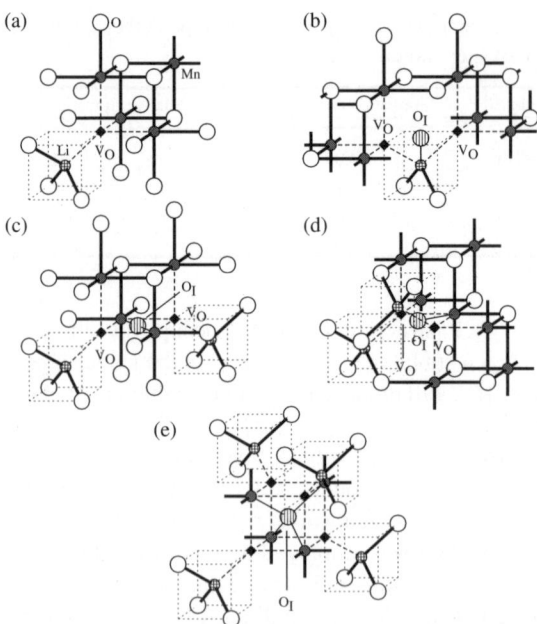

Fig. 7.5.1. Schematic defect structures of oxygen-vacancy type in LiMn$_2$O$_4$: (a) is a simple oxygen mono-vacancy. (b), (c) and (d) are "modified" oxygen mono-vacancies on the edge of a [LiO$_4$] tetrahedron, on the shared edge of [MnO$_6$] octahedron and on the unshared edge of [MnO$_6$] octahedron, respectively. (e) is a modified oxygen tri-vacancy. Li, Mn and O atoms at the normal positions are denoted by crossed-patterned, filled and open circles, respectively. Vacancy position and interstitial oxygen are denoted by filled diamonds and striped-patterned circles, respectively.

7.5.4 Defects of metal-interstitial type (I): simple interstitial atoms

Oxygen deficiency in oxides can be ascribed to either oxygen vacancy or interstitial metal in general. In this subsection, the formation of interstitial Li and Mn in the oxygen deficient LiMn$_2$O$_{4-\delta}$ is examined. Because of the limitation of computable size of the supercell, separated calculations of each interstitial defect were made using a [(Li$_8$)^{8a} (Li$_1$)^{16c} (Mn$_{16}$)^{16d} (O$_{32}$)^{32e}] supercell and a [(Li$_8$)^{8a} (Mn$_1$)^{16c} (Mn$_{16}$)^{16d} (O$_{32}$)^{32e}] supercell. This corresponds to the situation that the two interstitial atoms are located with an infinitive separation. In the present case the formation energy can be given by

$$E^F = 1/4\{E_T[\text{Li}_9\text{Mn}_{16}\text{O}_{32}] + 2E_T[\text{Li}_8\text{Mn}_{17}\text{O}_{32}] - 25E_T[\text{LiMn}_2\text{O}_4]\} + \mu_O \quad . \tag{7.5.6}$$

Three kinds of models were considered for both of interstitial Li and Mn. The first one is a simple mono-interstitial occupying an empty 16c position (Fig. 7.5.2-a). The octahedron centered by the 16c position faces to two tetrahedrons centered by the 8a positions that are already occupied by Li atoms. It may be locally tight for interstitial

7.5 First principles calculations of formation energies and electronic structures of defects 263

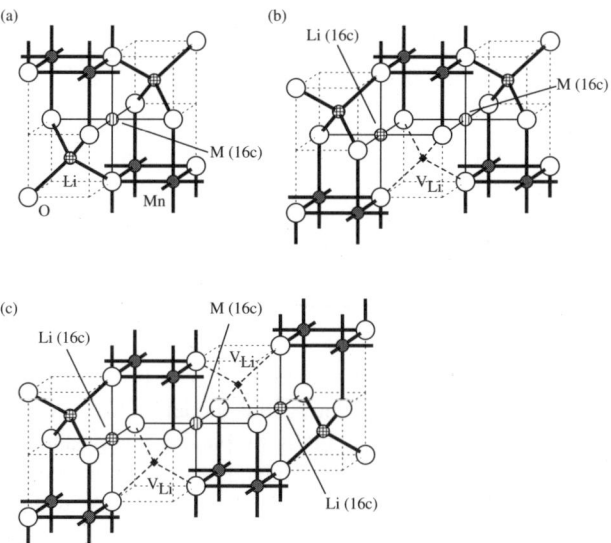

Fig. 7.5.2. Schematic defect structures of metal-interstitial type: (a) simple interstitial metal (Li or Mn) occupying a 16c position, (b) an interstitial metal with the move of one Li at the neighboring 8a position to the next 16c position, and (c) an interstitial metal with the move of two neighboring Li atoms. Notations of atoms are the same as in Fig.7.5.1. Interstitial metal atoms are denoted by striped-patterned circles.

Li or Mn at the 16c position, thereby making the formation of the simple mono-interstitial Li or Mn expensive in energy. If this is the case, the formation energy may be decreased by the removal of the Li at the neighboring 8a position. The idea is reflected in other two models shown in Figs.7.5.2-b and 7.5.2-c. In the former model, one of two Li at the 8a positions was moved to the neighboring empty 16c position. In the later model, both of the two Li were moved. From a different viewpoint, the model shown in Fig.7.5.2-b can be interpreted as two interstitial atoms at the 16c with one Li vacancy at the 8a. The one in Fig.7.5.2-c can be three interstitial atoms at the 16c with two Li vacancies at the 8a.

Relative total energies of interesting defects shown in Fig.7.5.2 are summarized in Table 7.5.2. The relative energies for the interstitial Li and Mn correspond to the first and second term of the right-hand side of the Eq. 7.5.6, respectively. Contrary to the case of the oxygen mono-vacancy, the energy of the defect decreases by the modification of atomic arrangement around the interstitial Li or Mn of the model in Fig.7.5.2-b. This is consistent with the idea as described above. However, relative energy increases when two Li were put into the 16c positions as in Fig.7.5.2-c. In the spinel-type Mn_2O_4 framework, the tetrahedral 8a position may be more stable for Li than the octahedral 16c position. The decrease in Fig.7.5.2-b and increase in Fig.7.5.2-c in the total energy suggest that local stress introduced by the interstitial atom is almost released by the first step. The second step simply results in an increase in energy, because of the increase in the number of Li at the interstitial 16c positions.

Table 7.5.2. Relative energies of metal-interstitial type defects.

Defect model	Relative energy (eV)
Li-interstitial type: [Li$_9$Mn$_{16}$O$_{32}$]-cell	
Fig. 7.5.2-a	0
Fig. 7.5.2-b	–0.07
Fig. 7.5.2-c	+0.17
Mn-interstitial type: [Li$_8$Mn$_{17}$O$_{32}$]-cell	
Fig. 7.5.2-a	0
Fig. 7.5.2-b	–0.70
Fig. 7.5.2-c	–0.43

The absolute value of the formation energy E^F can be calculated by Eq. 7.5.6. The formation energy of the simple interstitial model is 1.83 eV per one oxygen deficiency, which is 0.34 eV smaller than the simple oxygen-vacancy model. Further reduction in the formation energy can be obtained when both of the interstitial Li and Mn show the configurations given in Fig.7.5.2-b. It is 1.46 eV, which is the lowest energy configuration among the models discussed in the present subsection.

7.5.5 Defects of metal-interstitial type (II): with occupation of Mn at the 8a position

In the previous subsection, all of the 16d positions were made to occupy by Mn atoms. Either Li atoms or vacancies were put into the 8a positions. Occupation of Mn atoms at the 8a positions was not considered. Occupation of Li atoms at the 16d position was not allowed either. In this subsection, the excess Mn is put onto the 8a position to substitute for Li. Firstly the exchange energy of the interstitial Mn at the 16c position and its neighboring Li at the 8a position is evaluated. The exchange energy using two models shown in Figs.7.5.3-a and 7.5.3-b is calculated to be +0.02 eV, which is slightly positive. Another way to put Mn at the 8a position is shown in Fig.7.5.3-c. This model can be made by moving the interstitial Li in the model of Fig.7.5.3-b toward the infinitive separation. In other words, the excess Li made by the Mn substitution of the 8a position was taken into account as a mono-interstitial. In this case, the formation energy per one oxygen deficiency can be obtained by

$$E^F = 1/4\{2E_T[Li_7Mn_{17}O_{32}] + 3E_T[Li_9Mn_{16}O_{32}] - 41E_T[LiMn_2O_4]\} + \mu_O \quad . \tag{7.5.7}$$

When the lowest energy model of the interstitial Li is taken as shown in Fig.7.5.2-b, E^F of this model is 1.46 eV, which is as low as the smallest value obtained in the previous subsection. The formation energies are summarized in Table 7.5.3. On the basis of these results, the occurrence of the oxygen-vacancy type defects can be clearly ruled out in the oxygen deficient LiMn$_2$O$_{4-\delta}$ with the spinel-type structure. Among models examined in the present study, two kinds of models with extra Li at the interstitial 16c positions show smallest formation energy. According to these results, the excess Mn occupies either 8a or 16c position with the move of the

7.5 First principles calculations of formation energies and electronic structures of defects

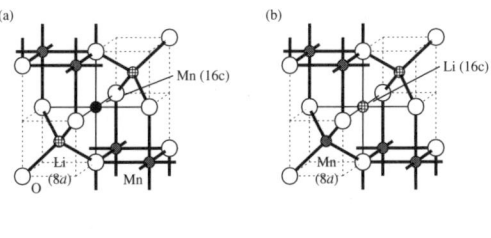

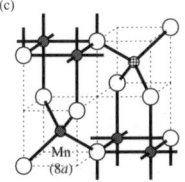

Fig. 7.5.3. Schematic defect structure of substituted Mn: (a) Simple interstitial Mn at the 16c position, which is the same model as Fig.7.5.2-a. (b) The substituted Mn with the exchange of Mn at the 16c position and its neighboring Li at the 8a position. (c) The substitution of Mn with the absence of an atom at the 16c position. Notations of atoms are the same as in Fig.7.5.1.

Table 7.5.3. Summary of the formation energies of defects that occurs in oxygen deficient LiMn$_2$O$_4$.

Defect type		Formation energy (eV)
Oxygen-vacancy type		
Fig. 7.5.1-a		2.17
Metal-interstitial type		
Li	Mn	
Fig. 7.5.2-a	Fig. 7.5.2-a	1.83
Fig. 7.5.2-b	Fig. 7.5.2-b	1.46
Fig. 7.5.2-b	Fig. 7.5.3-c	1.46

neighboring Li. This result is consistent with the experimental results of density measurements and diffusion experiments on LiMn$_2$O$_{4-\delta}$ and LiMg$_y$Mn$_{2-y}$O$_{4-\delta}$ [27, 31].

7.5.6 Local electronic structures around defects

Local electronic structures of interstitial Mn atom at the 16c position and substituted one at the 8a position in LiMn$_2$O$_4$ will be discussed. Because of the assumption that all Mn ions exhibit high spin states, only electronic states of majority spin, which are related to the redox reaction associated with Li insertion/extraction, will be discussed.

Figure 7.5.4-a shows total density of states (DOS) of the perfect LiMn$_2$O$_4$ and projected DOS for Mn in it. Fermi energies are set to zero and DOS are broadened

266   7 First principles calculations of lithium battery materials

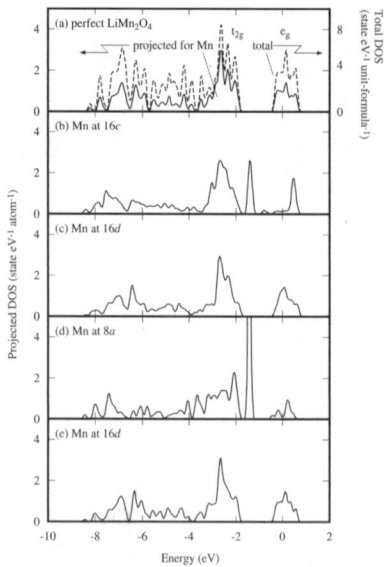

Fig. 7.5.4. Densities of states (DOS) of majority spin in the perfect and the defective $LiMn_2O_4$: (a) Total and projected DOS in the perfect $LiMn_2O_4$. (b) Projected DOS for the interstitial Mn at the 16cc position. (c) Projected DOS for normal Mn at the 16d in $LiMn_2O_4$ with interstitial Mn as in Fig.7.5.2-b. (d) Projected DOS for the substituted Mn at the 8a. (e) Projected DOS for normal Mn at d in $LiMn_2O_4$ with substitution of Mn as in Fig.7.5.3-c.

using a Gaussian function with a full width at half maximum (FWHM) of 0.1 eV. Because Mn is located at the octahedral position in the spinel-type structure, the Mn-3d band is split into the t_{2g}-like and e_g-like bands. In the perfect $LiMn_2O_4$, the t_{2g}-like band and a quarter of e_g-like band are occupied. In other words, formal charge of Mn is +3.5. Projected DOS for Mn atoms in the defective $LiMn_2O_4$ are shown in Figs.7.5.4-b to 7.5.4-e. The former two are for Mn atoms in the interstitial Mn model as shown in Fig.7.5.2-b: one is for the excess Mn at the interstitial 16c position (Fig.7.5.4-b) and the other is for the Mn that is located at the normal 16d position and farthest from the excess Mn (Fig.7.5.4-c). The later two are for Mn atoms in the substituted defect model as shown in Fig. 7.5.3-c: one is for the substituted Mn at the tetrahedral 8a position (Fig.7.5.4-d) and the other is for the Mn that is located at the 16d position and farthest from the excess Mn (Fig.7.5.4-e).

The projected DOS for the Mn at the 16d position in the interstitial defect model (Fig.7.5.4-c) is similar to that for the perfect $LiMn_2O_4$. It is also the case for the Mn at the 16d position in the substituted defect model (Fig.7.5.4-e). These results mean that the influence of the interstitial and substituted Mn atoms upon the electronic structures is well localized within the supercells used in the present study. In contrast to them, the projected DOS for the interstitial and substituted Mn are much different from that for Mn at the 16d position. As for the interstitial Mn, a new and fully occupied band is found at around −1.4 eV. The new band is found to be the e_g-like states of the interstitial Mn by the examination of the shape of the wave function. As a matter of fact, two states per one interstitial Mn are included in this band that are mainly composed of interstitial Mn orbitals. Therefore, the formal charge of the interstitial Mn can be concluded to be +2. The band around the Fermi energy is mainly composed of other Mn atoms at the normal 16d positions with a small contribution of the interstitial Mn.

In the projected DOS of substituted Mn (Fig.7.5.4-e), a new, narrow and fully occupied band is also located at around -1.4 eV. Since the substituted Mn at the 8a position is tetrahedrally coordinated, the Mn-3d band is split into e-like and t_2-like bands. The new band is found to correspond to the t_2-like band. There are three states per one substituted Mn in this band that is mainly composed of the substituted Mn orbitals. Therefore, the formal charge of the substituted Mn is also +2. The formal charge of the interstitial or substituted Mn atoms is smaller by 1.5 as compared to those at normal positions. This suggests that these Mn atoms suffer different electrostatic potential than those at normal positions. Excess electrons are generally introduced when neutral anion deficiency occurs in an ionic compound. In the present case, a large part of the excess electrons is localized to the extra Mn atoms.

7.5.7 Discussion

Theoretical results in the present study imply that the oxygen deficiency in $LiMn_2O_4$ occurs not by the oxygen vacancies but by the combination of interstitial Li at the 16c position and excess Mn occupying either 16c or 8a position. Although the conclusion is consistent with the experimental results of density and diffusion measurements [27, 31], one may feel uncomfortable by the conclusion. As a matter of fact, the same theoretical tool was employed for the study of the energetics of native defects in ZnO to find that the formation energy of neutral oxygen vacancy was more than 1 eV smaller than that of the interstitial Zn [32].

Other experimental results on $LiMn_2O_4$ that are supportive of the present theoretical results will be provided here. It is well known by experiments that tetragonal $Li_2Mn_2O_4$ appears when Li is electrochemically inserted into $LiMn_2O_4$. The structure of $Li_2Mn_2O_4$ can be made from the structure of $LiMn_2O_4$ by the full insertion of Li to the 16c positions together with the extraction of Li from the 8a positions. The experimental fact is suggestive of the stability of the defect structure shown in Fig.7.5.2-b. It was found experimentally that heating of $LiMn_2O_4$ introduced oxygen deficiency and eventual decomposition into Mn_3O_4 and orthorhombic $LiMnO_2$ [20, 21, 25]. The structure of Mn_3O_4 can be made by repeated substitution of Li by Mn at the 8a positions of $LiMn_2O_4$ if a distortion of the Mn_3O_4 structure due to the cooperative Jahn-Teller distortion of Mn^{3+} ion is ignored. The fact suggests that the defect structure shown in Fig.7.5.3-c is stable. In order to examine the relative stability of the tetragonal $Li_2Mn_2O_4$ and the orthorhombic $LiMnO_2$, an extra calculation was made. The two phases are found to show an indistinguishable energy difference. The above discussion based on the $Li_2Mn_2O_4$ is valid although the formation of LiMnO$_2$ has often been reported by high temperature experiments.

7.5.8 Conclusion

In order to find out defect structures in oxygen deficient LiMn$_2$O$_{4-\delta}$, systematic first principles calculations by the plane-wave pseudopotential method have been made. Among defects of oxygen-vacancy type, a simple oxygen mono-vacancy shows the smallest formation energy of 2.17 eV when the chemical potential of oxygen is chosen at the equilibrium condition of Mn$_2$O$_3$ and MnO$_2$. Some defects of metal-interstitial type show significantly smaller formation energies than that of the simple oxygen mono-vacancy. The smallest value, 1.46 eV, is obtained for two models: In both of the models, an excess Li occupies an empty 16c position. An excess Mn is present either at the 8a position substituting for Li or at the empty 16c position with moving of one neighboring Li into the next 16c position. From detailed examination of the local electronic structures of the defects, the formal charges of excess Mn either at the 8a or 16c positions are found to be +2. The preference of the metal-excess type is consistent with experimental results of density measurements, diffusion experiments and decomposition reaction at high temperatures.

References

1. M. M. Thackeray, P. J. Johnson, L. A. de Picciotto, P. G. Bruece and J. B. Goodenough, *Mater. Res. Bull.*, **19** (1984) 179.
2. J. M. Tarascon, E. Wang, F. K. Shokoohi, W. R. McKinnon and S. Colson, *J. Electrochem. Soc.*, **138** (1991) 2859.
3. M. M. Thackeray, A. de Kock, M. H. Rossouw, D. Eiles, R. Bittihn and D. Hoge, *J. Electrochem. Soc.*, **139** (1992) 363.
4. Y. Gao and J. R. Dahn, *J. Electrochem. Soc.*, **143** (1996) 100.
5. Y. Xia, T. Sakai, T. Fujieda, X. Q. Yang, X. Sun, Z. F. Ma, M. McBreen and M. Yoshio, *J. Electrochem. Soc.*, **148** (2001) A723.
6. C. Sigala, D. Guyomard, A. Verbaere, Y. Piffard and M. Tournoux, *Solid State Ionics*, **81** (1995) 167.
7. Y. Gao, K. Myrtle, M. Zhang, J. N. Reimers and J. R. Dahn, *Phys. Rev. B*, **54** (1996) 16670.
8. Q. Zhong, A. Bonakdarpour, M. Zhang, Y. Gao and J. R. Dahn, *J. Electrochem. Soc.*, **144** (1997) 205.
9. Y. Ein-Eli and W. F. Howard, Jr., *J. Electrochem. Soc.*, **144** (1997) L205.
10. M. N. Obrovac, Y. Gao and J. R. Dahn, *Phys. Rev. B*, **57** (1998) 5728.
11. H. Kawai, M. Nagata, H. Tukamoto and A. R. West, *J. Mater. Chem.*, **8** (1998) 837.
12. T. Ohzuku, S. Takeda and M. Iwanaga, *J. Power Sources*, **81–82** (1999) 90.
13. M. Okada, Y.-S. Lee and M. Yoshio, *J. Power Sources*, **90** (2000) 196.
14. A. Yamada and M. Tanaka, *Mater. Res. Bull.*, **30** (1995) 715.

15. M. Tabuchi, C. Masquelier, H. Kobayashi, R. Kanno, Y. Kobayashi, T. Akai, Y. Maki, H. Kageyama and O. Nakamura, *J. Power Sources*, **68** (1997) 623.
16. G. G. Amatucci, C. N. Schmutz, A. Blyr, C. Sigala, A. S. Gozdz, D. Larcher and J. M. Tarascon, *J. Power Sources*, **69** (1997) 11.
17. K. Oikawa, T. Kamiyama, F. Izumi, B. C. Chakoumakos, H. Ikuta, M. Wakihara, J. Li and Y. Matsui, *Solid State Ionics*, **109** (1998) 35.
18. J. Rodríguez-Carvajal, G. Rousse, C. Masquelier and M. Hervieu, *Phys. Rev. Lett.*, **81** (1998) 4660.
19. G. Rousse, C. Masquelier, J. Rodríguez-Carvajal and M. Hervieu, *Electrochem. Solid-State Lett.*, **2** (1999) 6.
20. J. M. Tarascon, W. R. McKinnon, F. Coowar, T. N. Bowmer, G. Amatucci and D. Guyomard, *J. Electrochem. Soc.*, **141** (1994) 1421.
21. J. M. Tarascon, F. Coowar, G. Amatucci, F. K. Shokoohi and D. Guyomard, *J. Power Sources*, **54** (1995) 103.
22. R. J. Gummow, A. de Kock and M. M. Thackeray, *Solid State Ionics*, **69** (1994) 59.
23. G. Pistoia, A. Antonini, R. Rosati, C. Bellitto and G. M. Ingo, *Chem. Mater.*, **9** (1997) 1443.
24. J.-H. Lee, J.-K. Hong, D.-H. Jang, Y.-K. Sun and S.-M. Oh, *J. Power Sources*, **89** (2000) 7.
25. J. Sugiyama, T. Atsumi, T. Hioki, S. Noda and N. Kamegashira, *J. Power Sources*, **68** (1997) 641.
26. R. Kanno, A. Kondo, M. Yonemura, R. Gover, Y. Kawamoto, M. Tabuchi, T. Kamiyama, F. Izumi, C. Masquelier and G. Rousse, *J. Power Sources*, **81–82** (1999) 542.
27. M. Hosoya, H. Ikuta, T. Uchida and M. Wakihara, *J. Electrochem. Soc.*, **144** (1997) L52.
28. V. Milman, B. Winkler, J. A. White, C. J. Pickard, M. C. Payne, E. V. Akhmatskaya and R. H. Nobes, *Int. J. Quantum Chem.*, **77** (2000) 895; The CASTEP program code was used in the present study. [Accelrys Inc., San Diego, CA.]
29. D. Vanderbilt, *Phys. Rev. B*, **41** (1990) 7892.
30. J. P. Perdew and A. Zunger, *Phys. Rev. B*, **23** (1981) 5048.
31. N. Hayashi, H. Ikuta and M. Wakihara, *J. Electrochem. Soc.*, **146** (1999) 1351.
32. F. Oba, S. R. Nishitani, S. Isotani, H. Adachi and I. Tanaka, *J. Appl. Phys.*, **90** (2001) 824.

7.6 Summary and conclusions

In order to understand the electronic mechanism behind various properties of electrode materials for lithium batteries, systematic and theoretical investigations are performed using the first principles molecular orbital calculations with model clusters and also the first principles band-structure calculations. The present study is successful to obtain fundamental knowledge about the difference in the electrode properties. On the basis of the knowledge obtained about the voltage, new positive electrode materials for high voltage lithium batteries are proposed and the validity of the proposal is quantitatively examined through the first principles calculation. The obtained results are summarized as follows:

In Section 7.2, the electronic and bonding states of $LiMO_2$ and MO_2 (M = V, Cr, Co and Ni) are investigated by the first principles molecular orbital calculation employing the discrete variational (DV)-Xα method with model clusters in order to understand the electronic mechanism behind the difference in the intercalation capability among four $LiMO_2$. Special attention is paid to understanding the changes in chemical bondings by the extraction of Li. Li is found to be nearly completely ionized in $LiMO_2$. Strong covalent interaction between M and O is noted. The extraction of Li significantly increases the interaction between M and O. This results in the "oxidation" associated with the extraction of Li not of M but of O. The formal redox notation for the intercalation, i.e., M(III)/M(IV), is thus far from reality. The difference in the intercalation capability among four $LiMO_2$ oxides is ascribed to the difference in the magnitude of the electrostatic repulsion between the oxygen layers in MO_2. The net charges of oxygen in CoO_2 and NiO_2 are found to be close to zero, resulting in weak electrostatic repulsion and stabilization of the MO_2 structure.

In Section 7.3, first principles molecular orbital calculations using model clusters by the DV-Xα method and by band-structure calculations of the full potential linearized augmented plane wave (FLAPW) method are combined in order to understand the electronic mechanisms that determine the voltages of $LiMO_2$ (M = Ti - Ni). The voltages are quantitatively evaluated using the FLAPW method. The electronic structures are then analyzed using molecular orbital calculations. Electrostatic potential (Madelung potential) at the Li position is evaluated using the net charges by the Ewald summation technique. The effective electrostatic potential should be related to the energy required to remove a Li ion from $LiMO_2$. It is smaller by approximately 7 V than the potential deduced by the formal charges because of the smaller ionicity. Besides the absolute values, the dependence on the atomic number of transition-metal element is different. The effective potential monotonically increases with increase in the atomic number of transition-metal elements. The potential deduced by formal charges should be overly sensitive to the variation of the interatomic distances. The energy required to remove an electron from the system is estimated from the energy of the highest occupied molecular orbital. It shows a minimum value at $LiFeO_2$, which can be explained by the balance of two factors: 1) the decrease in energy of the M-3d band with an atomic number of transition-metal element, and 2) the spin configuration of the M-3d band. The value V^*_{AVE}, which may be a good approximation of theoretical voltage, reproduces experimental voltage within an error of 1 V. The sum of the two factors as described above is found to explain the experimental dependence of the voltage on the atomic number of transition-metal element. The sum also agrees well with the V^*_{AVE} except for $LiFeO_2$ and $LiNiO_2$ which show discrepancies of about 1 V.

In Section 7.4, on the basis of the discussion about the factors determining voltage, cobalt fluorides are proposed as the new positive electrode materials for high voltage lithium batteries. Even when the same redox reaction of Co(III)/Co(IV)

7.6 Summary and conclusions

is used, higher voltages by 2 V in the fluorides than in $LiCoO_2$ can be expected. The average voltage of CoF_3 in the whole range is expected to be somewhat higher than that of $LiCoO_2$, even though the redox reaction of Co(II)/Co(III) is used in the former instead of Co(III)/Co(IV) in the later. Higher voltages of the fluorides may be simply ascribed to the greater electronegativity of fluorine than oxygen. The changes in volume due to the extraction of Li of $LiCaCoF_6$ and $LiCdCoF_6$ are very small of only +0.4 % and –2.5 %, respectively. In contrast, the volume change from $Li_{1/2}CoF_3$ to CoF_3 is –6.0 %, though the density of Li is the same. The degree of change in volume depends not only on the density of Li but also on the balance of the opposite changes in interatomic distances between Li-anion and Co-anion due to the extraction of Li together with the arrangement of Li and Co.

In Section 7.5, systematic first principles calculations on defects in oxygen deficient $LiMn_2O_{4-\delta}$ by the PW-PP method are made in order to find out the desirable defect structures. Among defects of oxygen-vacancy type, a simple oxygen mono-vacancy shows the smallest formation energy of 2.17 eV when the chemical potential of oxygen is chosen at the equilibrium condition of Mn_2O_3 and MnO_2. Some defects of metal-interstitial type show significantly smaller formation energies than that of the simple oxygen mono-vacancy. The smallest value, 1.46 eV, is obtained for two models: In both of the models, an excess Li occupies an empty $16c$ position in the spinel-type structure. An excess Mn is present either at the $8a$ position substituting for Li or at the empty $16c$ position with moving of one neighboring Li into the next $16c$ position. From detailed examination of the local electronic structures of the defects, the formal charges of excess Mn either at the $8a$ or $16c$ positions are found to be +2. The preference of the metal-excess type is consistent with experimental results of density measurements, diffusion experiments and decomposition reaction at high temperatures.

Index

Ab initio   112
ab initio calculations   116, 119, 122
AC conductivity   178
acid dissociation constant   74
activation energies   137
alkaline manganese batteries   19
alkaline secondary batteries   134
all-solid-state cell   54, 59, 126, 127, 177
all-solid-state energy devices   31
all-solid-state lithium ion batteries   104
all-solid-state Ni/MH batteries   173
amorphous   187
amorphous Si   26
apparent activation energy   165, 166
Arrhenius plots   115, 137

BET surface areas   88
binding energies   122
biocompatibility   222
branched poly(oxyethylene)   191
bridging oxygen   34

C-rate   19
capacity retention   148, 154, 157
carrier ions   114
cell voltage *vs.* current density   84
charge transfer   104, 105, 108, 119, 124
charge transfer resistance   177
charge-discharge cycle performance   174
chemical bonding   230
clustering   210
composite   80
conductive additives   59
conductivity   171

corrosion   177
coulomb efficiency   181
coulombic potential calculation   100
crosslinked poly(acrylate)   134
crosslinked polymer   197
crystallization   38
CuS   50
cyclability   57
cycle life   179
cyclic voltammetry   139
cyclic voltammogram   204

Debye relaxation time   124
decoupled systems   31
defects   258
density functional theory (DFT)   112
desolvation/solvation process   120
differential scanning calorimetry   213
diffusion path   100, 101
dimethylsiloxane polymers   11
direct methanol fuel cell   19
discharge efficiency   179
disordered spinel   101
disordered(Fd$\bar{3}$m)   100
dissociation   119, 123
dissociation constant   114
dry polymer electrolytes   16
DV-Xα 232
dynamic modulus   197
dynamic solvent effect   106

EDLC   167
elastic polymers   77
elastomer   187, 206, 209

electric double layer capacitors 74, 134, 159
electrochemical window 33
electrode-electrolyte interface 104, 119, 181
electrode/solid electrolyte 126
electrolyte creepage 143, 145
electrolyte leakage 143
electrolyte/electrode interface 105, 111, 124
electron conducting path 59
electron transfer rates 106
energy density 19
epoxycyclohexylethyltrimethoxysilane 91
Evans equation 218
exchange current density 109, 111, 119, 121

Fd$\bar{3}$m 98
first principles calculations 225
5V class 95, 96
FLAPW method 240
fluorides 250
free volume theory of polymer 112
functionality 221
functionality elastomer 221

gas-permeability 221
gel electrolyte 15, 165
gelled polymer electrolytes 7
genuine polymer electrolytes 11
GeSe$_{5.5}$ 67
Gibbs activation energies 111, 120, 125
Gibbs activation energy 106, 107, 124
glass-ceramics 38, 55
glass-transition temperature 114, 187
glassy electrolyte 128
3-glycidoxypropyltrimethoxysilane 88
graphite 25
graphite anode 27
GSM 9

$H_3PMo_{12}O_{40} \cdot 29H_2O$ 74
H_3PO_4-doped poly(ethylene oxide)-modified poly(methacrylate) polymeric gel 166
$H_3PW_{12}O_{40} \cdot 29H_2O$ (WPA) 74
Hard and Soft Acids and Bases (HSAB) theory 111, 117

Hartree-Fock (HF) 112
heteropolyacid hydrates 134, 177
high- and low-temperature characteristics 21
high-rate chargeability 153
high-rate dischargeability 153, 162
highest energy occupied molecular orbital (HOMO) 117
Hittorf's method 138
hybrid electric vehicles 20
hybrid electrochemical capacitors 134
hydrated water 74
hydration number 178
hydrogen storage alloy 133, 136
hydrogen-absorbing alloy 24
Hydromecanique et Frottements (HEF) 68

In_2Se_3 67
inorganic glass electrolytes 104
inorganic solid electrolytes 177
inorganic-organic hybrid 87
inter-metallic phases 43
interfacial product, SEI 130
interfacial reaction rate 104
ion transport number 138
ionene 209
ionic aggregation 210
ionic conductivity 104, 111, 133, 163
IR drop 160
irreversible capacity loss 157

Lewis acid 105, 111, 119, 123, 124
Lewis acidity 111, 115, 116, 119
Lewis basicity 111, 116, 119
$Li_{1-x}CoO_2$ 54
$Li_{3+5x}P_{1-x}S_4$ 39
$Li_{4-x}Ge_{1-x}P_xS_4$ 39, 40
$Li_yCo_{0.3}Ni_{0.7}O_2$ 48
$Li_xV_2O_5$ 67
$Li_{3.4}V_{0.6}Si_{0.4}O_4$ 67
$Li_{3.6}Si_{0.6}P_{0.4}O_4$ glass 66
Li_3PO_4-P_2S_5 67
$Li_{4.4}Si$ 43
$Li_{4/3}Ti_{5/3}O_4$ 56
$Li_4P_2S_7$ 67
$LiCo_{0.3}Ni_{0.7}O_2$ 43
$LiCoO_2$ 5, 48, 56, 67, 230
$LiCoPO_4$ 57
$LiM_yMn_{2-y}O_4$ 96

Index   275

LiMn$_2$O$_4$ 67, 95, 258
LiMO$_2$ 230
LiNi$_{0.5}$Mn$_{1.5}$O$_4$ 95, 96, 99, 102
LiNi$_{1/2}$Mn$_{1/2}$O$_2$ 56
LiNi$_{1/3}$Mn$_{1/3}$Co$_{1/3}$O$_2$ 13
LiPF$_6$ 26
LiPON   17, 53, 67
lithium ion batteries   12
lithium ion conductors   32
lithium ion secondary batteries   5
lithium poly-sulfides   50
lithium polymer batteries   8
lithium silicide   43
locking-chair type   67
longitudinal relaxation time   107, 108, 123, 124
lowest energy unoccupied molecular orbital (LUMO)   117

manganese dioxide   179
Marcus microscopic theories   124, 125
Marcus microscopic theory   105
mechanical milling   35
mechanical property   217
mechanochemical synthesis   35
medium temperature   77, 87
melt-quenching method   32
membrane/electrode assemblies (MEAs)   81
mesoporous silica gel   76
meta-stable new phase   43
metal excess structure   99
metal-interstitial   262, 264
metastable superionic crystal phases   38
misch metal-based alloy   178
MnO$_2$/MH solid-state battery   184
mobility   111, 114
molecular beam deposition   68
12-molybdophosphoric acid hydrate   179
monolithic cell   59
morphology   215

Nafion$^®$ 89
NASICON   17
Nd:YAG laser   68
negative electrode   42
net charge   235
Ni/MH battery   178
nickel oxyhydroxide   179

nickel-cadmium   19
nickel-metal hydride (Ni-MH)   19, 133
nickel/metal hydride (Ni/MH) batteries   146, 171
NMR   35, 78
non-sintered nickel electrode   21
nonbridging oxygen   35
nonstoichiometry   258
notebook-type personal computers   19

open circuit voltage   84
open structure   31
ordered spinel   101
ordered(P4$_3$32)   100
organic/inorganic composite   201
orientation   207
oxygen defect   99
oxygen evolution   21
oxygen permeability   141
oxygen reduction current   141, 144, 145
oxygen-vacancy   260
oxysulfide glass   32, 201

P-doping   171
P/Si ratio   174
P4$_3$32   98
PEG-borate ester   105, 119, 125
perovskite   17
phosphoric acid-doped (P-doped) silica gels   134, 171
phosphosilicate   77
plane-wave pseudopotential method   252, 260
plasma CVD   26
poly(ethylene glycol dimethylether   166
poly(ethylene glycol) dimethylether   163
Poly(ethylene oxide)   189
poly(ethylene oxide) dimethacrylate   163
poly(ethylene oxide) monomethacrylate   163
poly(ethylene oxide)-modified poly(methacrylate)   165
poly(oxyethylene)   188
poly(oxytetramethylene)   210
polyhexafluoropropylene   8
polyimide   81
polyisoprene (PI)   77
polymer electrolyte fuel cells (PEFCs)   80, 87

polymer electrolytes   105
polymer gel electrolyte   134
polymer hydrogel electrolyte   134, 136
polymer ionic conductors   134
polymer network   196
polymer solid electrolyte   187
polymeric membrane   163
polymeric proton conductor   134
polyvinylidene fluoride   8
pore structure   75
positive electrode   47
potential window   140
power density   19
precursor sol   89
processability   199
proton   73
proton conducting polymers   134
proton conductivity   135, 171, 173
proton-conducting electrolyte   134
proton-conducting polymeric gel electrolytes   162
pseudocapacitance   161

rate capability   169
rate performance   100
rate property   102
reversible capacity loss   156
RF magnetron sputtering   26
rubber elasticity   187
rubbery state   187

sealed-type Ni/MH cell   140
secondary battery   19
segmental motion   114, 115
SEI   130
self-discharge   154
shear dynamic modulus   215
Si alloy   13
silanol   73
silicon   25
sintered nickel electrode   21
SiS_2-Li_2S-Li_4SiO_4 glass electrolyte   104, 126
Sn-based glassy materials   57
SnO   68
soft composite   200
sol-gel method   73
solid electrolyte   104, 126
solid polymer electrolyte   104

solid-solid interfaces   59
solvation state   122
specific capacitance   169
spinel   95, 258
styrene-ethylene-butylene-styrene (SEBS)   77
styrene-isoprene-styrene block copolymer (SIS)   77
sulfide solid electrolytes   16
sulfonated polypropylene separators   150
sulfur   50
surface tension   144
surfactants   75

TeO_2 67
terpolymer   196
thermoplastically   77
thin film battery   64
thio-LISICON   39
TiN   60
tin   25
TiOS   67
TiS_2 43, 66, 67
transfer coefficient   120
transference number   203, 218
transport number   104, 111, 115, 116, 119
12-tungstophosphoric acid hydrate   179
twin-roller apparatus   32

uniaxial stretching   206

V_2O_5 56, 67
VGCF   60
viscosity   111, 199
Vogel-Tamman-Fulcher (VTF)   74

water content   171
water-absorbing capacity   137
water-holding capacity   143
wide-angle X-ray diffraction (WAXD)   193, 206
Williams-Landel-Ferry (WLF) equation   112, 113, 115, 194, 214
WLF parameter   113, 115

X-ray absorption spectroscopy (XAS)   126
XeCl Eximer laser   68

zero-shear viscosity   199
zirconium-based alloy   178